Current Topics in Microbiology and Immunology

Volume 376

For further volumes:
http://www.springer.com/series/82

Hubert Hilbi
Editor

Molecular Mechanisms in *Legionella* Pathogenesis

Responsible Series Editor: Klaus Aktories

Springer

Editor
Hubert Hilbi
Max von Pettenkofer Institute
Ludwig-Maximilians University München
Munich
Germany

ISSN 0070-217X ISSN 2196-9965 (electronic)
ISBN 978-3-642-40590-7 ISBN 978-3-642-40591-4 (eBook)
DOI 10.1007/978-3-642-40591-4
Springer Heidelberg New York Dordrecht London

Library of Congress Control Number: 2013948854

Printed on acid-free paper

Springer is part of Springer Science+Business Media (www.springer.com)

Preface

Legionnaires' disease is a potentially fatal pneumonia primarily affecting elderly and immunocompromised persons. The disease is caused by the ubiquitous environmental bacterium *Legionella pneumophila*, which was first identified more than 35 years ago in the aftermath of a pneumonia epidemic that swept through a convent of the American Legion in Philadelphia, USA. The water-borne bacteria are inhaled via contaminated aerosols, resist degradation by alveolar macrophages, and trigger a fulminant pneumonia. Direct inhalation represents the sole route of infection with *L. pneumophila*; person-to-person transmission does not occur.

Macrophage resistance of *L. pneumophila* is a prerequisite for its virulence. This trait has likely been acquired through long-standing evolutionary cross-talk with free-living protozoa. Accordingly, the genome of *L. pneumophila* encodes a number of eukaryotic-like genes presumably acquired by horizontal transkingdom gene transfer. Thus, the adaptation of *L. pneumophila* to bactericidal protozoa did select for virulence traits required for growth in mammalian cells. Many aspects of pathogen–phagocyte interactions seem to be mechanistically conserved between protozoan natural hosts and mammalian "accidental" target cells. Given these similarities, protozoa such as *Dictyostelium, Acanthamoeba, Hartmanella,* or *Tetrahymena* spp. are powerful models to dissect cell-autonomous aspects of *L. pneumophila* infection.

The interactions of *L. pneumophila* with phagocytes are regulated by at least four different two-component systems (PmrAB, CpxRA, LetAS, and LqsRS). These networks involve and converge on small regulatory RNAs, as well as RNA-binding proteins. *L. pneumophila* survives intracellularly in macrophages and amoebae by forming a specific replication-permissive compartment, the *Legionella*-containing vacuole (LCV). LCVs communicate with the endocytic, secretory, and retrograde vesicle trafficking pathways, but do not fuse with lysosomes. To gain insights into the composition of LCVs, intact pathogen vacuoles have recently been purified and analyzed by proteomics.

L. pneumophila governs the formation of LCVs and other pathogen–host interactions through distinct protein secretion systems, such as the Lsp type II secretion system (T2SS) and the Icm/Dot type IV secretion system (T4SS).

Whereas the Lsp T2SS secretes at least 25 proteins, the Icm/Dot T4SS translocates the astonishing number of ~300 different "effector" proteins into host cells. The function of most of these proteins is not understood, but they are thought to subvert host signal transduction and vesicle trafficking pathways.

Some Icm/Dot substrates are exceptionally intriguing, since they catalyze novel biochemical reactions. The eukaryotic small GTPase Rab1, which is implicated in secretory vesicle trafficking, is targeted by no fewer than six different *L. pneumophila* effectors. Whereas SidM (*alias* DrrA) activates Rab1 through its guanine nucleotide exchange factor (GEF) activity, LepB functions as a Rab1 GTPase activating protein (GAP). Furthermore, SidM and AnkX covalently modify Rab1 by attaching an AMP or a phosphocholine moiety, respectively. The reverse deadenylylation or dephosphocholination reactions are catalyzed by the effector proteins SidD or Lem3. Finally, the Icm/Dot substrate LidA assists SidM by binding with high affinity to activated Rab1.

Another interesting aspect of *L. pneumophila* host cell subversion is how translocated effectors localize to the cytoplasmic face of LCVs. Whereas the Icm/Dot substrate LegG1 is lipidated by the host prenylation machinery, the Rab1 GEF SidM and the ER interactor SidC anchor to LCVs through the phosphoinositide (PI) lipid phosphatidylinositol-4-phosphate (PtdIns(4)*P*). In contrast, the Rab1 deadenylylase SidD as well as the Arf1 GEF RalF, bind to the LCV membrane via unknown receptors apparently without targeting lipids. The host cell lipid pattern is directly modified by approximately 20 *L. pneumophila* T2SS or T4SS substrates, which act as phospholipases or PI phosphatases, respectively, thereby destroying host membranes and/or modulating host signaling pathways.

An important class of *L. pneumophila* effectors interferes with host cell ubiquitination. Icm/Dot substrates such as LubX and AnkB are functional mimics of eukaryotic E3 ubiquitin ligases that mark bacterial and host proteins for proteasomal degradation or modification of activity. Moreover, cytotoxic *L. pneumophila* glucosyltransferases modify the ribosome, thereby inhibiting protein synthesis (Lgt1-3), or subvert endosomal vesicle trafficking (SetA). Finally, *L. pneumophila* and host cell kinases, as well as the protein phosphorylation pattern and corresponding signal transduction pathways define pathogen–host interactions.

Whereas many aspects of *L. pneumophila* virulence can be satisfactorily analyzed using uni-cellular (protozoan) models, the study of inflammation and immune responses relies on mouse models of Legionnaireś disease, which faithfully mimick human pathology. To this end, the A/J strain of mice proved instrumental, as macrophages with this genetic background fail to restrict *L. pneumophila* replication. This is due to a Naip5 (*alias* Birc1e) protein that does not recognize flagellin, and consequently, does not trigger flagellin-dependent inflammasome activation.

In summary, this book contributes to an in-depth understanding of Legionnaireś disease by comprising comprehensive reviews about different facets of *L. pneumophila* pathogenesis. Topics covered include comparative phagocyte infection, virulence gene regulation, biochemical functions of effector proteins, cellular pathogen–host interactions, as well as host responses and immunity against *L. pneumophila*. Thus, this compilation provides a state-of-the-art overview of current insights into the molecular pathogenesis of an opportunistic but potentially fatal bacterial respiratory pathogen.

Hubert Hilbi

Contents

From Amoeba to Macrophages: Exploring the Molecular Mechanisms of *Legionella pneumophila* Infection in Both Hosts 1
Pedro Escoll, Monica Rolando, Laura Gomez-Valero and Carmen Buchrieser

The *Legionella pneumophila* Two-Component Regulatory Systems that Participate in the Regulation of Icm/Dot Effectors 35
Gil Segal

Facets of Small RNA-Mediated Regulation in *Legionella pneumophila* 53
Hana Trigui, Nilmini Mendis, Laam Li, Mariam Saad and Sebastien P. Faucher

Type II Secretion and *Legionella* Virulence 81
Nicholas P. Cianciotto

Effector Translocation by the *Legionella* Dot/Icm Type IV Secretion System 103
Jiazhang Qiu and Zhao-Qing Luo

Modulation of Small GTPases by *Legionella* 117
Roger S. Goody and Aymelt Itzen

Host Lipidation: A Mechanism for Spatial Regulation of *Legionella* Effectors 135
Stanimir S. Ivanov and Craig Roy

Phosphoinositide Lipids and the *Legionella* Pathogen Vacuole 155
Ina Haneburger and Hubert Hilbi

***Legionella* Phospholipases Implicated in Virulence** 175
Katja Kuhle and Antje Flieger

Cytotoxic Glucosyltransferases of *Legionella pneumophila* 211
Yury Belyi, Thomas Jank and Klaus Aktories

Modulation of the Ubiquitination Machinery by *Legionella* 227
Andree Hubber, Tomoko Kubori and Hiroki Nagai

Host Signal Transduction and Protein Kinases Implicated in *Legionella* Infection 249
Andrew D. Hempstead and Ralph R. Isberg

Mouse Models of Legionnaires' Disease 271
Andrew S. Brown, Ian R. van Driel and Elizabeth L. Hartland

Index 293

From Amoeba to Macrophages: Exploring the Molecular Mechanisms of *Legionella pneumophila* Infection in Both Hosts

Pedro Escoll, Monica Rolando, Laura Gomez-Valero and Carmen Buchrieser

Abstract *Legionella pneumophila* is a Gram-negative bacterium and the causative agent of Legionnaires' disease. It replicates within amoeba and infects accidentally human macrophages. Several similarities are seen in the *L. pneumophila*-infection cycle in both hosts, suggesting that the tools necessary for macrophage infection may have evolved during co-evolution of *L. pneumophila* and amoeba. The establishment of the *Legionella*-containing vacuole (LCV) within the host cytoplasm requires the remodeling of the LCV surface and the hijacking of vesicles and organelles. Then *L. pneumophila* replicates in a safe intracellular niche in amoeba and macrophages. In this review we will summarize the existing knowledge of the *L. pneumophila* infection cycle in both hosts at the molecular level and compare the factors involved within amoeba and macrophages. This knowledge will be discussed in the light of recent findings from the *Acanthamoeba castellanii* genome analyses suggesting the existence of a primitive immune-like system in amoeba.

Contents

1 Introduction 2
2 Molecular Pathways Activated Upon Attachment and Phagocytosis of Legionella pneumophila 8
  2.1 Legionella pneumophila Attachment to Host Cells 8
  2.2 Legionella pneumophila Phagocytosis by the Host Cell 10
  2.3 Molecular Pathways Immediately Activated After Legionella pneumophila *Uptake* 11
3 Evasion from the Endocytic Pathway and Avoidance of Vacuole Acidification 13
4 Remodeling of the LCV 15

P. Escoll · M. Rolando · L. Gomez-Valero · C. Buchrieser (✉)
Institut Pasteur, Biologie des Bactéries Intracellulaires and CNRS UMR, 3525 Paris, France
e-mail: cbuch@pasteur.fr

Current Topics in Microbiology and Immunology (2013) 376: 1–34
DOI: 10.1007/82_2013_351

Published Online: 15 August 2013

4.1 Recruitment of Secretory Vesicles, Endoplasmic Reticulum and Polyubiquitinated Proteins 15
4.2 Recruitment of Mitochondria 17
4.3 Recruitment of Ribosomes and Rough Endoplasmic Reticulum 18
5 Modulation of Host Immune Processes 18
6 Bacterial Egress 23
7 Conclusions and Future Perspectives 23
References 24

1 Introduction

When *Legionella pneumophila*, a Gram-negative bacterium and the causative agent of Legionnaires' disease, is engulfed by free-living amoeba or lung alveolar macrophages, complex pathogen-host interactions lead to its intracellular replication within a sophisticated *Legionella*-containing vacuole (LCV) (Isberg et al. 2009). Free-living amoeba in aquatic environments are the natural reservoir and protecting niche for these bacteria (Lau and Ashbolt 2009), however, the development of aerosolized systems, such as air conditioning systems, cooling towers or showers, has allowed *L. pneumophila* to encounter also human alveolar macrophages when *Legionella*-containing water droplets are inhaled (Lau and Ashbolt 2009).

L. pneumophila is a pathogen that infects and replicates within a wide range of hosts (Table 1), including numerous amoeba species, ciliate protozoa and different mammalian cells like macrophages and epithelial cells (Fields 1996; Lau and Ashbolt 2009). *L. pneumophila* is responsible for one third of the cases of community-acquired pneumonia, with a mortality rate of 8–10 % (Falcó et al. 1991; Lau and Ashbolt 2009), but most of the patients are able to effectively eliminate the infection (Shin 2012). In fact, mainly immunocompromised and elderly person have a high risk to develop a severe disease with respiratory failure, but in healthy persons the innate immune response is thought to allow to control *L. pneumophila* infection (Falcó et al. 1991; Shin 2012). Moreover, there are no cases of transmission from humans to humans reported. Collectively these facts seems to indicate that *L. pneumophila* is not well adapted to infect humans (Shin 2012).

At the cellular level, many similarities in the *L. pneumophila*-infection cycle in amoeba and macrophages are observed: both hosts engulf *L. pneumophila* by phagocytosis and the LCV is rapidly formed within the host cytoplasm avoiding fusion with lysosomes. In both hosts the establishment of the LCV requires the remodeling of the LCV surface by recruiting endoplasmic reticulum (ER) vesicles, ribosomes and mitochondria (Isberg et al. 2009), thereby creating a "friendly and safe" niche for replication. These common strategies used by *L. pneumophila* to infect and replicate in amoeba and macrophages, combined with the lack of reported transmission between humans, have led to the hypothesis that the interaction of *L. pneumophila* with amoeba has provided the selective pressure to supply the bacteria with the factors allowing also successful replication within accidentally encountered mammalian macrophages (Al-Quadan et al. 2012;

Table 1 List of cells and organisms supporting intracellular replication of *Legionella pneumophila*

Class	Specie	Type	Reference
Unicellular organisms	*Acanthamoeba castellanii*	Amoeba	(Bozue and Johnson 1996)
	Acanthamoeba polyphaga	Amoeba	(Gao et al. 1997)
	Acanthamoeba hatchetti	Amoeba	(Breiman et al. 1990)
	Naegleri fowleri	Amoeba	(Newsome et al. 1985)
	Comandonia operculata	Amoeba	(Breiman et al. 1990)
	Hartmannella cantabrigiensis	Amoeba	(Breiman et al. 1990)
	Hartmannella vermiformis	Amoeba	(King et al. 1991)
	Paratetramitus jugosis	Amoeba	(Breiman et al. 1990)
	Vahlkampfia ustiana	Amoeba	(Breiman et al. 1990)
	Dictyostelium discoideum	Social amoeba	(Solomon et al. 2000)
	Tetrahymena pyriformis	Ciliated protozoa	(Fields et al. 1986)
	Tetrahymena thermophila	Ciliated protozoa	(Kikuhara et al. 1994)
Experimental infection	*Mus musculus*	Mouse A/J strain	(Fujio et al. 1992)
	Cavia porcellus	Guinea pig Hartley strain	(Fitzgeorge et al. 1983)
	Macaca mulatta	Rhesus monkey	(Fitzgeorge et al. 1983)
	Galleria mellonella	Honeycomb moth	(Harding CR et al. 2012)
	Caenorhabditis elegans	Roundworm	(Brassinga 2013)
Cell lines	*Mus musculus*	RAW 264.7 (Macrophage)	(Cirillo et al. 1994)
	Mus musculus	L929 (Fibroblast)	(Fernandez et al. 1989)
	Homo sapiens	THP-1 (Macrophage-like)	(Cirillo et al. 1994)
	Homo sapiens	U-937 (Macrophage-like)	(Fields 1996)
	Homo sapiens	HeLa (Epithelial)	(Garduno et al. 1998b)
	Homo sapiens	A549 (Epithelial)	(Gao et al. 1998)
Primary cells	*Mus musculus*	Bone Marrow-derived Macrophages (BMMs)	(Kagan and Roy 2002)
	Homo sapiens	Monocyte-Derived Macrophages (MDMs)	(Hilbi et al. 2001)

Cianciotto and Fields 1992; Franco et al. 2009; Newsome et al. 1985). Thus, environmental amoeba are thought to be the "training grounds" where *L. pneumophila* has acquired its capacity to replicate intracellularly also in mammalian macrophages as both, free-living amoeba and human macrophages, are eukaryotic cells that share conserved molecular pathways targeted by *L. pneumophila* (Al-Quadan et al. 2012; Molmeret et al. 2005; Richards et al. 2013).

One of the most important *L. pneumophila* virulence factor is the Dot/Icm type IVB secretion system (T4BSS) that translocates an exceptional high number of nearly 300 effectors in the host cell allowing *L. pneumophila* to modulate many signaling and metabolic pathways of the host to its benefit (Table 2)(Burstein et al. 2009; Campodonico et al. 2005; de Felipe et al. 2005, 2008; Heidtman et al. 2009; Lifshitz et al. 2013; Shohdy et al. 2005; Zhu et al. 2011). In both, macrophages and amoeba, the establishment of the LCV and the escape from lysosomal fusion requires the quick translocation of these effectors to the host cytoplasm (Isberg et al. 2009). The Dot/Icm secretion system, initially described in *L. pneumophila*, is quite unique. A similar system with a considerable degree of sequence similitarity is present only in the zoonotic pathogen *Coxiella burnetii* and in the arthropod pathogen *Rickettsiella grylli* (Nagai and Kubori 2011). It is a multiprotein apparatus encoded by a set of genes highly conserved among *Legionella* species, called *dot/icm* (*dot*: defective in organelle trafficking; *icm*: intracellular multiplication) (Berger and Isberg 1993; Brand et al. 1994). Most surprisingly, about 10 % of *L. pneumophila* genome code for these translocated effectors (Al-Quadan et al. 2012). Furthermore, many of them harbor eukaryotic domains that mediate the interaction with host proteins and organelles to modulate its functions (Cazalet et al. 2004; de Felipe et al. 2008; Hubber and Roy 2010; Nora et al. 2009; Rolando and Buchrieser 2012). Mutants, like a *dotA* mutant, that lack a functional T4BSS are unable to remodel the LCV and to escape from the phagosome-lysosome fusion (Berger et al. 1994; Tilney et al. 2001). Indeed, this secretion machinery is essential for replication in both, amoeba and macrophages (Segal and Shuman 1999). The crucial role of the Dot/Icm system and its translocated effectors for replication of *L. pneumophila* within amoeba and macrophages, suggests that its acquisition may have allowed *Legionella* to infect and replicate in eukaryotic cells. Furthermore, its particular large repertoire of effectors seems to be the basis for the broad host range of *L. pneumophila*.

The question whether *L. pneumophila* triggers and targets similar molecular mechanisms subverting common cellular processes in macrophages and amoeba, is intriguing and only partially solved. We will thus hereafter compare at the molecular level some of the reported cellular processes exploited by *L. pneumophila* during infection to draw a clear picture of the similarities and differences that are known between these two hosts. We will put special focus on the Dot/Icm T4BSS effectors as these are known to subvert cellular functions in both hosts and are essential for a successful intracellular replication of *L. pneumophila*. An in depth comparison may help to better understand the appearance of *L. pneumophila* in human communities and will lead to new insights on the virulence strategies of *L. pneumophila* when infecting human macrophages.

Table 2 Dot/Icm-dependent translocated effectors with known function in amoeba and/or macrophages

Effector	Paris	Phila	Amoebal model	Function in amoeba	IGD	Mammal cell model	Function in mammal cells	IGD
AnkX	*lpp0750*	*lpg0695*	ND	ND	ND	BMMs (Pan et al. 2008)	Interference with fusion of the LCV with late endosomes	–
Ceg19	*lpp1121*	*lpg1121*	ND	ND	ND	U937 (Heidtman et al. 2009)	Manipulation of host cell vesicle trafficking factors	ND
Ceg9	*lpp0316*	*lpg0246*	ND	ND	ND	U937 (Heidtman et al. 2009)	Manipulation of host cell vesicle trafficking factors	ND
LaiA/ SdeA	*lpp2096*	*lpg2157*	*A. castellanii* (Bardill et al. 2005)	ND	+	A549 (Chang et al. 2005), BMMs (Bardill et al. 2005)	Promotes bacterial adherance and uptake	–
LegA9/ ankY	–	*lpg0402*	ND	ND	ND	BMMs (Khweek et al. 2013)	Targets the LCV for autophagy uptake	–
LegAU13/ AnkB/ Ceg27	*lpp2082*	*lpg2144*	*A. castellanii* (Lomma et al. 2010), *A. polyphaga* (Price et al. 2010a)	Acquisition of poliubiquitinated proteins by the LCV	+	THP1 and 549 (Lomma et al. 2010), U937 and hMDMs (Price et al. 2010a)	Acquisition of poliubiquitinated proteins by the LCV	+
LegC2/ YlfB	*lpp1848*	*lpg1884*	*A. castellanii* (Campodonico et al. 2005), *D. discoideum* (de Felipe et al. 2008)	Alteration of normal vacuolar trafficking	–	BMMs (Campodonico et al. 2005), CHO- FcεRII (de Felipe et al. 2008)	Alteration of normal vacuolar trafficking	–
LegC3	*lpp1666*	*lpg1701*	*D. discoideum* (de Felipe et al. 2008)	Alteration of normal vacuolar trafficking	ND	CHO- FcεRII (de Felipe et al. 2008)	Alteration of normal vacuolar trafficking	ND
LegC7/ YlfA	*lpp2246*	*lpg2298*	*A. castellanii* (Campodonico et al. 2005), *D. discoideum* (de Felipe et al. 2008)	Alteration of normal vacuolar trafficking	–	BMMs (Campodonico et al. 2005), CHO- FcεRII (de Felipe et al. 2008)	Alteration of normal vacuolar trafficking	–
LegK1	*lpp1439*	*lpg1483*	*A. castellanii* (Hervet et al. 2011)	ND	–	U937 (Ge et al. 2009), BMMs and U937 (Losick et al. 2010)	Contribution to NF-κB activation by phosphorylating the IκB family of inhibitors	–
LegK2	*lpp2076*	*lpg2137*	*A. castellanii* and *D. discoideum* (Hervet et al. 2011)	ER recruitment into the LCV	+	J774 (Hervet et al. 2011)	ND	ND
LegS2	*lpp2128*	*lpg2176*	*A. castellanii* (Degtyar et al. 2009)	ND	–	U937 and HL-60 (Degtyar et al. 2009)	Target mitochondria	–

(continued)

Table 2 (continued)

Effector	Paris	Phila	Amoebal model	Function in amoeba	IGD	Mammal cell model	Function in mammal cells	IGD
LegU1	*lpp0233*	*lpg0171*	*A. castellanii* (Ensminger and Isberg 2010)	ND	–	BMMs, HEK-293T (Ensminger and Isberg 2010)	Interferes with ubiquitin signaling	–
LegU2/LubX	*lpp2887*	*lpg2830*	ND	ND	–	BMMs and CHO- FcγRII (Kubori et al. 2008)	Interferes with ubiquitin signaling	–
Lem3	*lpp0751*	*lpg0696*	*D. discoideum* (Tan et al. 2011)	ND	–	HEK-293T, BMMs (Tan et al. 2011)	Regulation of AnkX	–
LepA	*lpp2839*	*lpg2793*	*A. castellanii* and *D. discoideum* (Chen 2004)	Promotion of nonlytic release from protozoa	–	ND	ND	ND
LepB	*lpp2555*	*lpg2490*	*A. castellanii* and *D. discoideum* (Chen 2004)	Promotion of nonlytic release from protozoa	–	U937(Chen 2004), CHO-FcγRII (Ingmundson et al. 2007)	Regulation of the activation state of Rab1	–
Lgt1	*lpp1322*	*lpg1368*	*A. castellanii* (Belyi et al. 2008)	ND	ND	Embryonic bovine lung (Belyi et al. 2008)	Blockage of host protein synthesis	ND
Lgt2	–	*lpg2862*	ND	ND	–	Embryonic bovine lung (Belyi et al. 2008)	Blockage of host protein synthesis	ND
Lgt3	*lpp1444*	*lpg1488*	*A. castellanii* (Belyi et al. 2008)	ND	ND	Embryonic bovine lung (Belyi et al. 2008)	Blockage of host protein synthesis	ND
LidA	*lpp1002*	*lpg0940*	ND	ND	ND	BMMs (Conover et al. 2003), BMMs (Machner and Isberg 2006)	Recruitment of Rab1 on the LCV membrane	+
LnaB	*lpp2592*	*lpg2527*	ND	ND	ND	BMMs and U937 (Losick et al. 2010)	Contribution to NF-κB activation	–
LncP	*lpp2972*	*lpg2905*	*A. castellanii* (Dolezal et al. 2012)	ND	–	THP-1 (Dolezal et al. 2012)	ATP transporter in mitochondrial membrane	–
PieA	–	*lpg1963*	*A. castellanii* (Ninio et al. 2009)	ND	–	CHO- FcγRII, BMMs and U937 (Ninio et al. 2009)	ND	–
RavZ	–	*lpg1683*	ND	ND	ND	BMMs (Choy et al. 2012)	Inhibition of host autophagy	–

(continued)

Table 2 (continued)

Effector	Paris	Phila	Amoebal model	Function in amoeba	IGD	Mammal cell model	Function in mammal cells	IGD
SdcA	*lpp2578*	*lpg2510*	*D. discoideum* (Ragaz et al. 2008)	ER recruitment into the LCV	–	RAW264.7 (Ragaz et al. 2008)	ER recruitment into the LCV	–
SdhA	*lpp0443*	*lpg0376*	*D. discoideum* (Laguna et al. 2006)	ND	–	BMMs (Creasey and Isberg 2012)	Maintenance of LCV integrity	+
SetA	*lpp1961*	*lpg1978*	*D. discoideum* (Heidtman et al. 2009)	ND	–	BMMs (Heidtman et al. 2009), RAW 264.7 (Jank et al. 2012)	Manipulation of host cell vesicle trafficking factors	–
SidC	*lpp2579*	*lpg2511*	*D. discoideum* (Ragaz et al. 2008)	ER recruitment into the LCV	–	RAW264.7 (Ragaz et al. 2008)	ER recruitment into the LCV	–
SidD	–	*lpg2465*	ND	ND	ND	BMMs (Tan and Luo 2011), BMMs (Neunuebel et al. 2011)	Removal of Rab1 from the LCV	ND
SidF	*lpp2637*	*lpg2584*	ND	ND	ND	U937 (Banga et al. 2007), BMMs (Hsu et al. 2012)	Apoptosis resistance, anchoring of PI(4)P binding effectors to the LCV	–
SidJ	*lpp2094*	*lpg2155*	*D. discoideum* (Liu and Luo 2007)	ER recruitment into the LCV	+	BMMs (Liu and Luo 2007)	ER recruitment into the LCV	+
SidK	*lpp1030*	*lpg0968*	*D. discoideum* (Xu et al. 2010)	ND	–	BMMs and U937 (Xu et al. 2010)	Blockage of LCV acidification	–
SidM/ DrrA	–	*lpg2464*	*D. discoideum* (Brombacher et al. 2009)	Recruitment of Rab1 on the LCV membrane	ND	BMMs (Machner and Isberg 2006)	Recruitment of Rab1 on the LCV membrane	–
VipA	*lpp0457*	*lpg0390*	*A. castellanii* (Franco et al. 2012)	ND	–	THP-1 (Franco et al. 2012)	Actin nucleator	–
VipD	*lpp2888*	*lpg2831*	*D. discoideum* (VanRheenen et al. 2006)	ND	–	BMMs (VanRheenen et al. 2006), RAW264.7 (Ku et al. 2012)	Interferes with endosomal trafficking	–

Corresponding genes from *L. pneumophila* strain Paris and strain Phila are shown. *ND* not determined. *IGD* Intracellular Growth Defect

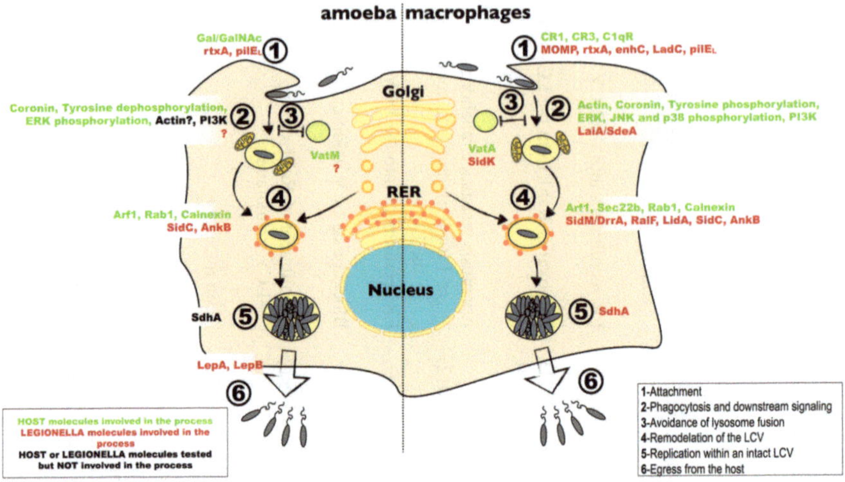

Fig. 1 Host factors (*green*) and *L. pneumophila* effectors (*red*) involved in uptake and intracellular replication of *L. pneumophila* in amoeba or macrophages. Host or *L. pneumophila* molecules tested but not involved in the analyzed processes are presented in *black*. Only molecules discussed in this chapter are shown. For a more complete list, please refer to Table 2

2 Molecular Pathways Activated Upon Attachment and Phagocytosis of Legionella pneumophila

2.1 Legionella pneumophila Attachment to Host Cells

The attachment of *L. pneumophila* to the host cell surface can be considered as the first step of the infection cycle. Bacterial factors identified till now that are involved in the entry of *L. pneumophila* into host cells are RtxA, $PilE_L$, EnhC, MOMP, LadC and Lcl. RtxA and $PilE_L$ are involved in the attachment of the extracellular bacteria to the amoeba surface (Fig. 1). RtxA appears to be involved in the attachment and entry of *L. pneumophila* into *Acanthamoeba castellanii* but the detailed mechanism of RtxA function remains still unknown (Cirillo et al. 2002). The $pilE_L$ gene, responsible for the expression of a long pili in *L. pneumophila*, plays a role in the attachment of *L. pneumophila* to *Acanthamoeba polyphaga* (Stone and Abu Kwaik 1998). Moreover, $PilE_L$ is similar to pili from other Gram-negative bacteria which are involved in the adhesion of the bacteria to the host cell surface (Strom and Lory 1993). The proteins RtxA and $PilE_L$ have also been shown to mediate *L. pneumophila* attachment to human cells, as *rtxA* or $pilE_L$ mutants displayed a diminished adherence and entry into human epithelial and monocytic cell lines (Cirillo et al. 2000; Stone and Abu Kwaik 1998). EnhC seems also important in the adherence of *L. pneumophila* to *A. castellanii*, and for attachment and entry into human epithelial and monocytic cell lines (Cirillo et al. 2000). However, the role of EnhC in adherence seems controversial as these results were not confirmed later (Liu et al. 2008). In addition, other *L. pneumophila* proteins implicated in adhesion to macrophages have been identified. These include the major outer membrane protein (MOMP), a *L. pneumophila* collagen-like protein named Lcl, and a putative *L. pneumophila*-specific adenylate cyclase named LadC that is present in the bacterial inner membrane (Bellinger-Kawahara and Horwitz 1990; Krinos et al. 1999; Newton et al. 2008; Vandersmissen et al. 2010). However, to our knowledge, the detailed functional role of these proteins in the attachment of *L. pneumophila* to amoeba has not been reported.

The host-specific factors that are exploited by *L. pneumophila* for its attachment to the eukaryotic cells seem to depend on the cell infected (macrophages or amoeba). Among the amoebal surface molecules involved in *L. pneumophila* attachment, a 170 kDa galactose/N-acetylgalactosamine-inhibitable lectin (Gal/GalNAc) has been identified as an amoebal receptor for *L. pneumophila* adherence to the protozoan species *Hartmannella vermiformis* (Venkataraman et al. 1997). Interestingly, while the use of Gal or GalNAc sugars completely blocked adherence of *L. pneumophila* to *H. vermiformis*, these sugars display just a weak effect, if any, in the adherence to *A. polyphaga*, suggesting different mechanisms for the attachment of *L. pneumophila* to different amoebal species (Harb et al. 1998).

In contrast, in macrophages, the attachment of *L. pneumophila* is mediated by the complement receptors CR1 (CD35) and CR3 (CD18/CD11b) that are exposed on the macrophage surface, as shown by the adherence of complement-coated bacteria. Furthermore, monoclonal antibodies against either CR1 or CR3 receptors inhibited *L. pneumophila* attachment (Payne and Horwitz 1987). However, the presence of specific antibodies generated during the adaptative immune response of the host, and a role of corresponding Fc-receptors for recognizing these antibodies, seem to be necessary for the complement-mediated adherence of *L. pneumophila* (Fields 1996; Husmann and Johnson 1992; Lau and Ashbolt 2009). Non-complement-mediated adherence of *L. pneumophila* was also reported (Elliott and Winn 1986; Falcó et al. 1991; Gibson et al. 1994; Lau and Ashbolt 2009; Rodgers and Gibson 1993). It might be responsible for the bacterial attachment to macrophages at early stages of infection, before a specific antibody-mediated response is mounted. However, the host cell receptor involved in the non-complement-mediated attachment of *L. pneumophila* to macrophages remains unknown. Finally, there seem to be a link between the *L. pneumophila*-proteins MOMP and Lcl and the macrophage-receptors involved in the adhesion of the bacteria to the host cell surface. MOMP, a porin (Gabay et al. 1985; Shin 2012), binds specifically to the complement component C3b and C3bi (that further bind to CR1 and CR3, respectively). It seems to be the main bacterial molecule implicated in the complement-mediated attachment to macrophages (Bellinger-Kawahara and Horwitz 1990; Falcó et al. 1991; Shin 2012). However, MOMP has also been shown to play a role in the binding of *L. pneumophila* to U937 macrophage-like cells in complement-independent assays (Krinos et al. 1999; Shin 2012). Furthermore, Lcl has been shown to be implicated in adherence to A549 human epithelial cells and U937 macrophage-like cells through the interaction with C1qR

(CD93), another complement receptor (Isberg et al. 2009; Vandersmissen et al. 2010).

In summary, attachment of *L. pneumophila* to the host cell seems to be a host-specific process. However, there is still an important lack of detailed knowledge about the molecular mechanisms allowing attachment and uptake of *L. pneumophila*, in particular for the interaction of *L. pneumophila* with its different amoebal hosts.

2.2 *Legionella pneumophila* Phagocytosis by the Host Cell

After attachment to the host cell surface, *L. pneumophila* is phagocytosed by macrophages trough a unique uptake process called "coiling phagocytosis" (Horwitz 1984). Coiling phagocytosis, in contrast to the common symmetrical and circumferential (zipper-like, "conventional") uptake of pathogens, consists in the asymmetrical engulfment of the bacteria by unilateral pseudopods encircling extracellular bacterial prior to entry (Rittig et al. 1998). However, the functional importance of this process in virulence and pathogen fate remains elusive as heat-killed and formalin-fixed *L. pneumophila* are also engulfed by coiling phagocytosis (Horwitz 1984). Moreover, uptake of other *Legionella* strains and species by conventional phagocytosis has also been reported independently of their specific virulence status (Al-Quadan et al. 2012; Elliott and Winn 1986; Molmeret et al. 2005; Rechnitzer and Blom 1989). It seems that differences regarding the bacterial strain, as well as the mammalian phagocyte and the experimental methods used may affect the uptake and may thus lead to the differences observed in how *Legionella* is phagocytosed.

Importantly, uptake of *L. pneumophila* by *A. castellanii* is mediated by the same coiling pseudopods as those seen in human macrophages (Bozue and Johnson 1996). In addition to occasional coiling phagocytosis, engulfment of *L. pneumophila* by *H. vermiformis* occurs mainly by zipper-like conventional phagocytosis (Abu Kwaik 1996), and uptake by *D. discoideum* seems to occur by macropinocytosis, a receptor-independent mechanism of endocytosis (Peracino et al. 2010). Importantly, even if uptake of *L. pneumophila* occurs mainly by host-driven phagocytosis, the Dot/Icm system enhances endocytic events in macrophage-like cells as well as in *A. castellanii* (Hilbi et al. 2001; Khelef et al. 2001). Moreover, Dot/Icm effectors translocated very early in the infection, like LaiA/SdeA, have been suggested to play a role in adherence and uptake by macrophages (Bardill et al. 2005; Chang et al. 2005).

At the molecular level, the formation of the nascent phagosome after *L. pneumophila* phagocytosis by mammalian cells is strongly actin-dependent, as treatment of cells with cytochalasin-D (an actin polymerization inhibitor (Flanagan and Lin 1980)) impairs *L. pneumophila* uptake by macrophages (Charpentier et al. 2009; Elliott and Winn 1986; Hayashi et al. 2008; King et al. 1991; Welsh et al. 2004) and by lung epithelial cells (Prashar et al. 2012). In contrast several studies report that *L. pneumophila* phagocytosis by the amoeba *H. vermiformis*, *A. castellanii* and *A. polyphaga* is an actin-independent process as it is unaffected by cytochalasin-D treatment (Harb et al. 1998; King et al. 1991; Köhler et al. 2000; Moffat and Tompkins 1992). However, Lu and coworkers, suggested that as these groups did not demonstrate that cytochalasin-D causes depolymerization of actin filaments in the amoebal hosts, they might have drawn a wrong conclusion, and thus suggests that *L. pneumophila* uptake in amoeba might well be actin-dependent (Lu and Clarke 2005). Furthermore, *L. pneumophila* uptake by the amoeba *D. discoideum* was reported to be an actin-dependent process, that, surprisingly, is insensitive to cytochalasin-D but sensitive to cytochalasin-A (Lu and Clarke 2005; Peracino et al. 2010; Weber et al. 2006). Recently, a *L. pneumophila* Dot/Icm-dependent translocated effector called VipA has been described as being an actin nucleator that directly polymerizes microfilaments without the requirement of additional proteins. However, although VipA alters host cell organelle trafficking it is not essential for entry or replication in *A. castellanii* or THP-1 macrophage-like cells (Franco et al. 2012).

In addition to actin, other proteins called coronins are conserved from amoeba to mammals (Yan et al. 2005), and transient recruitment of coronin (in fact an actin-binding protein) to the phagocytic cup has been shown in *D. discoideum* and U937 macrophage-like cells after *L. pneumophila* infection at very early times, disappearing then quickly from the nascent phagosome in both host cells (Hayashi et al. 2008; Lu and Clarke 2005).

In summary, phagocytosis of *L. pneumophila* is a process that, despite some differences at the cellular level, seems to be conserved at the molecular level between macrophages and amoeba, as coronin and probably actin are implicated in *L. pneumophila* phagocytosis by macrophages and amoeba (Fig. 1).

2.3 Molecular Pathways Immediately Activated After Legionella pneumophila Uptake

After *L. pneumophila* uptake several host cell pathways are activated in response to bacterial invasion. For example, the addition of a phosphate group to a tyrosine residue on a protein is a key step in signal transduction. Indeed, tyrosine phosphorylation of host proteins is modulated in amoeba and macrophages after *L. pneumophila* uptake differently. *L. pneumophila* attachment and invasion to *H. vermiformis* were associated with a time-dependent tyrosine dephosphorylation of multiple host cell proteins (including the 170 kDa Gal/GalNAc lectin involved in attachment), while invasion of macrophages was reported to lead to tyrosine phosphorylation of multiple host proteins trough a tyrosine kinase-mediated pathway (Coxon et al. 1998; Venkataraman et al. 1997). Thus the activation of this pathway does not seem to be conserved in these two hosts.

Mitogen-activated protein kinase (MAPK) cascades have also been analyzed after *L. pneumophila* uptake by human macrophages, showing that they are activated upon *L. pneumophila* infection (Welsh et al. 2004). MAPKs are serine/threonine protein kinases that are key signaling proteins in eukaryotic cells. They are conserved from amoeba to mammals. MAPK activation is an early and quick signaling event in response to a wide variety of cellular stimuli, such as mitogens, osmotic stress, heat shock, and microbial detection. They regulate a large number of cellular activities, including gene expression, cellular differentiation and proliferation, and cell survival and death (Shin 2012). Mammalian MAPKs comprise three types of kinases called p38, c-Jun N-terminal kinase (JNK) and extracellular signal-regulated kinase (ERK) that are all phosphorylated by upstream kinases for activation. Once activated they phosphorylate downstream proteins of the kinase cascade. The only members of the MAPK pathway present in *A. castellani* and *D. discoideum* are ERK proteins (Clarke et al. 2013; Li et al. 2009). Similar to what is seen in human macrophages, ERK1 is phosphorylated rapidly after contact with *L. pneumophila* during *D. discoideum* infection (Li et al. 2009). Human and murine macrophages show a significantly increased level of phosphorylation and MAPK activity after wild-type infection with *L. pneumophila* when compared to the avirulent *dotA* mutant or to the *E. coli* controls (Shin et al. 2008). Moreover, chemical inhibitors demonstrated the need of both the JNK and p38 signals for intracellular growth of virulent *L. pneumophila* in human macrophages (Welsh et al. 2004). Whether *L. pneumophila* can effectively modulate host MAPKs signaling for its own benefit through translocated Dot/Icm-dependent effectors is not known. However, it has been shown recently that *L. pneumophila* effectors that inhibit host protein synthesis may account for MAPK activation in murine macrophages. This might suggest that the regulation of host MAPKs after *L. pneumophila* infection is an indirect consequence of subversion of other host cellular processes (Fontana et al. 2012).

L. pneumophila also exploits phosphoinositide (PI) lipids during the establishment of the replicative vacuole. Phosphatidylinositol 3-kinases (PI(3)Ks), that catalyzes the phosphorylation of the 3' position on the inositol ring, has been shown to regulate the early steps of phagosome maturation downstream of diverse phagocytic receptors. This involves the activation and recruitment of multiple small Rab and Rac guanosine-triphosphatases (GTPases) that modulate endosomal trafficking and remodel the actin cytoskeleton, respectively (Thi and Reiner 2012). The activity of PI(3)K has been linked to actin cytoskeleton rearrangements by a PKCζ-dependent pathway (Chou et al. 1998; Nakanishi et al. 1993; Uberall et al. 1999).

A role of PI3 K in the phagocytosis of avirulent *L. pneumophila* mutants by macrophages, but not wild-type *L. pneumophila* was reported (Harada et al. 2012; Khelef et al. 2001), whereas others saw a role of PI(3)K also after wild-type *L. pneumophila* uptake (Charpentier et al. 2009; Peracino et al. 2010; Tachado et al. 2008). To explain these conflicting results, the authors suggested that these might be due to differences between the infected hosts cells used (human U937 and murine J774 macrophage-like cell lines, respectively) (Harada et al. 2012;

Tachado et al. 2008). However, recent findings showed the recruitment of class I PI(3)K to the nascent phagosome after uptake of wild-type or the avirulent *L. pneumophila dotA* mutant also by murine macrophages (Harada et al. 2012). Class I PI(3)K mediates the production of PI(3,4,5)P_3 from PI(4,5)P_2. Although the recruitment of class I PI(3)K occurs after uptake of both, wild-type or avirulent strains, PI(3)K activation and production of PI(3,4,5)P_3 around the bacteria occurs only after the phagocytosis of the avirulent *dotA* mutant (Harada et al. 2012), suggesting a role of Dot/Icm translocated effectors for the lack of PI(3,4,5)P_3 production by class I PI(3)K after uptake of wild-type *L. pneumophila*. A recent study reported that the *L. pneumophila* translocated effector SidF functions as a phosphoinositide 3-phosphatase that specifically hydrolyzes PI(3,4,5)P_3 to generate PI(4)P (Hsu et al. 2012), suggesting that it is implicated in this process.

The PI(3)K pathway has also been investigated in the amoeba *D. discoideum* after *L. pneumophila* infection. A recent study reported that *L. pneumophila* uptake by *Dictyostelium* is partially dependent on PI(3)K (Peracino et al. 2010). A second study showed that inhibiting PI(3)K has only a modest if any effect on the uptake of wild-type *L. pneumophila* by *Dictyostelium* (Weber et al. 2006). These results suggest that this pathway is not conserved between macrophages and amoeba. Furthermore, upon PI(3)K inactivation *L. pneumophila* growth was enhanced in *Dictyostelium*, and PI(3)K activity was restricted to early infection. It is involved mainly in intracellular replication and trafficking of wild-type *L. pneumophila* and not in phagocytosis (Peracino et al. 2010; Weber et al. 2006) (See also Sect. 4.1.)

3 Evasion from the Endocytic Pathway and Avoidance of Vacuole Acidification

A common feature of the infection of macrophages and amoeba by *L. pneumophila* is its ability to evade the endocytic pathway and immediately following uptake to prevent phagosome-lysosome fusion. Following phagocytosis, phagosomes containing both inert particles and non-pathogenic bacteria follow an intracellular pathway that mirrors the stages of the endosomal-lysosomal pathway known as endocytic pathway. The internalized material is delivered to early endosomes that follow endocytic maturation, with fusion/fission processes gradually transforming the phagosome into a phagolysosome. Whereas the last sorting step takes place in the late endosomes, lysosomes are generally believed to represent the end stage (Gruenberg 2001; Vieira et al. 2002). The microenvironment of phagolysosomes is acidic and filled with hydrolases that induce bacterial killing (Kornfeld and Mellman 1989). A key feature of phagosome maturation is luminal acidification mediated primarily by ATP-dependent proton transporters known as the vacuolar H^+-ATPases or v-ATPases (Forgac 2007).

Interestingly, *L. pneumophila* maintains a neutral luminal pH during infection, particularly within the first about six hours after uptake (Horwitz 1983a; Sturgill-

Koszycki and Swanson 2000). Whereas nascent phagosomes containing live *L. pneumophila* keep a neutral pH, phagosomes containing heat killed *L. pneumophila* or *E. coli* show an acidic pH. Blocking acidification and maturation of the LCV by treating infected macrophages with the proton ATPase inhibitor bafilomycin A1 arrests bacterial replication (Sturgill-Koszycki and Swanson 2000). During *L. pneumophila* replication, a significant proportion of the LCVs acquire lysosomal characteristics. At about 18 h after infection, the vacuoles have an acidic pH and have acquired endosomal markers like lysosomal-associated membrane protein 1 (LAMP-1). Furthermore, replicating bacteria obtained from macrophages, but not broth, are acid resistant (Sturgill-Koszycki and Swanson 2000). This led to the hypothesis that, although maintaining a proper phagosomal pH is important at early stages of infection, final fusion of the replicating LCV with the lysosomal compartment promotes, rather than inhibits, *L. pneumophila* growth in macrophages (Swanson and Hammer 2000).

The importance of avoidance of vacuole acidification at early stages of infection is generally recognized, but almost nothing is known about the molecular mechanisms involved (Xu et al. 2010). Recently, a *L. pneumophila* translocated effector, SidK that targets a proton transporter has been identified. SidK interacts with VatA, a key component of the proton pump, thus inhibiting ATP hydrolysis, proton translocation and subsequently vacuole acidification. When delivered into macrophages, SidK inhibits vacuole acidification and impairs the ability of the cells to digest non-pathogenic *E. coli* (Xu et al. 2010).

In the amoeba *D. discoideum*, VatM (a transmembrane subunit of the v-ATPase) is delivered to phagosomes containing the *L. pneumophila icmT* mutant that is defective in the T4SS causing vacuole acidification, but phagosomes containing wild-type *L. pneumophila* seem to avoid VatM recruitment and thus subsequent acidification of the LCV (Chen 2004). However, later proteome studies found VatM on isolated LCVs from PI(3)K-infected cells harboring wild-type *L. pneumophila* (Shevchuk et al. 2009; Urwyler et al. 2009b). Although these differing results have been reported, it seems that absence of v-ATPase activity at early times of infection is a key feature for successful intracellular replication of *L. pneumophila* within macrophages and amoeba. Thus, different host-specific mechanisms may be acting: avoidance of v-ATPase recruitment to the LCV in amoeba or regulation of v-ATPase activity in macrophages.

In addition, also isolated vesicles shed from *L. pneumophila*, known as outer membrane vesicles (OMVs), have been shown to cause a specific blockage of phagosome-lysosome fusion in murine macrophages (Fernandez-Moreira et al. 2006). Moreover, isolation of non-vesicular lipopolysaccharide (LPS) from OMVs suggested that LPS contributes to arrest the phagosome-lysosome fusion in *A. castellanii* and human macrophages (Seeger et al. 2010). In agreement with the results described above, this effect is more pronounced at 1 h than 5 h post-infection (Seeger et al. 2010), highlighting the importance of the avoidance of acidification at early time points of infection for both hosts.

4 Remodeling of the LCV

Immediately after phagocytosis of *L. pneumophila* by a host cell the bacterium is internalized into a compartment that escapes the endocytic pathway (Derré and Isberg 2004; Horwitz 1983b; Horwitz and Maxfield 1984). The LCV membrane is surrounded by mitochondria and small vesicles derived from the endoplasmic reticulum (ER) (Horwitz 1983b; Tilney et al. 2001). Redecoration of the LCV with ER-derived vesicles, host proteins and mitochondria precedes the recruitment of ribosomes to the LCV, establishing then a remodeled LCV that permits intracellular replication of *L. pneumophila* (Tilney et al. 2001). At the molecular level, these events are highly similar in amoeba and macrophages, and similar proteins are involved.

4.1 Recruitment of Secretory Vesicles, Endoplasmic Reticulum and Polyubiquitinated Proteins

After uptake, the LCV membrane lacks plasma membrane or endocytic markers and hijacks ER-derived vesicles (Horwitz 1983b; Tilney et al. 2001). This process has been well described in macrophages, showing that the LCV intercepts early secretory vesicles that transit between the ER and the Golgi, incorporate their luminal content into the LCV and create in this way an ER-like organelle that supports replication of *L. pneumophila* (Derré and Isberg 2004; Horwitz 1983b; Kagan and Roy 2002; Tilney et al. 2001). In macrophages, the acquisition of proteins involved in vesicle budding into the LCV, like ADP-ribosylation factor 1 (Arf1), Sec22b or Rab1, precedes the incorporation of resident ER proteins like calnexin or glucose-6-phosphatase, suggesting that in macrophages the LCV fuses with secretory vesicles exiting from the ER (Arasaki and Roy 2010; Derré and Isberg 2004; Kagan and Roy 2002; Kagan et al. 2004; Robinson and Roy 2006). In macrophages, recruitment of Rab1 and Arf1 to the LCV is mediated by the Dot/Icm-dependent translocated effectors SidM/DrrA and RalF, respectively (Machner and Isberg 2006; Murata et al. 2006; Nagai et al. 2002), but other Dot/Icm translocated substrates also participate in the recruitment of ER vesicles. LidA attaches to the cytoplasmic face of the LCV and synergizes with SidM/DrrA in the recruitment of Rab1 (Conover et al. 2003; Machner and Isberg 2006). The effector SidC localizes also on the cytoplasmic face of the LCV (Luo and Isberg 2004). It possesses a PI(4)P-binding domain, allowing its attachment to the LCV membrane, and a N-terminal domain involved in the recruitment of ER vesicles to the LCV (Ragaz et al. 2008). The discovery that *L. pneumophila* uses the host PI metabolism for the attachment of effectors to the LCV membrane and the subsequent recruitment of ER-derived vesicles allowed to explain the mechanism used by other effectors to be attached. SidM/DrrA and LidA are also PI-binding proteins that attach in the same way to the LCV (Brombacher et al. 2009; Schoebel et al. 2010).

L. pneumophila also hijacks secretory vesicles and ER when infecting amoebal hosts like *H. vermiformis* and *D. discoideum* (Abu Kwaik 1996; Fajardo et al. 2004; Lu and Clarke 2005; Urwyler et al. 2009b). ER-derived calnexin is attached to the LCV of *D. discoideum* (Fajardo et al. 2004; Lu and Clarke 2005). Thus magnetic-purified calnexin^{+} LCVs were isolated from infected amoeba cells to analyze the proteome of the LCV content, revealing the existence of 566 host proteins including among others Arf1 and Rab1, as well as other Rab GTPases (Urwyler et al. 2009b). A second proteome analysis of the LCV, using cell fractionation, two-dimensional gel electrophoresis and MALDI-TOF MS combined with genomic data identified 157 phagosome host proteins (Shevchuk et al. 2009). These results suggest a high degree of similarity in the infection between macrophages and amoeba at this step. However, although SidM/DrrA has been shown to be attached to the LCV in *D. discoideum*-infected cells (Brombacher et al. 2009), to our knowledge whether SidM/DrrA or RalF recruit host GTPases also in amoeba has not been tested. Importantly, the PI(4)P-binding effector SidC was found in the LCV of *D. discoideum*-infected cells, similarly to what was found in murine RAW 264.7 macrophage-like cells, showing that anchoring of this effector to a phosphatidylinositol-derived molecule in the LCV is a common mechanism within both hosts, amoeba and murine macrophages (Ragaz et al. 2008; Weber et al. 2006).

In addition to the recruitment of secretory vesicles and ER, *L. pneumophila* also hijacks polyubiquitinated (polyUb) proteins that accumulate on the vacuole (Dorer et al. 2006; Lomma et al. 2010; Price et al. 2011). The translocated effector AnkB is partly responsible for the recruitment of ubiquitinated proteins. It encodes an ankyrin and a F-box domain, which are necessary for the attachment of polyUb host proteins to the LCV, and a CaaX motif that allows anchoring of the effector to the LCV membrane through the action of a host farnesyltransferase (Ivanov et al. 2010; Lomma et al. 2010; Price et al. 2010b). Remarkably, AnkB/Lpp2082 of *L. pneumophila* strain Paris lacks the CaaX motif. Thus the effector localizes to the host cytoplasmic membrane (Lomma et al. 2010). Besides anchoring of polyUb proteins on the LCV, it was suggested that the proteasomal degradation of ubiquitinated host proteins at the LCV leads to the increase of cellular levels of amino acids, providing therewith a source of carbon and energy for the replicating bacteria (Price et al. 2011). Importantly, the function of AnkB recruiting polyUb proteins into the LCV has been shown in amoeba and macrophages and the *ankB* mutant is defective in intracellular replication within both hosts (Al-Quadan and Kwaik 2011; Lomma et al. 2010; Price et al. 2011). Thus exploitation of the host ubiquitination/proteasome machinery is a conserved strategy used by *L. pneumophila* during infection of macrophages and amoeba.

4.2 Recruitment of Mitochondria

One hour after infection the LCV is surrounded by smooth vesicles and the majority of vacuoles is also surrounded by at least one mitochondrion close to the vacuolar membrane (Horwitz 1983b). How *L. pneumophila* recruits mitochondria and what the advantage is for the bacterium remains unknown. A role of translocated Dot/Icm effectors has been suggested for mitochondria recruitment to the LCV, as *dot/icm* mutants deficient of a functional T4BSS do not recruit mitochondria (Berger et al. 1994; Chong et al. 2009; Tilney et al. 2001). Although several Dot/Icm translocated effectors have been shown to be targeted to mitochondria, like LncP (Dolezal et al. 2012) or LegS2/Spl (Degtyar et al. 2009), what may suggest that interaction with host mitochondria provides a certain benefit to the pathogen, no Dot/Icm effector has been identified yet that has a role in recruitment of mitochondria to the LCV. Interestingly, when testing mitochondria recruitment in *Drosophila melanogaster* cells, the density of mitochondria near vacuoles formed by infection with wild type *L. pneumophila* was not different from that found in *dotA* mutant-infected cells during the first 4 h after infection (Sun et al. 2013). The *L. pneumophila* chaperonin HtpB is upregulated upon contact with host cells and accumulates in the lumen of the LCV during the course of infection (Fernandez et al. 1996; Garduno et al. 1998a). It was reported that HtpB alters mitochondrial trafficking of U937-derived macrophages (Chong et al. 2009), but HtpB is not a Dot/Icm translocated effector and remains in the LCV lumen. Thus this chaperonin may not be responsible for the recruitment defect of *dot/icm* mutants, but the existence of translocated effectors that, directly or indirectly, participates in mitochondria recruitment to the LCV is expected.

Mitochondria recruitment around the LCV has also been reported after *L. pneumophila* infection of the amoeba *D. discoideum, N. fowleri* and *H. vermiformis* (Abu Kwaik 1996; Francione et al. 2009; Newsome et al. 1985). Importantly, in *D. discoideum* genetically diseased for mitochondria by disruption of the mitochondrial large ribosomal RNA gene *rnl* or by antisense inhibition of expression of an essential nuclear-encoded mitochondrial protein, enhanced intracellular replication of *L. pneumophila* was observed. It was independent from bacterial uptake but related to the fact that AMP-activated kinase (AMPK) is chronically activated (Francione et al. 2009; Francione and Fisher 2011). Since *L. pneumophila* infection upregulates the transcription of AMPK in *D. discoideum* (Farbrother et al. 2006), and increased levels of AMPK activation correlate with increased proliferation of *L. pneumophila* (Francione et al. 2009), it can be hypothesized that *L. pneumophila* modulates mitochondrial function in *D. discoideum*-infected cells to its benefit. Supporting this hypothesis, *L. pneumophila* infection of *D. discoideum* results in dramatic decrease of mitochondrial RNAs and in the specific cleavage of mitochondrial rRNA, suggesting that *L. pneumophila* specifically disrupts mitochondrial protein synthesis in *D. discoideum* during infection (Zhang and Kuspa 2009).

4.3 Recruitment of Ribosomes and Rough Endoplasmic Reticulum

Disappearance of the LCV-attached ER-derived vesicles and mitochondria has been associated with the increased appearance of ribosomes on the LCV. After the formation of this rough ER-like compartment, bacterial replication was observed in macrophages (Horwitz 1983b; Tilney et al. 2001). Replication of *L. pneumophila* within a ribosome-studded LCV has been reported in the amoeba *H. vermiformis*, *N. fowleri* and in the ciliated protozoa *Tetrahymena pyriformis* (Abu Kwaik 1996; Fields et al. 1986; Newsome et al. 1985). Since 67 ribosomal components were found to be associated with purified LCVs from *Dictyostelium*-infected cells, it seems plausible that the recruitment of ribosomes to the LCV occurs also in this amoebal species (Urwyler et al. 2009b).

Although recruitment of ribosomes to the LCV is a conserved infection feature in macrophages and amoeba, and *dot/icm* mutants lacking a functional T4BSS translocation apparatus are unable to attract ribosomes around the vacuole (Berger et al. 1994; Tilney et al. 2001), no Dot/Icm-dependent translocated effectors has yet been described with a clear role in ribosome recruitment to the LCV. The molecular mechanism of ribosome recruitment is unknown, however one could speculate that the LCV becomes sufficiently ER-like during infection to recruit ribosome spontaneously.

5 Modulation of Host Immune Processes

Although it is generally thought that amoeba lack an immune system, some molecular pathways are starting to be discovered showing that amoeba may have a primitive immune-like system that might help to discriminate between pathogenic and non-pathogenic bacterial preys (Chen et al. 2007). In fact, MAPK activation and subsequent regulation by dual-specific phosphatases (DUSPs), a well-known pathway involved in the innate immune response elicited by macrophages and triggered by *L. pneumophila* (Losick and Isberg 2006), participate also in the amoebal response to *L. pneumophila*. This pathway appears to be directly involved in the modulation of certain immune-like genes like TirA, a protein with similarities to mammalian Toll-like receptors (TLRs) that allows *Dictyostelium* to survive killing by *L. pneumophila* (Chen et al. 2007; Li et al. 2009). TLRs are eukaryotic pattern-recognition receptors (PRRs) activated by microbial-associated molecules that trigger conserved signaling cascades leading to the modulation of inflammatory host responses dependent on microbial pathogenicity (Shin 2012; Takeda et al. 2003). Discovery of the TLR-like protein TirA and its function in protecting *Dictyostelium* from *L. pneumophila* infection led to search for primitive immune processes that may allow amoeba to deal with pathogenic bacteria (Cosson and Soldati 2008). Certain structural components of *L. pneumophila* are TLR ligands: in murine macrophages LPS and flagellin activate TLR2 and TLR5, respectively (Akamine et al. 2005; Girard et al. 2003; Hawn et al. 2006, 2007). In addition, *L. pneumophila* can be sensed also by cytoplasmic PRRs known as Nod-like receptors (NLRs), that activate subsequently the inflammasome platform and the secretion of the proinflammatory cytokine Interleukin-1β (IL-1β) in murine macrophages (Lightfield et al. 2008; Pereira et al. 2011; Zamboni et al. 2006; Zhao et al. 2011). A role of NLRs sensing of *L. pneumophila* has also been suggested in infection of human macrophages (Vinzing et al. 2008).

Although no genes coding homologues of TirA have been found in amoebal hosts, like the recently sequenced amoeba *A. castellanii* (Clarke et al. 2013), other genes coding proteins with homology to known PRRs are present (Cosson and Soldati 2008). Here we have undertaken an in silico search in all unicellular *L. pneumophila* hosts whose genome has been completely sequenced (*D. discoideum*, *A. castellanii*, *N. gruberi* and the ciliated protozoa *T. thermophila*) to analyze which conserved domains of proteins involved in mammalian immune processes are present in these organisms. Indeed, as shown in Table 3 some of the conserved domains can also be found in unicellular *Legionella*-hosts.

The CD36 domain, a motif present in mammalian Scavenger receptor class B member 1 (SCARB1) and protein acts as a PRR for bacterial components in humans and mice is present in unicellular *L. pneumophila*-hosts (Baranova et al. 2008; Hoebe et al. 2005). In addition, a large number of proteins containing the LBP_BPI_CETP domain can be found in *A. castellanii* and other protists (Table 3). Since the LBP_BPI_CETP motif is involved in the recognition of LPS from the outer membrane of Gram-negative bacteria in mammalian PRRs (Krasity et al. 2011), this large number of proteins in unicellular *Legionella*-hosts might highlight that these organisms are well equipped to recognize bacterial pathogens. On the other hand, *L. pneumophila* seems to be able to modulate its own production of bacterial products that stimulate PRRs, to alter host cell recognition by mammalian cells (Liu et al. 2008, 2012). Moreover, the Dot/Icm translocated effector SdhA actively promotes the stability of the LCV to hide the replicating bacteria from cytosolic host defenses, as disruption of the vacuole membrane during intracellular replication in macrophages is fatal to both the host cell and the bacterium (Creasey and Isberg 2012). Therefore it seems that *L. pneumophila* has developed diverse strategies to counteract the recognition processes not only in macrophages, but also in their natural amoebal hosts.

In order to control intracellular proliferation of pathogenic bacteria, macrophages modulate intracellular iron homoeostasis, thus depriving the phagosome of the iron flux needed for bacterial replication. Natural resistance-associated macrophage protein 1 (Nramp1) is a proton/divalent cation antiporter that has a well-established, unique role in innate resistance to intraphagosomal pathogens in human and mice. Iron is a key requirement for growth of *L. pneumophila*. Thus

Table 3 Proteins of unicellular *Legionella* hosts, containing domains commonly related to the immune system

Protein domain	Dictyostelium discoideum	Acanthamoeba castellanii	Tetrahymena thermophila	Naegleria gruberi	Examples of proteins in higher eukaryotes containing the corresponding described domain	Role in immunity of example proteins
WRKY	DDB_G0275267	ACA1_216030 ACA1_291040	–	-	WRKY transcription factor protein 1	Pathogen defense (plants) (Eulgem et al. 1999)
TIR	DDB_G0289237	-	-	-	TLRs	PRR (Takeda et al. 2003)
CD36	DDB_G0267406 DDB_G0267440 DDB_G0287035	ACA1_126030	TTHERM_00371120 TTHERM_00238900 TTHERM_00046450	-	Scavenger receptor class B member 1 (SCARB1, CD36L1)	PRR (Baranova et al. 2008)
Caspase	DDB_G0277689 DDB_G0293196	ACA1_087710	-	-	Caspase-1	IL-1β maturation (Thornberry et al. 1992)
MATH-TRAF	DDB_G0272340 DDB_G0274899 DDB_G0280369 DDB_G0285149	-	TTHERM_00476630	NAEGRDRAFT _72663	TNF receptor-associated factor-6 (TRAF6)	TLR/IL-1 signaling (Takeda et al. 2003)
Death Domain	-	-	-	-	MyD88	TLR/IL-1 signaling (Takeda et al. 2003)
IPT_NFKB	-	-	-	-	p65 subunit of transcription factor NF-κB	Transcription of immune response genes (Gilmore and Wolenski 2012)
Uban	-	-	-	-	NF-κB modulator NEMO	NF-κB signaling (Gilmore and Wolenski 2012)
RHD-n_c-Rel	-	-	-	-	NF-κB c-Rel	NF-κB signaling (Gilmore and Wolenski 2012)
LBP_BPI_CETP	DDB_G0271242 DDB_G0288097	ACA1_143930 ACA1_252520 ACA1_208310 ACA1_382630 ACA1_281820 ACA1_234080 ACA1_368250 ACA1_382580	TTHERM_00823420 TTHERM_01287960 TTHERM_00137710 TTHERM_00190810 TTHERM_00433940 TTHERM_00433910	NAEGRDRAFT_ 78340 NAEGRDRAFT_47514 NAEGRDRAFT_82253 NAEGRDRAFT_76153	Lipopolysaccharide-binding protein (LBP)	PRR (Jack et al. 1997)

(continued)

Table 3 (continued)

Protein domain	Dictyostelium discoideum	Acanthamoeba castellanii	Tetrahymena thermophila	Naegleria gruberi	Examples of proteins in higher eukaryotes containing the corresponding described domain	Role in immunity of example proteins
PG_binding_1	DDB_G0280311 DDB_G0286137	ACA1_152930	TTHERM_00220760	-	Matrix metalloproteinase-3 (Stromelysin 1, MMP3)	Immunity to intestinal bacteria (Li et al. 2004) Degradation of IL-1β (Ito et al. 1996)
Gal_Lectin	-	ACA1_126170 ACA1_184380 ACA1_030220 ACA1_201210 ACA1_318670 ACA1_029840 ACA1_047640 ACA1_047640 ACA1_033480 ACA1_059570	-	NAEGRDRAFT_ 72022	D-galactoside/L- rhamnose binding SUEL lectin protein	PRR (invertebrates) (Watanabe et al. 2009)
GILT	-	ACA1_184120 ACA1_116130	-	-	Gamma-interferon-inducible lysosomal thiol reductase (GILT)	MHC Class II-restricted antigen presentation (Hastings and Cresswell 2011)
NRAMP	DDB_G0275815 DDB_G0276973	ACA1_225890	TTHERM_00680650 TTHERM_00691720 TTHERM_00697030	-	Natural resistance-associated macrophage protein 1 (Nramp1)	Resistance against intraphagosomal pathogens (Alter-Koltunoff et al. 2008)

Only species whose genome has been completely sequenced have been analyzed: *Dictyostelium discoideum*, *Acanthamoeba castellanii*, *Naegleria gruberi* and the ciliated protozoa *Tetrahymena thermophila* patients with iron overload as well as smokers whose lungs also contain elevated iron levels are at increased risk for Legionnaires' disease (Cianciotto 2007). Expression of Nramp1 is reported to be restricted to monocyte/macrophage cells (Alter-Koltunoff et al. 2008). However, a homologue of Nramp1 is present in *D. discoideum* and it was shown to protect cells from *L. pneumophila* infection by regulating the iron flux into the phagosome (Peracino et al. 2006, 2010). In addition, other homologues of Nramp1 have been identified recently in *A. castellanii* (Clarke et al. 2013) and in *T. thermophila* (Table 3). Our results suggest that a high degree of conservation of this innate immune process between macrophages and different amoeba species exists. Since *L. pneumophila* uses multiple pathways for iron acquisition (Cianciotto 2007), perhaps the variety in these mechanisms counteracts at some degree the conserved antimicrobial mechanism of Nramp1 in a wide range of susceptible hosts.

Finally, other immune pathways that have a well-established importance in *L. pneumophila*-infection in macrophages are not conserved in amoeba. These are for example caspase-mediated apoptosis and the NF-κB pathway. Caspases are a family of cysteine proteases that play essential roles in apoptosis and inflammation that are activated after infection of mammalian cells by *L. pneumophila*. Whereas caspase-1 is involved in the maturation of IL-1β to its immunologically active form, caspase-3 and caspase-7 play a role in the activation of apoptosis in macrophages (Amer 2010). It has been suggested that a delicate balance of apoptotic signals facilitates intracellular replication of *L. pneumophila* in murine macrophages. The requirement of the Dot/Icm system to modulate apoptotic processes in macrophages seems to suggest that *L. pneumophila* is adaptated to exploit caspase-activation and to delay apoptosis until replication within the host cell has been started (Amer 2010). However, where this adaptation took place remains unknown, because although homologues to the mammalian caspase domain can be found in *Dictyostelium* and *Acanthamoeba* proteins (Table 3), their function in amoeba seems not related to programmed cell death (Roisin-Bouffay et al. 2004).

NF-κB is a transcription factor that controls gene expression of multiple cellular processes including inflammation, innate immunity and apoptosis. Transcriptional profiling demonstrates that *L. pneumophila* infection of macrophages upregulates a number of pro-inflammatory proteins, many of which are targets of the NF-κB pathway, and also certain NF-κB-controlled anti-apoptotic genes, rendering infected macrophages resistant to apoptotic stimuli (Abu-Zant et al. 2007; Bartfeld et al. 2009; Losick and Isberg 2006; Shin et al. 2008). Among the Dot/Icm translocated effectors, LegK1 and LnaB are capable of activating an anti-apoptotic and protective NF-κB response (Ge et al. 2009; Losick and Isberg 2006). The recently discovered presence of NF-κB-related domains in the unicellular holozoan *Capsaspora owczarzaki* (Sebé-Pedrós et al. 2011) suggest that the NF-κB system evolved about 1000 million years ago. The amoebozoa branch separated earlier, and proteins harboring NF-κB-related domains are indeed not present in the unicellular *Legionella*-hosts that we analyzed (Table 3).

6 Bacterial Egress

When the infection cycle is completed and the host cell is spent, the bacterial progeny egresses from the host cell. Although several mutants showing defective egress from amoebal and mammalian cells have been identified (Alli et al. 2000; Gao and Kwaik 2000), how *L. pneumophila* exits the host cell after intracellular replication is not well understood. Formation of a cytolysin/egress pore required for host cell lysis has been proposed in U937-macrophage-like cells and in *A. polyphaga* (Alli et al. 2000; Gao and Kwaik 2000). In addition, the T4BSS-dependent effectors LepA and LepB seem to be involved in a non-lytic process that allows *L. pneumophila* to egress from amoeba, but not from macrophages (Chen 2004).

7 Conclusions and Future Perspectives

Many cellular processes observed during *L. pneumophila* infection in amoeba are similar to what can be seen in infection of mammalian macrophages (Fig. 1). *L. pneumophila* establishes a safe, intracellular niche within both hosts, the LCV. It evades the endocytic pathway and the subsequent phagosome-lysosome fusion, delays its acidification and allows intracellular replication of *L. pneumophila*. The Dot/Icm T4BSS and their translocated effectors are essential in macrophages and amoeba for establishing this LCV and for modulating host cellular processes to allow successful intracellular replication (Segal and Shuman 1999). However, from over 300 identified T4BSS translocated effectors, just a few have a known function. Moreover, among those with a known role in infection, their function in amoeba remains largely unstudied (Fig. 1 and Table 2). Processes subverted by *L. pneumophila* using highly similar strategies in macrophages and amoeba are the remodeling of the LCV, the use of the host's PI metabolism to anchor effectors in the LCV, acquisition of host secretory vesicles containing Arf1 and Rab1 and converting the LCV in an ER-derived niche for replication (Brombacher et al. 2009; Ingmundson et al. 2007; Urwyler et al. 2009a; Weber et al. 2006). Importantly, the effector SidC has been shown to play the same role in macrophages and amoeba (Ragaz et al. 2008; Weber et al. 2006). The benefit for the bacterium to replicate in an ER-like vacuole remains unknown, but it may be related to immune evasion in macrophages (Roy et al. 2006). Furthermore, the attachment of poly-Ub proteins to the LCV is highly conserved in macrophages and amoeba (Al-Quadan and Kwaik 2011; Lomma et al. 2010; Price et al. 2011). Other processes like phagocytosis or avoidance of phagosome acidification share only some common features in both hosts. Finally, the molecules implicated in the attachment to the host cell surface seem to be host-specific and some pathways like caspase-mediated apoptosis and activation of the NF-κB pathway are different in amoeba and macrophages. Interestingly, some of the features of *L. pneumophila* infection, can also be observed during the infection of other pathogens, like the manipulation of the phagosome maturation to form ER-rich vacuoles by *Brucella* (Campbell-Valois et al. 2012; Celli et al. 2003; Roy et al. 2006) or the evasion of phagosome-lysosome fusion triggered by *Mycobacterium tuberculosis* in human macrophages (Sturgill-Koszycki et al. 1994). A more in depth analysis and comparison at the molecular level of the shared processes in both evolutionary distant hosts might also help to better understand virulence of other intracellular pathogens that develop similar molecular strategies to infect human cells and for those which make transient association with amoeba, like many *Mycobacterium* spp., *Francisella tularensis,* or *Escherichia coli* O157.

Taken together, it is likely that the strategies used by *L. pneumophila* to infect human macrophages evolved mainly during its evolution within free-living protozoa, but we cannot exclude that the interaction with other susceptible hosts, closer to higher eukaryotes than amoeba, has also shaped the *Legionella*-host interactions. The broad host range of *L. pneumophila* might be due to the many different effectors this bacterium has acquired during evolution probably during interaction with distant hosts, helping it now to infect even humans. In fact, *L. pneumophila* can colonize and persist within the digestive tract of the nematode *Caenorhabditis elegans* (Brassinga et al. 2010), it has been identified in a phylogenetic analysis of microbial communities in the gastrointestinal tract of *Panaque nigrolineatus*, a tropical herbivorous freshwater fish (McDonald et al. 2012) and it can also naturally cause pneumonia in cattle (Fabbi et al. 1998). Thus, it is tempting to assume, that the large repertoire of T4BSS effectors contain also certain that might function either in amoeba or macrophages. Finally, the recent discovery of a primitive immune-like system in amoeba (Chen et al. 2007; Cosson and Soldati 2008) may suggest that some of the shared strategies used by *L. pneumophila* to infect macrophages and amoeba are a general strategy of intracellular pathogen to hide and evade from the eukaryotic host defense machinery.

Acknowledgments This work received financial support from the Institut Pasteur, the Centre National de la Recherche (CNRS), the Institut Carnot-Pasteur MI, the Fondation pour la Recherche Médicale (FRM) grant N° DEQ20120323697, the French region Ile-de-France (DIM Malinf) and the grant n°ANR-10-LABX-62-IBEID. M. Rolando is holder of a Roux contract financed by the Institut Pasteur.

References

Abu Kwaik Y (1996) The phagosome containing Legionella pneumophila within the protozoan Hartmannella vermiformis is surrounded by the rough endoplasmic reticulum. Appl Environ Microbiol 62(6):2022–2028

Abu-Zant A, Jones S, Asare R, Suttles J, Price C, Graham J et al (2007) Anti-apoptotic signalling by the Dot/Icm secretion system of L. pneumophila. Cell Microbiol 9(1):246–264

Akamine M, Higa F, Arakaki N, Kawakami K, Takeda K, Akira S et al (2005) Differential roles of Toll-like receptors 2 and 4 in in vitro responses of macrophages to Legionella pneumophila. Infect Immun 73(1):352–361

Al-Quadan T, Kwaik YA (2011) Molecular characterization of exploitation of the polyubiquitination and farnesylation machineries of dictyostelium discoideum by the AnkB F-box effector of Legionella pneumophila. Front Microbiol 2:23

Al-Quadan T, Price CT, Kwaik YA (2012) Exploitation of evolutionarily conserved amoeba and mammalian processes by Legionella. Trends Microbiol 20(6):299–306

Alli OA, Gao LY, Pedersen LL, Zink S, Radulic M, Doric M et al (2000) Temporal pore formation-mediated egress from macrophages and alveolar epithelial cells by Legionella pneumophila. Infect Immun 68(11):6431–6440

Alter-Koltunoff M, Goren S, Nousbeck J, Feng CG, Sher A, Ozato K et al (2008) Innate immunity to intraphagosomal pathogens is mediated by interferon regulatory factor 8 (IRF-8) that stimulates the expression of macrophage-specific Nramp1 through antagonizing repression by c-Myc. J Biol Chem 283(5):2724–2733

Amer AO (2010) Modulation of caspases and their non-apoptotic functions by Legionella pneumophila. Cell Microbiol 12(2):140–147

Arasaki K, Roy CR (2010) Legionella pneumophila promotes functional interactions between plasma membrane syntaxins and Sec22b. Traffic 11(5):587–600

Banga S, Gao P, Shen X, Fiscus V, Zong W-X, Chen L et al (2007) Legionella pneumophila inhibits macrophage apoptosis by targeting pro-death members of the Bcl2 protein family. Proc Natl Acad Sci U S A 104(12):5121–5126

Baranova IN, Kurlander R, Bocharov AV, Vishnyakova TG, Chen Z, Remaley AT et al (2008) Role of human CD36 in bacterial recognition, phagocytosis, and pathogen-induced JNK-mediated signaling. J Immunol 181(10):7147–7156

Bardill JP, Miller JL, Vogel JP (2005) IcmS-dependent translocation of SdeA into macrophages by the Legionella pneumophila type IV secretion system. Mol Microbiol 56(1):90–103

Bartfeld S, Engels C, Bauer B, Aurass P, Flieger A, Brüggemann H et al (2009) Temporal resolution of two-tracked NF-kappaB activation by Legionella pneumophila. Cell Microbiol 11(11):1638–1651

Bellinger-Kawahara C, Horwitz MA (1990) Complement component C3 fixes selectively to the major outer membrane protein (MOMP) of Legionella pneumophila and mediates phagocytosis of liposome-MOMP complexes by human monocytes. J Exp Med 172(4):1201–1210

Belyi Y, Tabakova I, Stahl M, Aktories K (2008) Lgt: a family of cytotoxic glucosyl transferases produced by Legionella pneumophila. J Bacteriol 190(8):3026–3035

Berger KH, Isberg RR (1993) Two distinct defects in intracellular growth complemented by a single genetic locus in Legionella pneumophila. Mol Microbiol 7(1):7–19

Berger KH, Merriam JJ, Isberg RR (1994) Altered intracellular targeting properties associated with mutations in the Legionella pneumophila dotA gene. Mol Microbiol 14(4):809–822

Bozue JA, Johnson W (1996) Interaction of Legionella pneumophila with Acanthamoeba castellanii: uptake by coiling phagocytosis and inhibition of phagosome-lysosome fusion. Infect Immun 64(2):668–673

Brand BC, Sadosky AB, Shuman HA (1994) The Legionella pneumophila icm locus: a set of genes required for intracellular multiplication in human macrophages. Mol Microbiol 14(4):797–808

Brassinga AKC, Kinchen JM, Cupp ME, Day SR, Hoffman PS, Sifri CD (2010) Caenorhabditis is a metazoan host for Legionella. Cell Microbiol 12(3):343–361

Brassinga AK, Sifri CD (2013) The Caenorhabditis elegans model of Legionella infection. Methods Mol Biol 954:439–461

Breiman RF, Fields BS, Sanden GN, Volmer L, Meier A, Spika JS (1990) Association of shower use with legionnaires-disease—possible role of amoebas. J Am Med Assoc 263(21): 2924–2926

Brombacher E, Urwyler S, Ragaz C, Weber SS, Kami K, Overduin M et al (2009) Rab1 guanine nucleotide exchange factor SidM is a major phosphatidylinositol 4-phosphate-binding effector protein of Legionella pneumophila. J Biol Chem 284(8):4846–4856
Burstein D, Zusman T, Degtyar E, Viner R, Segal G, Pupko T (2009) Genome-scale identification of Legionella pneumophila effectors using a machine learning approach. PLoS Pathog 5(7):e1000508
Campbell-Valois F-X, Trost M, Chemali M, Dill BD, Laplante A, Duclos S et al (2012) Quantitative proteomics reveals that only a subset of the endoplasmic reticulum contributes to the phagosome. Mol Cell Proteomics 11(7):M111.016378
Campodonico EM, Chesnel L, Roy CR (2005) A yeast genetic system for the identification and characterization of substrate proteins transferred into host cells by the Legionella pneumophila Dot/Icm system. Mol Microbiol 56(4):918–933
Cazalet C, Rusniok C, Brüggemann H, Zidane N, Magnier A, Ma L et al (2004) Evidence in the Legionella pneumophila genome for exploitation of host cell functions and high genome plasticity. Nat Genet 36(11):1165–1173
Celli J, de Chastellier C, Franchini D-M, Pizarro-Cerda J, Moreno E, Gorvel J-P (2003) Brucella evades macrophage killing via VirB-dependent sustained interactions with the endoplasmic reticulum. J Exp Med 198(4):545–556
Chang B, Kura F, Amemura-Maekawa J, Koizumi N, Watanabe H (2005) Identification of a novel adhesion molecule involved in the virulence of Legionella pneumophila. Infect Immun 73(7):4272–4280
Charpentier X, Gabay JE, Reyes M, Zhu JW, Weiss A, Shuman HA (2009) Chemical genetics reveals bacterial and host cell functions critical for type IV effector translocation by Legionella pneumophila. PLoS Pathog 5(7):e1000501 (Roy CR, editor)
Chen G, Zhuchenko O, Kuspa A (2007) Immune-like phagocyte activity in the social amoeba. Science 317(5838):678–681
Chen J (2004) Legionella effectors that promote nonlytic release from protozoa. Science 303(5662):1358–1361
Chong A, Lima CA, Allan DS, Nasrallah GK, Garduno RA (2009) The purified and recombinant Legionella pneumophila chaperonin alters mitochondrial trafficking and microfilament organization. Infect Immun 77(11):4724–4739
Chou MM, Hou W, Johnson J, Graham LK, Lee MH, Chen CS et al (1998) Regulation of protein kinase C zeta by PI 3-kinase and PDK-1. Curr Biol 8(19):1069–1077
Choy A, Dancourt J, Mugo B, O'Connor TJ, Isberg RR, Melia TJ et al (2012) The Legionella effector RavZ inhibits host autophagy through irreversible Atg8 deconjugation. Science 338(6110):1072–1076
Cianciotto NP (2007) Iron acquisition by Legionella pneumophila. Biometals. 20(3–4):323–331
Cianciotto NP, Fields BS (1992) Legionella pneumophila mip gene potentiates intracellular infection of protozoa and human macrophages. Proc Natl Acad Sci U S A 89(11):5188–5191
Cirillo JD, Falkow S, Tompkins LS (1994) Growth of Legionella pneumophila in Acanthamoeba castellanii enhances invasion. Infect Immun 62(8):3254–3261
Cirillo SL, Lum J, Cirillo JD (2000) Identification of novel loci involved in entry by Legionella pneumophila. Microbiology (Reading, Engl.) 146 (Pt 6):1345–1359
Cirillo SLG, Yan L, Littman M, Samrakandi MM, Cirillo JD (2002) Role of the Legionella pneumophila rtxA gene in amoebae. Microbiology (Reading, Engl.) 148(Pt 6):1667–1677
Clarke M, Lohan AJ, Liu B, Lagkouvardos I, Roy S, Zafar N et al (2013) Genome of Acanthamoeba castellanii highlights extensive lateral gene transfer and early evolution of tyrosine kinase signaling. Genome Biol 14(2):R11
Conover GM, Derré I, Vogel JP, Isberg RR (2003) The Legionella pneumophila LidA protein: a translocated substrate of the Dot/Icm system associated with maintenance of bacterial integrity. Mol Microbiol 48(2):305–321
Cosson P, Soldati T (2008) Eat, kill or die: when amoeba meets bacteria. Curr Opin Microbiol 11(3):271–276

Coxon PY, Summersgill JT, Ramirez JA, Miller RD (1998) Signal transduction during Legionella pneumophila entry into human monocytes. Infect Immun 66(6):2905–2913

Creasey EA, Isberg RR (2012) The protein SdhA maintains the integrity of the Legionella-containing vacuole. Proc Natl Acad Sci U S A 109(9):3481–3486

de Felipe KS, Glover RT, Charpentier X, Anderson OR, Reyes M, Pericone CD et al (2008) Legionella eukaryotic-like type IV substrates interfere with organelle trafficking. PLoS Pathog 4(8):e1000117

de Felipe KS, Pampou S, Jovanovic OS, Pericone CD, Ye SF, Kalachikov S et al (2005) Evidence for acquisition of Legionella type IV secretion substrates via interdomain horizontal gene transfer. J Bacteriol 187(22):7716–7726

Degtyar E, Zusman T, Ehrlich M, Segal G (2009) A Legionella effector acquired from protozoa is involved in sphingolipids metabolism and is targeted to the host cell mitochondria. Cell Microbiol 11(8):1219–1235

Derré I, Isberg RR (2004) Legionella pneumophila replication vacuole formation involves rapid recruitment of proteins of the early secretory system. Infect Immun 72(5):3048–3053

Dolezal P, Aili M, Tong J, Jiang J-H, Marobbio CM, Lee SF et al (2012) Legionella pneumophila secretes a mitochondrial carrier protein during infection. PLoS Pathog 8(1):e1002459 (Roy CR, editor)

Dorer MS, Kirton D, Bader JS, Isberg RR (2006) RNA interference analysis of Legionella in Drosophila cells: exploitation of early secretory apparatus dynamics. PLoS Pathog 2(4):e34

Elliott JA, Winn WC (1986) Treatment of alveolar macrophages with cytochalasin D inhibits uptake and subsequent growth of Legionella pneumophila. Infect Immun 51(1):31–36

Ensminger AW, Isberg RR (2010) E3 ubiquitin ligase activity and targeting of BAT3 by multiple Legionella pneumophila translocated substrates. Infect Immun 78(9):3905–3919

Eulgem T, Rushton PJ, Schmelzer E, Hahlbrock K, Somssich IE (1999) Early nuclear events in plant defence signalling: rapid gene activation by WRKY transcription factors. EMBO J 18(17):4689–4699

Fabbi M, Pastoris MC, Scanziani E, Magnino S, Di Matteo L (1998) Epidemiological and environmental investigations of Legionella pneumophila infection in cattle and case report of fatal pneumonia in a calf. J Clin Microbiol 36(7):1942–1947

Fajardo M, Schleicher M, Noegel A, Bozzaro S, Killinger S, Heuner K et al (2004) Calnexin, calreticulin and cytoskeleton-associated proteins modulate uptake and growth of Legionella pneumophila in Dictyostelium discoideum. Microbiology (Reading, Engl.) 150(Pt 9): 2825–2835

Falcó V, Fernández de Sevilla T, Alegre J, Ferrer A, Martínez Vázquez JM (1991) Legionella pneumophila. A cause of severe community-acquired pneumonia. Chest 100(4):1007–1011

Farbrother P, Wagner C, Na J, Tunggal B, Morio T, Urushihara H et al (2006) Dictyostelium transcriptional host cell response upon infection with Legionella. Cell Microbiol 8(3):438–456

Fernandez RC, Lee SH, Haldane D, Sumarah R, Rozee KR (1989) Plaque assay for virulent Legionella pneumophila. J Clin Microbiol 27(9):1961–1964

Fernandez RC, Logan SM, Lee SH, Hoffman PS (1996) Elevated levels of Legionella pneumophila stress protein Hsp60 early in infection of human monocytes and L929 cells correlate with virulence. Infect Immun 64(6):1968–1976

Fernandez-Moreira E, Helbig JH, Swanson MS (2006) Membrane vesicles shed by Legionella pneumophila inhibit fusion of phagosomes with lysosomes. Infect Immun 74(6):3285–3295

Fields BS (1996) The molecular ecology of Legionellae. Trends Microbiol 4(7):286–290

Fields BS, Barbaree JM, Shotts EB, Feeley JC, Morrill WE, Sanden GN et al (1986) Comparison of guinea pig and protozoan models for determining virulence of Legionella species. Infect Immun 53(3):553–559

Fitzgeorge RB, Baskerville A, Broster M, Hambleton P, Dennis PJ (1983) Aerosol infection of animals with strains of Legionella pneumophila of different virulence: comparison with intraperitoneal and intranasal routes of infection. J Hyg (Lond) 90(1):81–89

Flanagan MD, Lin S (1980) Cytochalasins block actin filament elongation by binding to high affinity sites associated with F-actin. J Biol Chem 255(3):835–838
Fontana MF, Shin S, Vance RE (2012) Activation of host MAP kinases by Legionella pneumophila secreted effectors that inhibit host protein translation. Infect Immun 80(10):3570–3575
Forgac M (2007) Vacuolar ATPases: rotary proton pumps in physiology and pathophysiology. Nat Rev Mol Cell Biol 8(11):917–929
Francione L, Smith PK, Accari SL, Taylor PE, Bokko PB, Bozzaro S et al (2009) Legionella pneumophila multiplication is enhanced by chronic AMPK signalling in mitochondrially diseased Dictyostelium cells. Dis Model Mech 2(9–10):479–489
Francione LM, Fisher PR (2011) Heteroplasmic mitochondrial disease in Dictyostelium discoideum. Biochem Pharmacol 82(10):1510–1520
Franco IS, Shohdy N, Shuman HA (2012) The Legionella pneumophila effector VipA is an actin nucleator that alters host cell organelle trafficking. PLoS Pathog 8(2):e1002546 (Stebbins CE, editor)
Franco IS, Shuman HA, Charpentier X (2009) The perplexing functions and surprising origins of Legionella pneumophila type IV secretion effectors. Cell Microbiol 11(10):1435–1443
Fujio H, Yoshida S, Miyamoto H, Mitsuyama M, Mizuguchi Y (1992) Investigation of the role of macrophages and endogenous interferon-gamma in natural resistance of mice against Legionella pneumophila infection. FEMS Microbiol Immunol 4(4):183–191
Gabay JE, Blake M, Niles WD, Horwitz MA (1985) Purification of Legionella pneumophila major outer membrane protein and demonstration that it is a porin. J Bacteriol 162(1):85–91
Gao LY, Harb OS, Abu Kwaik Y (1997) Utilization of similar mechanisms by Legionella pneumophila to parasitize two evolutionarily distant host cells, mammalian macrophages and protozoa. Infect Immun 65(11):4738–4746
Gao LY, Kwaik YA (2000) The mechanism of killing and exiting the protozoan host Acanthamoeba polyphaga by Legionella pneumophila. Environ Microbiol 2(1):79–90
Gao LY, Stone BJ, Brieland JK, Abu Kwaik Y (1998) Different fates of Legionella pneumophila pmi and mil mutants within macrophages and alveolar epithelial cells. Microb Pathog 25(6):291–306
Garduno RA, Faulkner G, Trevors MA, Vats N, Hoffman PS (1998a) Immunolocalization of Hsp60 in Legionella pneumophila. J Bacteriol 180(3):505–513
Garduno RA, Quinn FD, Hoffman PS (1998b) HeLa cells as a model to study the invasiveness and biology of Legionella pneumophila. Can J Microbiol 44(5):430–440
Ge J, Xu H, Li T, Zhou Y, Zhang Z, Li S et al (2009) A Legionella type IV effector activates the NF-kappaB pathway by phosphorylating the IkappaB family of inhibitors. Proc Natl Acad Sci U S A 106(33):13725–13730
Gibson FC, Tzianabos AO, Rodgers FG (1994) Adherence of Legionella pneumophila to U-937 cells, guinea-pig alveolar macrophages, and MRC-5 cells by a novel, complement-independent binding mechanism. Can J Microbiol 40(10):865–872
Gilmore TD, Wolenski FS (2012) NF-κB: where did it come from and why? Immunol Rev 246(1):14–35
Girard R, Pedron T, Uematsu S, Balloy V, Chignard M, Akira S et al (2003) Lipopolysaccharides from Legionella and Rhizobium stimulate mouse bone marrow granulocytes via Toll-like receptor 2. J Cell Sci 116(Pt 2):293–302
Gruenberg J (2001) The endocytic pathway: a mosaic of domains. Nat Rev Mol Cell Biol 2(10):721–730
Harada T, Tanikawa T, Iwasaki Y, Yamada M, Imai Y, Miyake M (2012) Phagocytic entry of Legionella pneumophila into macrophages through phosphatidylinositol 3,4,5-trisphosphate-independent pathway. Biol Pharm Bull 35(9):1460–1468
Harb OS, Venkataraman C, Haack BJ, Gao LY, Kwaik YA (1998) Heterogeneity in the attachment and uptake mechanisms of the Legionnaires' disease bacterium, Legionella pneumophila, by protozoan hosts. Appl Environ Microbiol 64(1):126–132

Harding CR, Schroeder GN, Reynolds S, Kosta A, Collins JW, Mousnier A, Frankel G (2012) Legionella pneumophila pathogenesis in the Galleria mellonella infection model. Infect Immun 80(8):2780–2790

Hastings KT, Cresswell P (2011) Disulfide reduction in the endocytic pathway: immunological functions of gamma-interferon-inducible lysosomal thiol reductase. Antioxid Redox Signal 15(3):657–668

Hawn TR, Berrington WR, Smith IA, Uematsu S, Akira S, Aderem A et al (2007) Altered inflammatory responses in TLR5-deficient mice infected with Legionella pneumophila. J Immunol 179(10):6981–6987

Hawn TR, Smith KD, Aderem A, Skerrett SJ (2006) Myeloid differentiation primary response gene (88)- and toll-like receptor 2-deficient mice are susceptible to infection with aerosolized Legionella pneumophila. J Infect Dis 193(12):1693–1702

Hayashi T, Miyake M, Fukui T, Sugaya N, Daimon T, Itoh S et al (2008) Exclusion of actin-binding protein p57/coronin-1 from bacteria-containing phagosomes in macrophages infected with Legionella. Biol Pharm Bull 31(5):861–865

Heidtman M, Chen EJ, Moy M-Y, Isberg RR (2009) Large-scale identification of Legionella pneumophila Dot/Icm substrates that modulate host cell vesicle trafficking pathways. Cell Microbiol 11(2):230–248

Hervet E, Charpentier X, Vianney A, Lazzaroni J-C, Gilbert C, Atlan D et al (2011) Protein kinase LegK2 is a type IV secretion system effector involved in endoplasmic reticulum recruitment and intracellular replication of Legionella pneumophila. Infect Immun 79(5):1936–1950

Hilbi H, Segal G, Shuman HA (2001) Icm/dot-dependent upregulation of phagocytosis by Legionella pneumophila. Mol Microbiol 42(3):603–617

Hoebe K, Georgel P, Rutschmann S, Du X, Mudd S, Crozat K et al (2005) CD36 is a sensor of diacylglycerides. Nature 433(7025):523–527

Horwitz MA (1983a) The Legionnaires' disease bacterium (Legionella pneumophila) inhibits phagosome-lysosome fusion in human monocytes. J Exp Med 158(6):2108–2126

Horwitz MA (1983b) Formation of a novel phagosome by the Legionnaires' disease bacterium (Legionella pneumophila) in human monocytes. J Exp Med 158(4):1319–1331

Horwitz MA (1984) Phagocytosis of the Legionnaires' disease bacterium (Legionella pneumophila) occurs by a novel mechanism: engulfment within a pseudopod coil. Cell 36(1):27–33

Horwitz MA, Maxfield FR (1984) Legionella pneumophila inhibits acidification of its phagosome in human monocytes. J Cell Biol 99(6):1936–1943

Hsu F, Zhu W, Brennan L, Tao L, Luo Z-Q, Mao Y (2012) Structural basis for substrate recognition by a unique Legionella phosphoinositide phosphatase. Proc Natl Acad Sci U S A 109(34):13567–13572

Hubber A, Roy CR (2010) Modulation of Host Cell Function by Legionella pneumophilaType IV Effectors. Annu Rev Cell Dev Biol 26(1):261–283

Husmann LK, Johnson W (1992) Adherence of Legionella pneumophila to guinea pig peritoneal macrophages, J774 mouse macrophages, and undifferentiated U937 human monocytes: role of Fc and complement receptors. Infect Immun 60(12):5212–5218

Ingmundson A, Delprato A, Lambright DG, Roy CR (2007) Legionella pneumophila proteins that regulate Rab1 membrane cycling. Nature 450(7168):365–369

Isberg RR, O'Connor TJ, Heidtman M (2009) The Legionella pneumophila replication vacuole: making a cosy niche inside host cells. Nat Rev Microbiol 7(1):13–24

Ito A, Mukaiyama A, Itoh Y, Nagase H, Thogersen IB, Enghild JJ et al (1996) Degradation of interleukin 1beta by matrix metalloproteinases. J Biol Chem 271(25):14657–14660

Ivanov SS, Charron G, Hang HC, Roy CR (2010) Lipidation by the host prenyltransferase machinery facilitates membrane localization of Legionella pneumophila effector proteins. J Biol Chem 285(45):34686–34698

Jack RS, Fan X, Bernheiden M, Rune G, Ehlers M, Weber A et al (1997) Lipopolysaccharide-binding protein is required to combat a murine gram-negative bacterial infection. Nature 389(6652):742–745

Jank T, Böhmer KE, Tzivelekidis T, Schwan C, Belyi Y, Aktories K (2012) Domain organization of Legionella effector SetA. Cell Microbiol 14(6):852–868

Kagan JC, Roy CR (2002) Legionella phagosomes intercept vesicular traffic from endoplasmic reticulum exit sites. Nat Cell Biol 4(12):945–954

Kagan JC, Stein M-P, Pypaert M, Roy CR (2004) Legionella subvert the functions of Rab1 and Sec22b to create a replicative organelle. J Exp Med 199(9):1201–1211

Khelef N, Shuman HA, Maxfield FR (2001) Phagocytosis of wild-type Legionella pneumophila occurs through a wortmannin-insensitive pathway. Infect Immun 69(8):5157–5161

Khweek AA, Caution K, Akhter A, Abdulrahman BA, Tazi M, Hassan H et al (2013) A bacterial protein promotes the recognition of the Legionella pneumophila vacuole by autophagy. Eur J Immunol 43(5):1333–1344

Kikuhara H, Ogawa M, Miyamoto H, Nikaido Y, Yoshida S (1994) Intracellular multiplication of Legionella pneumophila in Tetrahymena thermophila. J UOEH 16(4):263–275

King CH, Fields BS, Shotts EB, White EH (1991) Effects of cytochalasin D and methylamine on intracellular growth of Legionella pneumophila in amoebae and human monocyte-like cells. Infect Immun 59(3):758–763

Kornfeld S, Mellman I (1989) The biogenesis of lysosomes. Annu Rev Cell Biol 5:483–525

Köhler R, Bubert A, Goebel W, Steinert M, Hacker J, Bubert B (2000) Expression and use of the green fluorescent protein as a reporter system in Legionella pneumophila. Mol Gen Genet 262(6):1060–1069

Krasity BC, Troll JV, Weiss JP, McFall-Ngai MJ (2011) LBP/BPI proteins and their relatives: conservation over evolution and roles in mutualism. Biochem Soc Trans 39(4):1039–1044

Krinos C, High AS, Rodgers FG (1999) Role of the 25 kDa major outer membrane protein of Legionella pneumophila in attachment to U-937 cells and its potential as a virulence factor for chick embryos. J Appl Microbiol 86(2):237–244

Ku B, Lee K-H, Park WS, Yang C-S, Ge J, Lee S-G et al (2012) VipD of Legionella pneumophila targets activated Rab5 and Rab22 to interfere with endosomal trafficking in macrophages. PLoS Pathog 8(12):e1003082

Kubori T, Hyakutake A, Nagai H (2008) Legionella translocates an E3 ubiquitin ligase that has multiple U-boxes with distinct functions. Mol Microbiol 67(6):1307–1319

Laguna RK, Creasey EA, Li Z, Valtz N, Isberg RR (2006) A Legionella pneumophila-translocated substrate that is required for growth within macrophages and protection from host cell death. Proc Natl Acad Sci U S A 103(49):18745–18750

Lau HY, Ashbolt NJ (2009) The role of biofilms and protozoa in Legionella pathogenesis: implications for drinking water. J Appl Microbiol 107(2):368–378

Li CKF, Pender SLF, Pickard KM, Chance V, Holloway JA, Huett A et al (2004) Impaired immunity to intestinal bacterial infection in stromelysin-1 (matrix metalloproteinase-3)-deficient mice. J. Immunol. 173(8):5171–5179

Li Z, Dugan AS, Bloomfield G, Skelton J, Ivens A, Losick V et al (2009) The amoebal MAP kinase response to Legionella pneumophila is regulated by DupA. Cell Host Microbe. 6(3):253–267

Lifshitz Z, Burstein D, Peeri M, Zusman T, Schwartz K, Shuman HA et al (2013) Computational modeling and experimental validation of the Legionella and Coxiella virulence-related type-IVB secretion signal. Proc. Natl. Acad. Sci. U.S.A. 110(8):E707–E715

Lightfield KL, Persson J, Brubaker SW, Witte CE, von Moltke J, Dunipace EA et al (2008) Critical function for Naip5 in inflammasome activation by a conserved carboxy-terminal domain of flagellin. Nat Publ Group 9(10):1171–1178

Liu M, Conover GM, Isberg RR (2008) Legionella pneumophila EnhC is required for efficient replication in tumour necrosis factor α-stimulated macrophages. Cell Microbiol 10(9):1906–1923

Liu M, Haenssler E, Uehara T, Losick VP, Park JT, Isberg RR (2012) The Legionella pneumophila EnhC Protein Interferes with immunostimulatory muramyl peptide production to evade innate immunity. Cell Host Microbe 12(2):11

Liu Y, Luo Z-Q (2007) The Legionella pneumophila effector SidJ is required for efficient recruitment of endoplasmic reticulum proteins to the bacterial phagosome. Infect Immun 75(2):592–603

Lomma M, Dervins-Ravault D, Rolando M, Nora T, Newton HJ, Sansom FM et al (2010) The Legionella pneumophila F-box protein Lpp 2082 (AnkB) modulates ubiquitination of the host protein parvin B and promotes intracellular replication. Cell Microbiol 12(9):1272–1291

Losick VP, Haenssler E, Moy M-Y, Isberg RR (2010) LnaB: a Legionella pneumophila activator of NF-kappa B. Cell Microbiol 12(8):1083–1097

Losick VP, Isberg RR (2006) NF-kappaB translocation prevents host cell death after low-dose challenge by Legionella pneumophila. J Exp Med 203(9):2177–2189

Lu H, Clarke M (2005) Dynamic properties of Legionella-containing phagosomes in Dictyostelium amoebae. Cell Microbiol 7(7):995–1007

Luo Z-Q, Isberg RR (2004) Multiple substrates of the Legionella pneumophila Dot/Icm system identified by interbacterial protein transfer. Proc Natl Acad Sci U S A 101(3):841–846

Machner MP, Isberg RR (2006) Targeting of host Rab GTPase function by the intravacuolar pathogen Legionella pneumophila. Dev Cell 11(1):47–56

McDonald R, Schreier HJ, Watts JEM (2012) Phylogenetic analysis of microbial communities in different regions of the gastrointestinal tract in Panaque nigrolineatus, a wood-eating fish. PLoS ONE 7(10):e48018 (Ravel J, editor)

Moffat JF, Tompkins LS (1992) A quantitative model of intracellular growth of Legionella pneumophila in Acanthamoeba castellanii. Infect Immun 60(1):296–301

Molmeret M, Horn M, Wagner M, Santic M, Abu Kwaik Y (2005) Amoebae as training grounds for intracellular bacterial pathogens. Appl Environ Microbiol 71(1):20–28

Murata T, Delprato A, Ingmundson A, Toomre DK, Lambright DG, Roy CR (2006) The Legionella pneumophila effector protein DrrA is a Rab1 guanine nucleotide-exchange factor. Nat Cell Biol 8(9):971–977

Nagai H, Kagan JC, Zhu X, Kahn RA, Roy CR (2002) A bacterial guanine nucleotide exchange factor activates ARF on Legionella phagosomes. Science 295(5555):679–682

Nagai H, Kubori T (2011) Type IVB secretion systems of legionella and other gram-negative bacteria. Front Microbiol 2:136

Nakanishi H, Brewer KA, Exton JH (1993) Activation of the zeta isozyme of protein kinase C by phosphatidylinositol 3,4,5-trisphosphate. J Biol Chem 268(1):13–16

Neunuebel MR, Chen Y, Gaspar AH, Backlund PS, Yergey A, Machner MP (2011) De-AMPylation of the small GTPase Rab1 by the pathogen Legionella pneumophila. Science 333(6041):453–456

Newsome AL, Baker RL, Miller RD, Arnold RR (1985) Interactions between Naegleria fowleri and Legionella pneumophila. Infect Immun 50(2):449–452

Newton HJ, Sansom FM, Dao J, Cazalet C, Bruggemann H, Albert-Weissenberger C et al (2008) Significant role for ladC in initiation of Legionella pneumophila infection. Infect Immun 76(7):3075–3085

Ninio S, Celli J, Roy CR (2009) A Legionella pneumophila effector protein encoded in a region of genomic plasticity binds to Dot/Icm-modified vacuoles. PLoS Pathog 5(1):e1000278

Nora T, Lomma M, Gomez-Valero L, Buchrieser C (2009) Molecular mimicry: an important virulence strategy employed by Legionella pneumophila to subvert host functions. Future Microbiol 4(6):691–701

Pan X, Lührmann A, Satoh A, Laskowski-Arce MA, Roy CR (2008) Ankyrin repeat proteins comprise a diverse family of bacterial type IV effectors. Science 320(5883):1651–1654

Payne NR, Horwitz MA (1987) Phagocytosis of Legionella pneumophila is mediated by human monocyte complement receptors. J Exp Med 166(5):1377–1389

Peracino B, Balest A, Bozzaro S (2010) Phosphoinositides differentially regulate bacterial uptake and Nramp1-induced resistance to Legionella infection in Dictyostelium. J Cell Sci 123(Pt 23):4039–4051

Peracino B, Wagner C, Balest A, Balbo A, Pergolizzi B, Noegel AA et al (2006) Function and mechanism of action of Dictyostelium Nramp1 (Slc11a1) in bacterial infection. Traffic 7(1):22–38

Pereira MSF, Morgantetti GF, Massis LM, Horta CV, Hori JI, Zamboni DS (2011) Activation of NLRC4 by flagellated bacteria triggers caspase-1-dependent and -independent responses to restrict Legionella pneumophila replication in macrophages and in vivo. J Immunol 187(12):6447–6455

Prashar A, Bhatia S, Tabatabaeiyazdi Z, Duncan C, Garduño RA, Tang P et al (2012) Mechanism of invasion of lung epithelial cells by filamentous Legionella pneumophila. Cell Microbiol 14(10):1632–1655

Price CTD, Al-Khodor S, Al-Quadan T, Abu Kwaik Y (2010a) Indispensable role for the eukaryotic-like ankyrin domains of the ankyrin B effector of Legionella pneumophila within macrophages and amoebae. Infect Immun 78(5):2079–2088

Price CTD, Al-Quadan T, Santic M, Jones SC, Abu Kwaik Y (2010b) Exploitation of conserved eukaryotic host cell farnesylation machinery by an F-box effector of Legionella pneumophila. J Exp Med 207(8):1713–1726

Price CTD, Al-Quadan T, Santic M, Rosenshine I, Abu Kwaik Y (2011) Host proteasomal degradation generates amino acids essential for intracellular bacterial growth. Science 334(6062):1553–1557

Ragaz C, Pietsch H, Urwyler S, Tiaden A, Weber SS, Hilbi H (2008) The Legionella pneumophila phosphatidylinositol-4 phosphate-binding type IV substrate SidC recruits endoplasmic reticulum vesicles to a replication-permissive vacuole. Cell Microbiol 10(12):2416–2433

Rechnitzer C, Blom J (1989) Engulfment of the Philadelphia strain of Legionella pneumophila within pseudopod coils in human phagocytes. Comparison with other Legionella strains and species. APMIS 97(2):105–114

Richards AM, Dwingelo Von JE, Price CT, Abu Kwaik Y (2013) Cellular microbiology and molecular ecology of Legionella-amoeba interaction. Virulence 4(4):307–314

Rittig MG, Burmester GR, Krause A (1998) Coiling phagocytosis: when the zipper jams, the cup is deformed. Trends Microbiol 6(10):384–388

Robinson CG, Roy CR (2006) Attachment and fusion of endoplasmic reticulum with vacuoles containing Legionella pneumophila. Cell Microbiol 8(5):793–805

Rodgers FG, Gibson FC (1993) Opsonin-independent adherence and intracellular development of Legionella pneumophila within U-937 cells. Can J Microbiol 39(7):718–722

Roisin-Bouffay C, Luciani M-F, Klein G, Levraud J-P, Adam M, Golstein P (2004) Developmental cell death in dictyostelium does not require paracaspase. J Biol Chem 279(12):11489–11494

Rolando M, Buchrieser C (2012) Post-translational modifications of host proteins by Legionella pneumophila: a sophisticated survival strategy. Future Microbiol 7(3):369–381

Roy CR, Salcedo SP, Gorvel J-PE (2006) Pathogen-endoplasmic-reticulum interactions: in through the out door. Nat Rev Immunol 6(2):136–147

Schoebel S, Blankenfeldt W, Goody RS, Itzen A (2010) High-affinity binding of phosphatidylinositol 4-phosphate by Legionella pneumophila DrrA. EMBO Rep 11(8):598–604

Sebé-Pedrós A, de Mendoza A, Lang BF, Degnan BM, Ruiz-Trillo I (2011) Unexpected repertoire of metazoan transcription factors in the unicellular holozoan Capsaspora owczarzaki. Mol Biol Evol 28(3):1241–1254

Seeger EM, Thuma M, Fernandez-Moreira E, Jacobs E, Schmitz M, Helbig JH (2010) Lipopolysaccharide of Legionella pneumophila shed in a liquid culture as a nonvesicular fraction arrests phagosome maturation in amoeba and monocytic host cells. FEMS Microbiol Lett 307(2):113–119

Segal G, Shuman HA (1999) Legionella pneumophila utilizes the same genes to multiply within Acanthamoeba castellanii and human macrophages. Infect Immun 67(5):2117–2124

Shevchuk O, Batzilla C, Hägele S, Kusch H, Engelmann S, Hecker M et al (2009) Proteomic analysis of Legionella-containing phagosomes isolated from Dictyostelium. Int J Med Microbiol 299(7):489–508

Shin S (2012) Innate Immunity to Intracellular Pathogens: Lessons Learned from Legionella pneumophila. Adv Appl Microbiol 79:43–71

Shin S, Case CL, Archer KA, Nogueira CV, Kobayashi KS, Flavell RA et al (2008) Type IV secretion-dependent activation of host MAP kinases induces an increased proinflammatory cytokine response to Legionella pneumophila. PLoS Pathog 4(11):e1000220

Shohdy N, Efe JA, Emr SD, Shuman HA (2005) Pathogen effector protein screening in yeast identifies Legionella factors that interfere with membrane trafficking. Proc Natl Acad Sci U S A 102(13):4866–4871

Solomon JM, Rupper A, Cardelli JA, Isberg RR (2000) Intracellular growth of Legionella pneumophila in Dictyostelium discoideum, a system for genetic analysis of host-pathogen interactions. Infect Immun 68(5):2939–2947

Stone BJ, Abu Kwaik Y (1998) Expression of multiple pili by Legionella pneumophila: identification and characterization of a type IV pilin gene and its role in adherence to mammalian and protozoan cells. Infect Immun 66(4):1768–1775

Strom MS, Lory S (1993) Structure-function and biogenesis of the type IV pili. Annu Rev Microbiol 47:565–596

Sturgill-Koszycki S, Schlesinger PH, Chakraborty P, Haddix PL, Collins HL, Fok AK et al (1994) Lack of acidification in Mycobacterium phagosomes produced by exclusion of the vesicular proton-ATPase. Science 263(5147):678–681

Sturgill-Koszycki S, Swanson MS (2000) Legionella pneumophila replication vacuoles mature into acidic, endocytic organelles. J Exp Med 192(9):1261–1272

Sun EW, Wagner ML, Maize A, Kemler D, Garland-Kuntz E, Xu L et al (2013) Legionella pneumophila infection of Drosophila S2 cells induces only minor changes in mitochondrial dynamics. PLoS ONE 8(4):e62972

Swanson MS, Hammer BK (2000) Legionella pneumophila pathogesesis: a fateful journey from amoebae to macrophages. Annu Rev Microbiol 54:567–613

Tachado SD, Samrakandi MM, Cirillo JD (2008) Non-opsonic phagocytosis of Legionella pneumophila by macrophages is mediated by phosphatidylinositol 3-kinase. PLoS ONE 3(10):e3324

Takeda K, Kaisho T, Akira S (2003) Toll-like receptors. Annu Rev Immunol 21:335–376

Tan Y, Arnold RJ, Luo Z-Q (2011) Legionella pneumophila regulates the small GTPase Rab1 activity by reversible phosphorylcholination. Proc Natl Acad Sci U S A 108(52):21212–21217

Tan Y, Luo Z-Q (2011) Legionella pneumophila SidD is a deAMPylase that modifies Rab1. Nature 475(7357):506–509

Thi EP, Reiner NE (2012) Phosphatidylinositol 3-kinases and their roles in phagosome maturation. J Leukoc Biol 92(3):553–566

Thornberry NA, Bull HG, Calaycay JR, Chapman KT, Howard AD, Kostura MJ et al (1992) A novel heterodimeric cysteine protease is required for interleukin-1 beta processing in monocytes. Nature 356(6372):768–774

Tilney LG, Harb OS, Connelly PS, Robinson CG, Roy CR (2001) How the parasitic bacterium Legionella pneumophila modifies its phagosome and transforms it into rough ER: implications for conversion of plasma membrane to the ER membrane. J Cell Sci 114(Pt 24):4637–4650

Uberall F, Hellbert K, Kampfer S, Maly K, Villunger A, Spitaler M et al (1999) Evidence that atypical protein kinase C-lambda and atypical protein kinase C-zeta participate in Ras-mediated reorganization of the F-actin cytoskeleton. J Cell Biol 144(3):413–425

Urwyler S, Brombacher E, Hilbi H (2009a) Endosomal and secretory markers of the Legionella-containing vacuole. Commun Integr Biol 2(2):107–109

Urwyler S, Nyfeler Y, Ragaz C, Lee H, Mueller LN, Aebersold R et al (2009b) Proteome analysis of Legionella vacuoles purified by magnetic immunoseparation reveals secretory and endosomal GTPases. Traffic 10(1):76–87

Vandersmissen L, De Buck E, Saels V, Coil DA, Anné J (2010) A Legionella pneumophila-collagen-like protein encoded by a gene with a variable number of tandem repeats is involved in the adherence and invasion of host cells. FEMS Microbiol Lett 306(2):168–176

VanRheenen SM, Luo Z-Q, O'Connor T, Isberg RR (2006) Members of a Legionella pneumophila family of proteins with ExoU (phospholipase A) active sites are translocated to target cells. Infect Immun 74(6):3597–3606

Venkataraman C, Haack BJ, Bondada S, Abu Kwaik Y (1997) Identification of a Gal/GalNAc lectin in the protozoan Hartmannella vermiformis as a potential receptor for attachment and invasion by the Legionnaires' disease bacterium. J Exp Med 186(4):537–547

Vieira OV, Botelho RJ, Grinstein S (2002) Phagosome maturation: aging gracefully. Biochem J 366(Pt 3):689–704

Vinzing M, Eitel J, Lippmann J, Hocke AC, Zahlten J, Slevogt H et al (2008) NAIP and Ipaf control Legionella pneumophila replication in human cells. J Immunol 180(10):6808–6815

Watanabe Y, Tateno H, Nakamura-Tsuruta S, Kominami J, Hirabayashi J, Nakamura O et al (2009) The function of rhamnose-binding lectin in innate immunity by restricted binding to Gb3. Dev Comp Immunol 33(2):187–197

Weber SS, Ragaz C, Reus K, Nyfeler Y, Hilbi H (2006) Legionella pneumophila exploits PI(4)P to anchor secreted effector proteins to the replicative vacuole. PLoS Pathog 2(5):e46

Welsh CT, Summersgill JT, Miller RD (2004) Increases in c-Jun N-terminal kinase/stress-activated protein kinase and p38 activity in monocyte-derived macrophages following the uptake of Legionella pneumophila. Infect Immun 72(3):1512–1518

Xu L, Shen X, Bryan A, Banga S, Swanson MS, Luo Z-Q (2010) Inhibition of host vacuolar H + -ATPase activity by a Legionella pneumophila effector. PLoS Pathog 6(3):e1000822

Yan M, Collins RF, Grinstein S, Trimble WS (2005) Coronin-1 function is required for phagosome formation. Mol Biol Cell 16(7):3077–3087

Zamboni DS, Kobayashi KS, Kohlsdorf T, Ogura Y, Long EM, Vance RE et al (2006) The Birc1e cytosolic pattern-recognition receptor contributes to the detection and control of Legionella pneumophila infection. Nat Immunol 7(3):318–325

Zhang C, Kuspa A (2009) Transcriptional down-regulation and rRNA cleavage in Dictyostelium discoideum mitochondria during Legionella pneumophila infection. PLoS ONE 4(5):e5706

Zhao Y, Yang J, Shi J, Gong Y-N, Lu Q, Xu H et al (2011) The NLRC4 inflammasome receptors for bacterial flagellin and type III secretion apparatus. Nature 477(7366):596–600

Zhu W, Banga S, Tan Y, Zheng C, Stephenson R, Gately J et al (2011) Comprehensive identification of protein substrates of the Dot/Icm type IV transporter of Legionella pneumophila. PLoS ONE 6(3):e17638

The *Legionella pneumophila* Two-Component Regulatory Systems that Participate in the Regulation of Icm/Dot Effectors

Gil Segal

Abstract *Legionella pneumophila,* the causative agent of Legionnaires' disease, actively manipulates intracellular processes to establish a replication niche inside their host cells. The establishment of its replication niche requires a functional Icm/Dot type IV secretion system which translocates about 300 effector proteins into the host cells during infection. This enormous number of effectors should be coordinated at the level of gene expression, in order to be expressed and translocated at the correct time and appropriate amounts. One of the predominant ways in bacteria to regulate virulence gene expression is by the use of two-component systems (TCSs). To date, four TCSs have been shown to be involved in the regulation of Icm/Dot effector-encoding genes: The PmrAB and CpxRA TCSs that directly control, and the LetAS and LqsRS TCSs that indirectly control the level of expression of effector-encoding genes. According to our current knowledge, these four TCSs control the expression of about 70 effector-encoding genes. The regulation by different TCSs divides the effectors into groups of co-regulated effector-encoding genes that are probably co-expressed at a similar time during infection and might perform related functions. In addition, examples of interplay between these TCSs were already reported indicating that they form part of a regulatory network that orchestrates the expression of *L. pneumophila* effector-encoding genes during infection.

Abbreviations

GAP	GTPas activating protein
GEF	Guanine nucleotide exchange factor
Icm/Dot	Intracellular multiplication/Defect organelle trafficking
LCV	*Legionella* containing vacuole
TCS	Two-component system

G. Segal (✉)
The George S. Wise Faculty of Life Sciences, Department of Molecular Microbiology and Biotechnology, Tel-Aviv University, 69978 Ramat-Aviv, Tel-Aviv, Israel
e-mail: gils@tauex.tau.ac.il

Current Topics in Microbiology and Immunology (2013) 376: 35–52
DOI: 10.1007/82_2013_346

Published Online: 6 August 2013

Contents

1 Introduction 36
2 The CpxR-CpxA Two-Component System 37
3 The PmrA-PmrB Two-Component System 41
4 The LetA-LetS Two-Component System 42
5 The LqsR-LqsS Two-Component System 44
6 Interplay Between Different TCSs in The Regulation of *L. pneumophila* Effectors 45
6.1 Effectors Manipulating Rab1 are Regulated by Both the CpxRA and the PmrAB TCSs 45
6.2 The Regulatory Switch Between the PmrAB and the LetAS Regulated Effectors 46
6.3 Effectors Directly Regulated by Both the CpxRA and PmrAB TCSs 47
7 Conclusions 48
References 48

1 Introduction

Legionella pneumophila is an opportunistic human pathogen that multiplies within alveolar macrophages and causes a severe pneumonia known as Legionnaires' disease. In order to establish a replicative niche inside eukaryotic cells, *L. pneumophila* modulates host cell functions by the delivery of about 300 effector proteins through the Icm/Dot Type-IV secretion system (reviewed in Ensminger and Isberg 2009; Franco et al. 2009; Gomez-Valero et al. 2011; Shin and Roy 2008). The numerous effectors that take part in the establishment of the *L. pneumophila* containing vacuole (LCV); the many host cell pathways manipulated by *L. pneumophila* effectors (Dorer et al. 2006; O'Connor et al. 2012) and the stepwise process that occurs during the establishment of the LCV inside host cells (Horwitz 1983a); suggest that the effectors translocated by the Icm/Dot secretion system will most likely be regulated at the level of gene expression, in order to coordinate a successful infection (other levels of regulation of effectors translocation such as the recognition by components of the secretion complex and chaperons probably contribute to this stepwise process as well).

Two-component systems (TCSs) are widespread signal transduction devices in bacteria that enable them to respond to environmental stimuli mainly via changes in gene expression. These systems are used by many pathogenic bacteria that utilize multiple TCSs in order to control the expression of their virulence genes (Fass and Groisman 2009; Gooderham and Hancock 2009; Gotoh et al. 2010; Vogt and Raivio 2012). The TCSs are classically composed of a membrane-integrated sensor histidine kinase and a cytoplasmic transcriptional regulator containing an N-terminal receiver domain and a C-terminal DNA binding domain (helix-turn-helix domain). Generally, stimuli detected by the sensor histidine kinase lead to its autophosphorylation. Then, the phosphoryl group from the histidine residue is transferred to an aspartic acid residue in the receiver domain of the response regulator, thus leading to its activation (Jung et al. 2012; Laub and Goulian 2007).

To date, four TCSs have been found to regulate the expression of *L. pneumophila* effector-encoding genes (Table 1 and Fig. 1): (i) the CpxRA TCS was shown to directly activate or repress the expression of 12 effector-encoding genes and several *icm*/*dot* genes (Altman and Segal 2008; Gal-Mor and Segal 2003a); (ii) the PmrAB TCS was shown to directly activate the expression of 43 effector-encoding genes (Al-Khodor et al. 2009; Zusman et al. 2007); (iii) the LetAS TCS was shown to indirectly regulate the expression of four effector-encoding genes (Rasis and Segal 2009; Shi et al. 2006) and (iv) the LqsRS TCS that was shown to indirectly regulate the expression of 12 effector-encoding genes (Tiaden et al. 2010; Tiaden et al. 2007).

These four TCSs, the different functions mediated by the effectors they regulate and the interplay between them are the focus of this chapter.

2 The CpxR-CpxA Two-Component System

The *L. pneumophila* CpxR-CpxA TCS consists of the CpxR response regulator and the CpxA sensor histidine kinase (Gal-Mor and Segal 2003a). The CpxRA TCS has been studied in many bacteria, where CpxA was shown to sense misfolded proteins in the bacterial envelope and to activate (phosphorylate) CpxR. Phosphorylated CpxR was shown to regulate pilus assembly, adherence, and biofilm formation (Hunke et al. 2012; Vogt and Raivio 2012). Moreover, CpxR was shown to be required for host cell invasion in several species, including pathogenic *Escherichia coli* and *Salmonella enterica* (Humphreys et al. 2004; Nevesinjac and Raivio 2005).

The involvement of the CpxRA TCS in *L. pneumophila* virulence was first identified in a genetic screen looking for a direct regulator of the *icmR* gene (Gal-Mor and Segal 2003a). Later, by using bioinformatic approaches aimed at identifying additional genes that harbor the CpxR regulatory element (GTAAAnnnnnnGWAAA, W indicates T or A) this system was shown to participate in the regulation of two additional *icm/dot* genes (*icmV* and *icmW*), the *lvgA* gene and 11 effector-encoding genes (Altman and Segal 2008). The *L. pneumophila* CpxRA TCS was found to activate the expression of all the *icm/dot* genes it regulates, as well as five effector-encoding genes, and to repress the expression of six other effector-encoding genes (Altman and Segal 2008). However, deletion mutants in the genes coding for CpxR as well as CpxA were found to have no intracellular growth phenotype when examined in different host cells (Gal-Mor and Segal 2003a). The environmental stimuli that activate the *L. pneumophila* CpxA sensor kinase have not been discovered yet, but since CpxR-activated effectors were shown to translocate into host cells early during infection (see below), the CpxA activation might be related to *L. pneumophila* attachment to host cells like in the case of *E. coli* (Nevesinjac and Raivio 2005).

Table 1 *L. pneumophila* two-component systems involved in effectors regulation

Response regulator	Sensor kinase	Type of regulation	Additional components	Effectors regulated	Type of HTH[a]	References
CpxR-lpg1438	CpxA-lpg1437	Direct	–	CegC1, LegA10, CegC2, Ceg7, LegA11, Ceg18, CegC3, CegC4, SidH, SidM/DrrA, SidD, Ceg33	WHTH	Altman and Segal (2008); Gal-Mor and Segal (2003a)
PmrA-lpg1292	PmrB-lpg1291	Direct	–	Ceg2-7, SidE, Ceg8-10, SdhA, Ceg11, LegA9, Ceg14/SidL, Ceg15, Ceg17-21, SidG, Ceg22, LegC5, Ceg23, SidB, Ceg24-25, LegLC8, LegAU13, SdeC, SdeB, SdeA, Ceg28-30, LegA14, SdbB, LepB, Ceg32/SidI, SidF, Ceg33, VipE, Ceg34	WHTH	Al-Khodor et al. (2009); Zusman et al. (2007)
LetA-lpg2646	LetS-lpg1912	Indirect	RsmY, RsmZ, CsrA, LetE	VipA, YlfB/LegC2, RalF, YlfA/LegC7	HTH_LUXR	Bachman and Swanson, (2004); Edwards et al. (2010); Forsbach-Birk et al. (2004); Gal-Mor and Segal (2003b); Hammer et al. (2002); Lynch et al. (2003); Molofsky and Swanson (2003); Rasis and Segal (2009); Sahr et al. (2009)
LqsR-lpg2732	LqsS-lpg2734	Indirect	LqsA, LqsT	Lpg0081, lpg0294, lpg0634 lpg0967, SidK, RavN, lpg1453, MavB, CegC4, SdbB, lpg2844, lpg3000	No HTH	Kessler et al. (2013); Tiaden et al. (2008, 2007, 2010)

[a] *WHTH* winged helix-turn-helix, *HTH_LUXR* A LuxR type helix-turn-helix

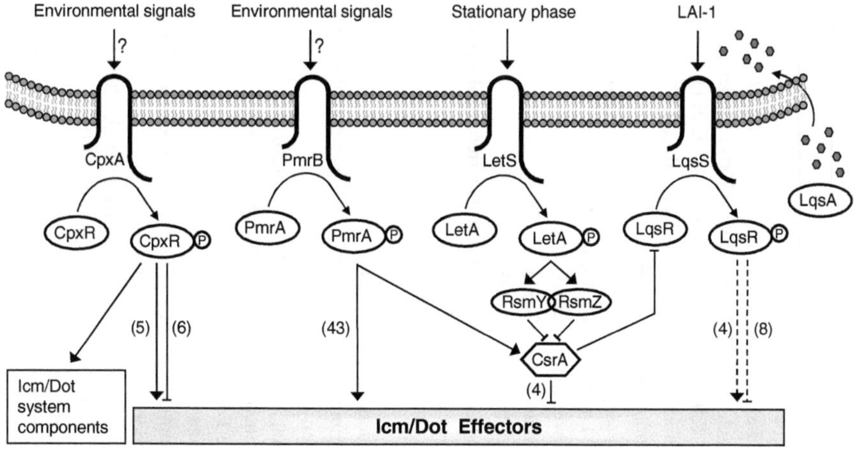

Fig. 1 Model of the TCSs that control the expression of the *L. pneumophila* Icm/Dot effector-encoding genes. The four TCSs (CpxRA, PmrAB, LetAS and LqsRS) and the components of the LetAS-RsmYZ-CsrA regulatory cascade as well as LqsA which is part of the Lqs system are schematically illustrated. The environmental signals sensed by CpxA and PmrB are currently not known, and the phosphorylation of these components is expected to be activated by transfer of the phosphate group to their cognate response regulators CpxR and PmrA respectively, which then directly activate or repress the transcription of their target effector-encoding genes. During stationary phase, the LetAS TCS activates the expression of the sRNAs RsmY and RsmZ that thus sequester CsrA from its target mRNAs and relieve the CsrA post-transcriptional repression. The *csr*A gene was also shown to be under the regulation of the PmrA transcriptional regulator and LqsR was shown to be repressed by CsrA. The quorum sensing system Lqs is being activated by binding of LAI-1 (*Legionella* auto inducer-1, generated by LqsA) to LqsS that thus activates the LqsR response regulator by phosphorylation. LqsR does not contain any DNA binding domain and it affects gene expression in an unknown way. The number of effector-encoding genes which were shown to be regulated by each of these TCSs is indicated in brackets. Solid lines and dashed lines indicate direct and indirect regulation, respectively. Solid arrows and T-shaped symbols indicate activation and repression, respectively

The function of most of the effectors regulated by the CpxRA TCS is currently not known. Two of the effectors that were found to be repressed by CpxR (LegA10 and LegA11) (Altman and Segal 2008) contain an Ankyrin domain (Table 2), which probably indicates that they interact with yet unknown host proteins. One effector (SidM/DrrA) that was shown to be strongly activated by CpxR (Altman and Segal 2008) was studied intensively. SidM/DrrA was found to recruit the host cell factor Rab1 to the LCV, it was shown to function as a guanine nucleotide exchange factor (GEF) for Rab1 as well as a GDP dissociation inhibitor (GDI) displacement factor (GDF) for Rab1 (Machner and Isberg 2006; Machner and Isberg 2007; Neunuebel et al. 2011). Besides these activities the amino-terminal domain of SidM/DrrA was shown to AMPylate Rab1, leading to its covalent modification. The AMPylation by SidM/DrrA limits the access of GTPase activating proteins (GAPs), thereby keeping Rab1 constitutively active (Muller et al. 2010; Murata et al. 2006). Interestingly, it was found that another effector-

Table 2 Effector proteins with known function which are regulated by TCSs

Lpg#	Gene product	Regulator	Motif[a]	Brief functional description[b]	References
lpg0038	LegA10/AnkQ	CpxR	AR		de Felipe et al. (2005)
lpg0436	LegA11/AnkJ	CpxR	AR		de Felipe et al. (2005); Habyarimana et al. (2008)
lpg2464	SidM/DrrA	CpxR		GEF and GDF for Rab1, AMPylates Rab1	Machner and Isberg (2006); (2007) Muller et al. (2006, 2010); Neunuebel et al. (2011)
lpg2465	SidD	CpxR		deAMPylates Rab1	Neunuebel et al. (2011); Tan and Luo (2011)
lpg0246	Ceg9	PmrA		Trafficking (yeast)	Heidtman et al. (2009)
lpg0376	SdhA	PmrA	GRIP	Vacuole integrity, anti-apoptotic	Creasey and Isberg (2012); Ge et al. (2012); Laguna et al. (2006)
lpg0402	LegA9	PmrA	AR	Preventing autophagy	Khweek et al. (2013)
lpg0437	Ceg14/SidL	PmrA		Inhibits eEF1A	Fontana et al. (2011)
lpg1121	Ceg19	PmrA		Trafficking (yeast)	Heidtman et al. (2009)
lpg1488	LegC5/Lgt3	PmrA		Glucosyltransferase of eEF1A	Belyi et al. (2008); Hurtado-Guerrero et al. (2010)
lpg2144	LegAU13/AnkB	PmrA	AR/F-box	Generates polyubiquitinated proteins on the LCV, supply amino acids for bacterial growth	Price et al. (2009, 2010, 2011)
lpg2452	LegA14	PmrA	AR	Trafficking (yeast)	Heidtman et al. (2009)
lpg2490	LepB	PmrA		GAP for Rab1	Ingmundson et al. (2007); Neunuebel et al. (2011)
lpg2504	Ceg32/SidI	PmrA		Inhibits eEF1A	Shen et al. (2009)
lpg2584	SidF	PmrA		Anti-apoptotic, phosphoinositide phosphatase	Banga et al. (2007); Hsu et al. (2012)
lpg0390	VipA	LetA		Trafficking, binds actin	Franco et al. (2012); Shohdy et al. (2005)
lpg1884	YlfB/LegC2	LetA		Trafficking	(de Felipe et al. (2008)
lpg1950	RalF	LetA	Sec7	GEF for ARF1	Nagai et al. (2002)
lpg2298	YlfA/LegC7	LetA		Trafficking	Campodonico et al. (2005); de Felipe et al. (2008)
lpg0634		LqsR		Trafficking (yeast)	Heidtman et al. (2009)
lpg0695	LegA8/AnkX	LqsA	AR	Phsphocholinates Rab1	Goody et al. (2012); Mukherjee et al. (2011); Pan et al. (2008); Tan et al. (2011)
lpg0968	SidK	LqsR		Inhibition of vacuolar ATPase	Xu et al. (2010)

[a] *AR* Ankyrin repeat, *F-box* F-box domain that mediates ubiquitination of proteins, *GRIP* A domain sufficient for targeting to the Golgi
[b] *Trafficking (yeast)* The involvement in trafficking was shown only in yeast using the CPY system

encoding gene sidD, which is located adjacent to *sidM/drrA*, harbors the opposite enzymatic activity and it functions as a deAMPylator for Rab1 (Neunuebel et al. 2011; Tan and Luo 2011). Examination of the effect of CpxR on the expression of *sidD* revealed that CpxR represses the level of expression of *sidD* as opposed to its activation of the level of expression of *sidM/drrA* (Zusman and Segal, unpublished results). The connection between the regulation and function of these two effectors will be discussed in Sect. 6.1.

3 The PmrA-PmrB Two-Component System

The *L. pneumophila* PmrA-PmrB TCS consists of the PmrA response regulator and the PmrB sensor histidine kinase (Zusman et al. 2007). The PmrAB TCS was studied extensively in *Salmonella enterica* serovar Typhimurium where it functions as the major regulator of lipopolysaccharide modification genes (Gunn 2008). The PmrA regulator was shown to be activated when its cognate sensor PmrB detects mildly acidic pH (Perez and Groisman 2007) or the presence of Fe^{3+} (Wosten et al. 2000). The *S. enterica* PmrAB TCS was found to be active when the bacteria are inside macrophages and during infection of mice (Merighi et al. 2005).

The involvement of the PmrAB TCS in *L. pneumophila* virulence was first identified in a bioinformatic analysis of several *L. pneumophila* effector-encoding genes that were found to contain a conserved regulatory element at their upstream regulatory region. This regulatory element consists of a tandem repeat sequence (cTTAATatT, lower case letters indicate less conserved nucleotides) with a spacer of two nucleotides (Zusman et al. 2007). Bioinformatic and literature searches of bacterial regulatory elements and regulatory factors revealed that a very similar sequence to the one described above has been previously identified in *S. enterica*, and it was shown to be recognized by the PmrA response regulator (Marchal et al. 2004; Wosten and Groisman 1999). Later, the *L. pneumophila* PmrAB TCS was found to activate the expression of many *L. pneumophila* effector-encoding genes (Al-Khodor et al. 2009; Zusman et al. 2007) and the identification of the PmrA regulatory element in the upstream regulatory region of many hypothetical proteins led to their validation as novel effector-encoding genes (Burstein et al. 2009; Zusman et al. 2007). In line with these results, the gene encoding for PmrA was shown to be required for intracellular growth of *L. pneumophila* in amoeba (Zusman et al. 2007). The environmental stimuli that activate the *L. pneumophila* PmrB sensor kinase are not known, but since *L. pneumophila* was shown to inhibit phagosome–lysosome fusion early during infection (Horwitz 1983b), PmrB activation might be related to the pH levels of the LCV, like in the case of *S. enterica* (Perez and Groisman 2007). Currently, the *L. pneumophila* PmrAB regulon consists of 43 effector-encoding genes, it is the largest effectors regulon, and it includes about 15 % of the known *L. pneumophila* effectors.

Of the large number of effectors that were shown to be regulated by the PmrAB TCS, the function of 11 effectors was uncovered (Table 2), and related functions among these effectors might indicate for the time during infection when the PmrAB TCS activates the expression of its target effectors. Two groups of PmrAB regulated effectors with related functions arise: (i) Three effectors (SidI, SidL, and Lgt3) were found to interact with components of the eukaryotic translation elongation machinery (eEF1A and eEF1B), interactions which lead to inhibition of host protein synthesis (Belyi et al. 2008; Fontana et al. 2011; Hurtado-Guerrero et al. 2010; Shen et al. 2009). Lgt3 was shown to function as a glucosyltransferase of eEF1A and the way by which the two other effectors (SidI and SidL) inhibit translation elongation is currently not known. Two additional effectors (Lgt1—Lpg1368 and Lgt2/LegC8—Lpg2862) were shown to function as glucosyltransferases of eEF1A (Belyi et al. 2008; Hurtado-Guerrero et al. 2010), but there is no information regarding their regulation. (ii) Three effectors (SdhA, SidF, and LegAU13/AnkB) seem to be involved in maintenance of the LCV in the host cell. Two of these effectors (SdhA and SidF) were shown to have anti-apoptotic activities (Banga et al. 2007; Laguna et al. 2006). SdhA was shown to actively stabilize the integrity of the LCV during intracellular replication (Creasey and Isberg 2012; Ge et al. 2012) and SidF was shown to contribute to apoptosis resistance of *L. pneumophila*-infected cells by specifically interacting with two proapoptotic members of the Bcl2 protein family (Banga et al. 2007). Beside these two effectors, the effector LegAU13/AnkB that harbors an ankyrin domain and an F-box motif, was shown to generate polyubiquitinated proteins on the LCV, and degradation of these proteins supply amino acids required for bacterial growth (Price et al. 2011). It is important to note that these three effectors (SdhA, SidF, and LegAU13/AnkB) are expected to perform their function after the establishment of the LCV (about 6 h post infection), when the bacteria grow exponentially inside the LCV (Horwitz 1983a). This result was indeed found with the effector SidI (described above) which was found to be expressed during exponential phase when examined in vitro (Shen et al. 2009). An additional PmrAB regulated effector with a known function is LepB which was shown to function as a GAP for Rab1, and was found to translocate into host cells and to perform its function several hours post-infection (Neunuebel et al. 2011) (see Sec. 6.1).

4 The LetA-LetS Two-Component System

The *L. pneumophila* LetA-LetS TCS consists of the LetA response regulator and the LetS sensor histidine kinase (Hammer et al. 2002). The LetAS TCS was found to be present in several γ-proteobacteria, in many of which it bears different names such as the *Pseudomonas aeruginosa*—GacS-GacA, *E. coli*—BarA-UvrY, *S. enterica*—BarA-SirA, and *Vibrio cholerae*—VarS-VarA. In most pathogenic bacteria that harbor this TCS was shown to be involved in virulence gene expression (Lapouge et al. 2008).

The involvement of the LetAS TCS in *L. pneumophila* virulence was first identified in a genetic screen looking for mutants that express the flagellin gene poorly (Hammer et al. 2002). Later, the signal transduction pathway from LetS to individual effector-encoding genes was found to comprise a regulatory cascade (Hovel-Miner et al. 2009; Rasis and Segal 2009; Sahr et al. 2009). During stationary phase LetS activates LetA by a four-step phosphorelay (Edwards et al. 2010). Activated (phosphorylated) LetA positively regulates the transcription of two small regulatory RNAs, RsmY, and RsmZ, which act in a redundant fashion to jointly antagonize CsrA, a small RNA-binding protein that negatively regulates the expression of effectors-encoding genes (Hovel-Miner et al. 2009; Rasis and Segal 2009; Sahr et al. 2009). Sequestration of CsrA by RsmY and RsmZ leads to derepression of CsrA-repressed mRNAs (Rasis and Segal 2009). A deletion mutant of LetA was found to be defective for intracellular growth of *L. pneumophila* in amoeba (Gal-Mor and Segal 2003b; Hammer et al. 2002; Lynch et al. 2003), and the gene encoding for CsrA was found to be essential for *L. pneumophila*, however, mutants containing a reduced level of this regulator were shown to be attenuated for intracellular multiplication in amoeba (Forsbach-Birk et al. 2004; Molofsky and Swanson 2003). In addition, the expression of the gene encoding for CsrA was shown to be activated by the PmrA response regulator described above (Rasis and Segal 2009). The current number of effector-encoding genes regulated by the LetAS-RsmYZ-CsrA regulatory cascade is rather small and includes only four effector-encoding genes (Rasis and Segal 2009; Shi et al. 2006), however, it is highly likely that additional *L. pneumophila* effectors will be found to be regulated by the LetAS-RsmYZ-CsrA regulatory cascade. The level of expression of additional effector-encoding genes was found to be affected by a deletion of *letA* or *letS* (Edwards et al. 2010; Shi et al. 2006), but it is not known if these effectors are regulated by the regulatory cascade described above.

Functional analyses of the four effectors regulated by the LetAS-RsmYZ-CsrA regulatory cascade were already performed and all these effectors were found to be involved in vesicular trafficking (Table 2). RalF, which was the first effector identified in *L. pneumophila*, functions as a GEF for Arf1 (ADP ribosylation factor) and it was shown to localize on the LCV early during infection (Nagai et al. 2002). The VipA effector was found to bind actin in vitro and directly polymerize actin microfilaments. During macrophage infection, VipA was found to be associated with actin patches and early endosomes indicating for its role in modulating organelle trafficking (Franco et al. 2012). The paralogous effectors YlfA and YlfB were also shown to be involved in vesicular trafficking and they were found within large structures that colocalized with anti-KDEL antibodies in mammalian cells (Campodonico et al. 2005; de Felipe et al. 2008). All the effectors known to be regulated by the LetAS-RsmYZ-CsrA regulatory cascade were found to be involved in vesicular trafficking that takes place during the establishment of the LCV. This result might indicate that effectors which are expressed at the end of an infection cycle (the equivalent of stationary phase) are translocated into host cells and perform their function early during the next infection cycle, when the bacteria actively modulate organelle trafficking.

5 The LqsR-LqsS Two-Component System

The *L. pneumophila* LqsR-LqsS TCS consists of the LqsR response regulator and the LqsS sensor histidine kinase (Spirig et al. 2008). The prototype of this system is the *V. cholerae* CqsAS (*Cholerae* quorum sensing) quorum sensing system which includes an autoinducer synthase (CqsA) and its cognate sensor (CqsS) (Miller et al. 2002). The CqsA and CqsS quorum sensing system in *Vibrio* promotes cell density-dependent regulation of virulence and biofilm formation (Miller et al. 2002; Henke and Bassler 2004).

The Lqs system was first identified in *L. pneumophila* by a bioinformatic analysis of the *L. pneumophila* genome with the aim of identifying a homologous system to the CqsAS system from *V. cholerae* and the corresponding *L. pneumophila* proteins were termed LqsA and LqsS (*Legionella* quorum sensing) (Tiaden et al. 2007). In *L. pneumophila*, a gene encoding for a putative response regulator (that lacks a DNA binding motif) is located between *lqsA* and *lqsS*, and this gene was termed *lqsR*. The autoinducer synthase LqsA was shown to catalyze the production of the diffusible signaling molecule 3-hydroxypentadecan-4-one (*Legionella* auto inducer-1—LAI-1, Fig. 1) that is presumably recognized by the sensor kinase LqsS, which in turn probably activates LqsR (Spirig et al. 2008; Tiaden et al. 2010). Recently an 'orphan' homologue of LqsS termed LqsT was identified which probably also respond to LAI-1 (Kessler et al. 2013). DNA microarray experiments revealed that LqsR affects the expression of genes involved in virulence including 12 effector-encoding genes (Tiaden et al. 2007). In addition, a transcriptome analysis of the Δ*lqsA*, Δ*lqsS,* and Δ*lqsT* mutants indicated that the level of expression of several other effector-encoding genes was changed in these mutants (Kessler et al. 2013; Tiaden et al. 2010). The expression of LqsR itself was found to require the RpoS sigma factor, and it was also found to be dependent to a smaller extent on the response regulator LetA (Tiaden et al. 2007). Moreover, the production of LqsR was found to be regulated at a post-transcriptional level by the sRNAs RsmY and RsmZ and by CsrA (Sahr et al. 2009). These results indicate that the Lqs system is involved in the regulation of gene expression during stationary phase, similarly to the LetAS TCS. However, since the LqsR response regulator lacks any known DNA binding motif it is currently not known how it affects gene expression.

The function of two effectors whose level of expression was changed by the Lqs system was determined (Table 2). The effector AnkX was found to catalyze the transfer of phosphocholine to Rab1, which like SidM/DrrA contributes to the activation of Rab1 on the LCV (Goody et al. 2012; Mukherjee et al. 2011; Pan et al. 2008; Tan et al. 2011). The function of the effector SidK was also uncovered and it was found to specifically target the host v-ATPase, a multisubunit complex responsible for organelle acidification in eukaryotic cells. SidK was found to specifically interact with VatA, a key component of the proton pump and this binding was shown to result with inhibition of ATP hydrolysis and proton translocation (Xu et al. 2010). The functions mediated by these two effectors are expected to be required early during infection which correlates well with their activation during stationary phase, similarly to effectors regulated by the LetAS TCS.

6 Interplay Between Different TCSs in The Regulation of *L. pneumophila* Effectors

The pioneering work performed during the early 1980 s by Marcus Horwitz demonstrated that the establishment of the LCV in host cells is a sequential process that includes several steps, a process which was more carefully described later on (Horwitz 1983a, b; Kagan and Roy 2002; Robinson and Roy 2006; Tilney et al. 2001). It is clear today that this stepwise process, which occurs similarly in both human macrophages and amoeba (Abu Kwaik 1996), is mediated by the numerous effectors translocated via the Icm/Dot secretion system (Gomez-Valero et al. 2011; Isberg et al. 2009; Segal et al. 2005). One way to accomplish these stepwise events that occur on the LCV is to translocate different sets of effectors in a timely fashion. The different sets of effectors might be regulated at the level of gene expression which should result with sequential expression and translocation of different sets of effectors which are likely to perform their functions group after group. Since the current knowledge about the regulatory factors that control the expression of effector-encoding genes is rather limited (the regulatory factors that control the expression of more than 200 effectors are not known) it is impossible to build a complete picture about *L. pneumophila* effectors regulation in relation to their function during infection. However, the interplay between different TCSs in relation to the function of different effectors during infection starts to uncover.

6.1 Effectors Manipulating Rab1 are Regulated by Both the CpxRA and the PmrAB TCSs

One of the best studied host factors manipulated by *L. pneumophila* is Rab1. Six effectors (SidM/DrrA, SidD, AnkX, Lem3, LidA, and LepB) were shown to target this host factor (reviewed in Neunuebel and Machner 2012), and the regulation of the genes encoding for some of these effectors is already known. The effector SidM/DrrA was found to activate Rab1 by functioning as a Rab1-GEF (Machner and Isberg 2006; Murata et al. 2006) and as an AMPylator for Rab1 early during infection (Muller et al. 2010; Neunuebel et al. 2011), and it was shown before to be activated by CpxR at the level of transcription (Altman and Segal 2008). SidD (which is located adjacent to *sidM/drrA*) was shown to deAMPylate Rab1, thus counteracting the activity of SidM/DrrA (Neunuebel et al. 2011; Tan and Luo 2011), and it was found to be repressed by CpxR at the level of transcription

(Zusman and Segal, unpublished results). In addition, LepB that functions as a Rab1-GAP (Ingmundson et al. 2007) was shown to be activated by PmrA (Zusman et al. 2007). Examination of the localization of these three effectors on the LCV during the first hours post-infection indicated that SidM/DrrA was localized first to the LCV, during the time when the LCV is being established, while SidD and LepB were found on the LCV several hours post-infection (Neunuebel et al. 2011), at the end of the LCV establishment. Thus, effectors activated by CpxR (SidM/DrrA) seem to be translocated very early during infection and to contribute to the LCV establishment, and effectors repressed by CpxR (SidD) as well as effectors activated by PmrA (LepB) are probably translocated into host cells several hours post-infection. This observation also fits the function of several other PmrA activated effectors (SdhA, SidF, and LegAU13/AnkB) which are likely to perform their function when the bacteria grow exponentially in the LCV.

6.2 The Regulatory Switch Between the PmrAB and the LetAS Regulated Effectors

Another type of interplay between regulators is the direct regulation of one regulator by another regulator. This type of regulation was found between the PmrAB TCS and the CsrA translational repressor, which is a part of the LetAS-RsmYZ-CsrA regulatory cascade (Rasis and Segal 2009). The direct activation of the CsrA encoding gene by PmrA suggests a regulatory switch between two groups of effectors. At the same time when the expression of the effector-encoding genes which are under the regulation of the PmrAB TCS will be activated, the expression of the CsrA-encoding gene will be activated as well, consequently this activation should lead to an increase in the amount of the CsrA post-transcriptional repressor in the bacterial cell that in turn will lead to repression of the group of effector-encoding genes that are under the repression of CsrA. Moreover, the function of the post-transcriptional repressor CsrA was found to be dependent on the LetAS TCS which is activated during stationary phase (Rasis and Segal 2009; Shi et al. 2006) and the PmrA regulated effectors seem to be expressed during exponential phase (Shen et al. 2009). Thus, the likely scenario is that PmrA activates the expression of its target effector-encoding genes during exponential phase at the same time when the level of the sRNAs RsmY and RsmZ anti-repressors in the bacterial cell is low, and then the increase in the amount of the CsrA repressor by PmrA will most likely result with a strong reduction in the level of expression of the effectors repressed by CsrA (Fig. 2). Thus, this interplay between two regulators of effector-encoding genes should result with two apparent groups of effectors: One group of effectors which are activated by the PmrAB TCS and are expressed during exponential phase and the second group of effectors which are de-repressed by the LetAS TCS and are expressed during stationary phase.

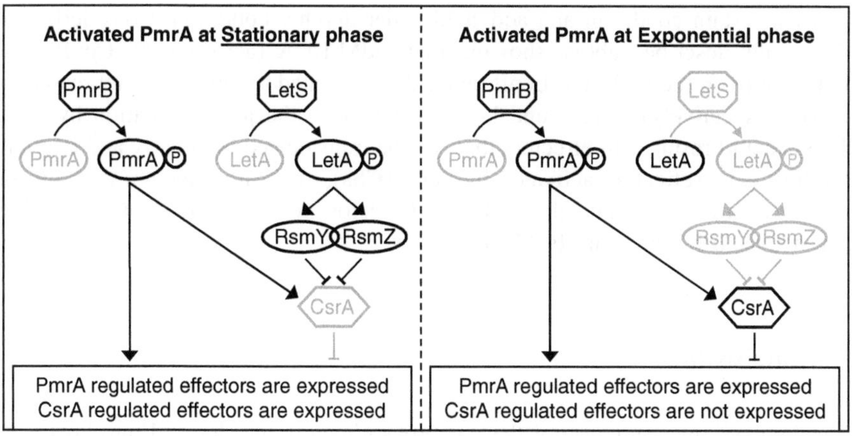

Fig. 2 The interplay between effectors regulated by PmrAB TCS and the LetAS-RsmYZ-CsrA regulatory cascade in relation to growth phase. The activation by the PmrA response regulator might occur during exponential or stationary phase of growth, when the LetAS-RsmYZ-CsrA regulatory cascade is inactivated or activated, respectively. If the PmrAB TCS activation occurs at stationary phase when the level of expression of the small sRNAs RsmY and RsmZ is very high, the activation of the level of expression of the gene encoding for CsrA by PmrA will be sequestered by the two sRNAs (*left side*). If the PmrAB TCS activation occurs at exponential phase (which is more likely, see text) when the level of expression of the small sRNAs RsmY and RsmZ is very low, the activation of the level of expression of the gene encoding for CsrA by PmrA will result with a reduction in the level of expression of effectors regulated by the LetAS-RsmYZ-CsrA regulatory cascade at the same time when effectors directly regulated by PmrA will be activated (*right side*). This interplay between these two regulatory systems will lead to a regulatory switch between the effectors regulated by the two systems. The regulatory components which are present and active at each growth phase are marked in *black*, the regulatory components which are absent or inactive are marked in *grey*

6.3 *Effectors Directly Regulated by Both the CpxRA and PmrAB TCSs*

An additional type of interplay between regulators is the direct regulation of a single effector by multiple TCSs. This situation was described thus far for three effectors (Ceg7, Ceg18, and Ceg33) which were shown to be directly regulated by both the CpxRA and the PmrAB TCSs (Altman and Segal 2008). The joint regulation by these TCSs includes two possible scenarios since CpxR was shown to function as a repressor or as an activator of effector-encoding genes and PmrA was shown to function only as an activator. Ceg07 and Ceg18 were found to be repressed by CpxR and activated by PmrA, and Ceg33 was found to be activated by both of these TCSs. Taking into account that these TCSs probably respond to different environmental stimuli (currently there is no information regarding the environmental stimuli that activate the CpxA or PmrB sensor histidine kinases) this type of regulation will result with effectors that will be expressed under conditions that activate both TCSs as well as with effectors that will be repressed under a certain condition and activated under another condition. Considering all the results described above showing that SidM/DrrA (activated by CpxR) was found early on the LCV and LepB and SidD (activated by PmrA and repressed by CpxR, respectively) were found later on the LCV, the joint regulation by these TCSs should result with effectors that will be translocated both early and late into the host cells (effectors activated by both TCSs), and with effectors that will be translocated only late into the host cells (effectors repressed by the CpxRA TCS and activated by the PmrAB TCS).

7 Conclusions

The study of the regulation of the numerous effectors translocated into host cells by *L. pneumophila* already uncovered four TCSs that participate in this process. However, the ways by which the majority of the effector-encoding genes are regulated is largely unknown. Further study of these TCSs and the environmental stimuli that activate them as well as identification of additional TCSs and other regulators of gene expression that coordinate the regulation of the *L. pneumophila* effector-encoding genes will deepen our understanding on the ways by which such a multicomponent pathogenesis system is controlled at the level of gene expression in order to result with a successful infection.

References

Abu Kwaik Y (1996) The phagosome containing *Legionella pneumophila* within the protozoan *Hartmannella vermiformis* is surrounded by the rough endoplasmic reticulum. Appl Environ Microbiol 62:2022–2028

Al-Khodor S, Kalachikov S, Morozova I, Price CT, Abu Kwaik Y (2009) The PmrA/PmrB two-component system of *Legionella pneumophila* is a global regulator required for intracellular replication within macrophages and protozoa. Infect Immun 77:374–386

Altman E, Segal G (2008) The response regulator CpxR directly regulates the expression of several *Legionella pneumophila icm/dot* components as well as new translocated substrates. J Bacteriol 190:1985–1996

Bachman MA, Swanson MS (2004) The LetE protein enhances expression of multiple LetA/LetS-dependent transmission traits by *Legionella pneumophila*. Infect Immun 72:3284–3293

Banga S, Gao P, Shen X, Fiscus V, Zong WX, Chen L, Luo ZQ (2007) *Legionella pneumophila* inhibits macrophage apoptosis by targeting pro-death members of the Bcl2 protein family. Proc Natl Acad Sci USA 104:5121–5126

Belyi Y, Tabakova I, Stahl M, Aktories K (2008) Lgt: a family of cytotoxic glucosyltransferases produced by *Legionella pneumophila*. J Bacteriol 190:3026–3035

Burstein D, Zusman T, Degtyar E, Viner R, Segal G, Pupko T (2009) Genome-scale identification of *Legionella pneumophila* effectors using a machine learning approach. PLoS Pathog 5:e1000508

Campodonico EM, Chesnel L, Roy CR (2005) A yeast genetic system for the identification and characterization of substrate proteins transferred into host cells by the *Legionella pneumophila* Dot/Icm system. Mol Microbiol 56:918–933

Creasey EA, Isberg RR (2012) The protein SdhA maintains the integrity of the *Legionella*-containing vacuole. Proc Natl Acad Sci USA 109:3481–3486

de Felipe KS, Glover RT, Charpentier X, Anderson OR, Reyes M, Pericone CD, Shuman HA (2008) *Legionella* eukaryotic-like type IV substrates interfere with organelle trafficking. PLoS Pathog 4:e1000117

de Felipe KS, Pampou S, Jovanovic OS, Pericone CD, Ye SF, Kalachikov S, Shuman HA (2005) Evidence for acquisition of *Legionella* type IV secretion substrates via interdomain horizontal gene transfer. J Bacteriol 187:7716–7726

Dorer MS, Kirton D, Bader JS, Isberg RR (2006) RNA interference analysis of *Legionella* in *Drosophila* cells: exploitation of early secretory apparatus dynamics. PLoS Pathog 2:e34

Edwards RL, Jules M, Sahr T, Buchrieser C, Swanson MS (2010) The *Legionella pneumophila* LetA/LetS two-component system exhibits rheostat-like behavior. Infect Immun 78:2571–2583

Ensminger AW, Isberg RR (2009) *Legionella pneumophila* Dot/Icm translocated substrates: a sum of parts. Curr Opin Microbiol 12:67–73

Fass E, Groisman EA (2009) Control of *Salmonella* pathogenicity island-2 gene expression. Curr Opin Microbiol 12:199–204

Fontana MF, Banga S, Barry KC, Shen X, Tan Y, Luo ZQ, Vance RE (2011) Secreted bacterial effectors that inhibit host protein synthesis are critical for induction of the innate immune response to virulent *Legionella pneumophila*. PLoS Pathog 7:e1001289

Forsbach-Birk V, McNealy T, Shi C, Lynch D, Marre R (2004) Reduced expression of the global regulator protein CsrA in *Legionella pneumophila* affects virulence-associated regulators and growth in *Acanthamoeba castellanii*. Int J Med Microbiol 294:15–25

Franco IS, Shohdy N, Shuman HA (2012) The *Legionella pneumophila* effector VipA is an actin nucleator that alters host cell organelle trafficking. PLoS Pathog 8:e1002546

Franco IS, Shuman HA, Charpentier X (2009) The perplexing functions and surprising origins of *Legionella pneumophila* type IV secretion effectors. Cell Microbiol 11:1435–1443

Gal-Mor O, Segal G (2003a) Identification of CpxR as a positive regulator of *icm* and *dot* virulence genes of *Legionella pneumophila*. J Bacteriol 185:4908–4919

Gal-Mor O, Segal G (2003b) The *Legionella pneumophila* GacA homolog (LetA) is involved in the regulation of *icm* virulence genes and is required for intracellular multiplication in *Acanthamoeba castellanii*. Microb Pathog 34:187–194

Ge J, Gong YN, Xu Y, Shao F (2012) Preventing bacterial DNA release and absent in melanoma 2 inflammasome activation by a *Legionella* effector functioning in membrane trafficking. Proc Natl Acad Sci USA 109:6193–6198

Gomez-Valero L, Rusniok C, Cazalet C, Buchrieser C (2011) Comparative and functional genomics of *Legionella* identified eukaryotic like proteins as key players in host-pathogen interactions. Front Microbiol 2:208

Gooderham WJ, Hancock RE (2009) Regulation of virulence and antibiotic resistance by two-component regulatory systems in *Pseudomonas aeruginosa*. FEMS Microbiol Rev 33:279–294

Goody PR, Heller K, Oesterlin LK, Muller MP, Itzen A, Goody RS (2012) Reversible phosphocholination of Rab proteins by Legionella pneumophila effector proteins. EMBO J 31:1774–1784

Gotoh Y, Eguchi Y, Watanabe T, Okamoto S, Doi A, Utsumi R (2010) Two-component signal transduction as potential drug targets in pathogenic bacteria. Curr Opin Microbiol 13:232–239

Gunn JS (2008) The *Salmonella* PmrAB regulon: lipopolysaccharide modifications, antimicrobial peptide resistance and more. Trends Microbiol 16:284–290

Habyarimana F, Al-Khodor S, Kalia A, Graham JE, Price CT, Garcia MT, Kwaik YA (2008) Role for the Ankyrin eukaryotic-like genes of *Legionella pneumophila* in parasitism of protozoan hosts and human macrophages. Environ Microbiol 10:1460–1474

Hammer BK, Tateda ES, Swanson MS (2002) A two-component regulator induces the transmission phenotype of stationary-phase *Legionella pneumophila*. Mol Microbiol 44:107–118

Heidtman M, Chen EJ, Moy MY, Isberg RR (2009) Large-scale identification of *Legionella pneumophila* Dot/Icm substrates that modulate host cell vesicle trafficking pathways. Cell Microbiol 11:230–248

Horwitz MA (1983a) Formation of a novel phagosome by the Legionnaires' disease bacterium (*Legionella pneumophila*) in human monocytes. J Exp Med 158:1319–1331

Horwitz MA (1983b) The Legionnaires' disease bacterium (*Legionella pneumophila*) inhibits phagosome-lysosome fusion in human monocytes. J Exp Med 158:2108–2126

Hovel-Miner G, Pampou S, Faucher SP, Clarke M, Morozova I, Morozov P, Russo JJ, Shuman HA, Kalachikov S (2009) SigmaS controls multiple pathways associated with intracellular multiplication of *Legionella pneumophila*. J Bacteriol 191:2461–2473

Hsu F, Zhu W, Brennan L, Tao L, Luo ZQ, Mao Y (2012) Structural basis for substrate recognition by a unique *Legionella* phosphoinositide phosphatase. Proc Natl Acad Sci USA 109:13567–13572

Humphreys S, Rowley G, Stevenson A, Anjum MF, Woodward MJ, Gilbert S, Kormanec J, Roberts M (2004) Role of the two-component regulator CpxAR in the virulence of *Salmonella enterica* serotype Typhimurium. Infect Immun 72:4654–4661

Hunke S, Keller R, Muller VS (2012) Signal integration by the Cpx-envelope stress system. FEMS Microbiol Lett 326:12–22

Hurtado-Guerrero R, Zusman T, Pathak S, Ibrahim AF, Shepherd S, Prescott A, Segal G, van Aalten DM (2010) Molecular mechanism of elongation factor 1A inhibition by a *Legionella pneumophila* glycosyltransferase. Biochem J 426:281–292

Ingmundson A, Delprato A, Lambright DG, Roy CR (2007) *Legionella pneumophila* proteins that regulate Rab1 membrane cycling. Nature 450:365–369

Isberg RR, O'Connor TJ, Heidtman M (2009) The *Legionella pneumophila* replication vacuole: making a cosy niche inside host cells. Nat Rev Microbiol 7:13–24

Jung K, Fried L, Behr S, Heermann R (2012) Histidine kinases and response regulators in networks. Curr Opin Microbiol 15:118–124

Kagan JC, Roy CR (2002) *Legionella* phagosomes intercept vesicular traffic from endoplasmic reticulum exit sites. Nat Cell Biol 4:945–954

Kessler A, Schell U, Sahr T, Tiaden A, Harrison C, Buchrieser C, Hilbi H (2013) The *Legionella pneumophila* orphan sensor kinase LqsT regulates competence and pathogen-host interactions as a component of the LAI-1 circuit. Environ Microbiol 15:646–662

Khweek AA, Caution K, Akhter A, Abdulrahman BA, Tazi M, Hassan H, Majumdar N, Doran A, Guirado E, Schlesinger LS, Shuman H, Amer AO (2013) A bacterial protein promotes the recognition of the *Legionella pneumophila* vacuole by autophagy. Eur J Immunol: doi. doi:10.1002/eji.201242835

Laguna RK, Creasey EA, Li Z, Valtz N, Isberg RR (2006) A *Legionella pneumophila*-translocated substrate that is required for growth within macrophages and protection from host cell death. Proc Natl Acad Sci USA 103:18745–18750

Lapouge K, Schubert M, Allain FH, Haas D (2008) Gac/Rsm signal transduction pathway of gamma-proteobacteria: from RNA recognition to regulation of social behaviour. Mol Microbiol 67:241–253

Laub MT, Goulian M (2007) Specificity in two-component signal transduction pathways. Annu Rev Genet 41:121–145

Lynch D, Fieser N, Gloggler K, Forsbach-Birk V, Marre R (2003) The response regulator LetA regulates the stationary-phase stress response in *Legionella pneumophila* and is required for efficient infection of *Acanthamoeba castellanii*. FEMS Microbiol Lett 219:241–248

Machner MP, Isberg RR (2006) Targeting of host Rab GTPase function by the intravacuolar pathogen *Legionella pneumophila*. Dev Cell 11:47–56

Machner MP, Isberg RR (2007) A bifunctional bacterial protein links GDI displacement to Rab1 activation. Science 318:974–977

Marchal K, De Keersmaecker S, Monsieurs P, van Boxel N, Lemmens K, Thijs G, Vanderleyden J, De Moor B (2004) In silico identification and experimental validation of PmrAB targets in *Salmonella typhimurium* by regulatory motif detection. Genome Biol 5:R9

Merighi M, Ellermeier CD, Slauch JM, Gunn JS (2005) Resolvase-in vivo expression technology analysis of the *Salmonella enterica* serovar Typhimurium PhoP and PmrA regulons in BALB/c mice. J Bacteriol 187:7407–7416

Miller MB, Skorupski K, Lenz DH, Taylor RK, Bassler BL (2002) Parallel quorum sensing systems converge to regulate virulence in *Vibrio cholerae*. Cell 110:303–314

Molofsky AB, Swanson MS (2003) *Legionella pneumophila* CsrA is a pivotal repressor of transmission traits and activator of replication. Mol Microbiol 50:445–461

Mukherjee S, Liu X, Arasaki K, McDonough J, Galan JE, Roy CR (2011) Modulation of Rab GTPase function by a protein phosphocholine transferase. Nature 477:103–106

Muller MP, Peters H, Blumer J, Blankenfeldt W, Goody RS, Itzen A (2010) The *Legionella* effector protein DrrA AMPylates the membrane traffic regulator Rab1b. Science 329:946–949

Murata T, Delprato A, Ingmundson A, Toomre DK, Lambright DG, Roy CR (2006) The *Legionella pneumophila* effector protein DrrA is a Rab1 guanine nucleotide-exchange factor. Nat Cell Biol 8:971–977

Nagai H, Kagan JC, Zhu X, Kahn RA, Roy CR (2002) A bacterial guanine nucleotide exchange factor activates ARF on *Legionella* phagosomes. Science 295:679–682

Neunuebel MR, Chen Y, Gaspar AH, Backlund PS Jr, Yergey A, Machner, MP (2011) De-AMPylation of the small GTPase Rab1 by the pathogen *Legionella pneumophila*. Science 333:453–456

Neunuebel MR, Machner MP (2012) The taming of a Rab GTPase by *Legionella pneumophila*. Small GTPases 3:28–33

Nevesinjac AZ, Raivio TL (2005) The Cpx envelope stress response affects expression of the type IV bundle-forming pili of enteropathogenic *Escherichia coli*. J Bacteriol 187:672–686

O'Connor TJ, Boyd D, Dorer MS, Isberg RR (2012) Aggravating genetic interactions allow a solution to redundancy in a bacterial pathogen. Science 338:1440–1444

Pan X, Luhrmann A, Satoh A, Laskowski-Arce MA, Roy CR (2008) Ankyrin repeat proteins comprise a diverse family of bacterial type IV effectors. Science 320:1651–1654

Perez JC, Groisman EA (2007) Acid pH activation of the PmrA/PmrB two-component regulatory system of *Salmonella enterica*. Mol Microbiol 63:283–293

Price CT, Al-Khodor S, Al-Quadan T, Santic M, Habyarimana F, Kalia A, Kwaik YA (2009) Molecular mimicry by an F-box effector of *Legionella pneumophila* hijacks a conserved polyubiquitination machinery within macrophages and protozoa. PLoS Pathog 5:e1000704

Price CT, Al-Quadan T, Santic M, Jones SC, Abu Kwaik Y (2010) Exploitation of conserved eukaryotic host cell farnesylation machinery by an F-box effector of *Legionella pneumophila*. J Exp Med 207:1713–1726

Price CT, Al-Quadan T, Santic M, Rosenshine I, Abu Kwaik Y (2011) Host proteasomal degradation generates amino acids essential for intracellular bacterial growth. Science 334:1553–1557

Rasis M, Segal G (2009) The LetA-RsmYZ-CsrA regulatory cascade, together with RpoS and PmrA, post-transcriptionally regulates stationary phase activation of *Legionella pneumophila* Icm/Dot effectors. Mol Microbiol 72:995–1010

Robinson CG, Roy CR (2006) Attachment and fusion of endoplasmic reticulum with vacuoles containing *Legionella pneumophila*. Cell Microbiol 8:793–805

Sahr T, Bruggemann H, Jules M, Lomma M, Albert-Weissenberger C, Cazalet C, Buchrieser C (2009) Two small ncRNAs jointly govern virulence and transmission in *Legionella pneumophila*. Mol Microbiol 72:741–762

Segal G, Feldman M, Zusman T (2005) The Icm/Dot type-IV secretion systems of *Legionella pneumophila* and *Coxiella burnetii*. FEMS Microbiol Rev 29:65–81

Shen X, Banga S, Liu Y, Xu L, Gao P, Shamovsky I, Nudler E, Luo ZQ (2009) Targeting eEF1A by a *Legionella pneumophila* effector leads to inhibition of protein synthesis and induction of host stress response. Cell Microbiol 11:911–926

Shi C, Forsbach-Birk V, Marre R, McNealy TL (2006) The *Legionella pneumophila* global regulatory protein LetA affects DotA and Mip. Int J Med Microbiol 296:15–24

Shin S, Roy CR (2008) Host cell processes that influence the intracellular survival of *Legionella pneumophila*. Cell Microbiol 10:1209–1220

Shohdy N, Efe JA, Emr SD, Shuman HA (2005) Pathogen effector protein screening in yeast identifies *Legionella* factors that interfere with membrane trafficking. Proc Natl Acad Sci USA 102:4866–4871

Spirig T, Tiaden A, Kiefer P, Buchrieser C, Vorholt JA, Hilbi H (2008) The *Legionella* autoinducer synthase LqsA produces an alpha-hydroxyketone signaling molecule. J Biol Chem 283:18113–18123

Tan Y, Arnold RJ, Luo ZQ (2011) *Legionella pneumophila* regulates the small GTPase Rab1 activity by reversible phosphorylcholination. Proc Natl Acad Sci USA 108:21212–21217

Tan Y, Luo ZQ (2011) *Legionella pneumophila* SidD is a deAMPylase that modifies Rab1. Nature 475:506–509

Tiaden A, Spirig T, Carranza P, Bruggemann H, Riedel K, Eberl L, Buchrieser C, Hilbi H (2008) Synergistic contribution of the *Legionella pneumophila lqs* genes to pathogen-host interactions. J Bacteriol 190:7532–7547

Tiaden A, Spirig T, Sahr T, Walti MA, Boucke K, Buchrieser C, Hilbi H (2010) The autoinducer synthase LqsA and putative sensor kinase LqsS regulate phagocyte interactions, extracellular filaments and a genomic island of *Legionella pneumophila*. Environ Microbiol 12:1243–1259

Tiaden A, Spirig T, Weber SS, Bruggemann H, Bosshard R, Buchrieser C, Hilbi H (2007) The *Legionella pneumophila* response regulator LqsR promotes host cell interactions as an element of the virulence regulatory network controlled by RpoS and LetA. Cell Microbiol 9:2903–2920

Tilney LG, Harb OS, Connelly PS, Robinson CG, Roy CR (2001) How the parasitic bacterium *Legionella pneumophila* modifies its phagosome and transforms it into rough ER: implications for conversion of plasma membrane to the ER membrane. J Cell Sci 114:4637–4650

Vogt SL, Raivio TL (2012) Just scratching the surface: an expanding view of the Cpx envelope stress response. FEMS Microbiol Lett 326:2–11

Wosten MM, Groisman EA (1999) Molecular characterization of the PmrA regulon. J Biol Chem 274:27185–27190

Wosten MM, Kox LF, Chamnongpol S, Soncini FC, Groisman EA (2000) A signal transduction system that responds to extracellular iron. Cell 103:113–125

Xu L, Shen X, Bryan A, Banga S, Swanson MS, Luo ZQ (2010) Inhibition of host vacuolar H + -ATPase activity by a *Legionella pneumophila* effector. PLoS Pathog 6:e1000822

Zusman T, Aloni G, Halperin E, Kotzer H, Degtyar E, Feldman M, Segal G (2007) The response regulator PmrA is a major regulator of the *icm/dot* type IV secretion system in *Legionella pneumophila* and *Coxiella burnetii*. Mol Microbiol 63:1508–1523

Facets of Small RNA-Mediated Regulation in *Legionella pneumophila*

Hana Trigui, Nilmini Mendis, Laam Li, Mariam Saad and Sebastien P. Faucher

Abstract *Legionella pneumophila* is a water-borne pathogen that causes a severe lung infection in humans. It is able to replicate inside amoeba in the water environment, and inside lung macrophages in humans. Efficient regulation of gene expression is critical for responding to the conditions that *L. pneumophila* encounters and for intracellular multiplication in host cells. In the last two decades, many reports have contributed to our understanding of the critical importance of small regulatory RNAs (sRNAs) in the regulatory network of bacterial species. This report presents the current state of knowledge about the sRNAs expressed by *L. pneumophila* and discusses a few regulatory pathways in which sRNAs should be involved in this pathogen.

The first two authors, Hana Trigui and Nilmini Mendis, have contributed equally to this work.

H. Trigui · N. Mendis · L. Li · M. Saad · S. P. Faucher (✉)
Faculty of Agricultural and Environmental Sciences,
Department of Natural Resource Sciences, McGill University,
Ste-Anne-de-Bellevue, QC H9X 3V9, Canada
e-mail: sebastien.faucher2@mcgill.ca

H. Trigui
e-mail: hana.trigui@mcgill.ca

N. Mendis
e-mail: nissanka.mendis@mail.mcgill.ca

L. Li
e-mail: laam.li@mail.mcgill.ca

M. Saad
e-mail: mariam.saad@mail.mcgill.ca

Current Topics in Microbiology and Immunology (2013) 376: 53–80 53
DOI: 10.1007/82_2013_347

Published Online: 6 August 2013

Contents

1 Introduction 54
2 Small Regulatory RNAs 56
2.1 Base-Pairing Small RNAs 57
2.2 Protein-Binding Small RNA 60
3 Small RNAs Expressed By *Legionella pneumophila* 61
3.1 Identification of sRNAs Expressed By L. pneumophila 61
3.2 RsmX/Y/Z and CsrA 61
3.3 6S RNA 64
3.4 Other sRNAs 66
3.5 Hfq-Regulated Genes 68
4 *Legionella pneumophila* Regulatory Systems That May Contain sRNAs 68
4.1 OxyR System and Oxidative Stress 69
4.2 RpoS-Regulating sRNA 71
4.3 Fur and Iron Uptake 72
5 Future Perspective 72
References 73

1 Introduction

A mysterious bout of illness amongst veterans during a meeting of the American Legion in 1976 shed light on a new pathogen causing, what is known today as, Legionellosis (Fraser et al. 1977). This umbrella term combines Legionnaires' disease (LD), a serious form of potentially life-threatening pneumonia and Pontiac fever, a milder flu-like illness (Fraser et al. 1977). There are over 50 species and >70 serogroups of this Gram-negative bacterium, but the most common etiological agent of Legionellosis is *Legionella pneumophila* serogroup 1, contributing to >85 % of disease (Jarraud et al. 2013; Reimer et al. 2010; Yu et al. 2002). The mortality rate ranges from 5 to 20 %, increasing to almost 50 % in some nosocomial outbreaks (Benin et al. 2002; Gomez-Valero et al. 2011; Marston et al. 1994; Swanson and Hammer 2000).

In the environment, *Legionella* is mainly an intracellular pathogen of amoeba but can also replicate in ciliate protozoans (Fields 1996). Upon entry into humans through inhalation of aerosols, *L. pneumophila* can infect and replicate inside human alveolar macrophages and potentially cause LD in susceptible individuals (Butler and Breiman 1998). The bacterium enters human macrophages through coiling phagocytosis after which it is found in a modified phagosome called a *Legionella* Containing Vacuole (LCV) (Horwitz 1983a). *L. pneumophila* prevents phagosome-lysosome fusion (Horwitz 1983b). Nascent LCVs are progressively associated with vesicles originating from the host endoplasmic reticulum, and replication of the pathogen is observed approximately 8 h postinfection (Horwitz 1983a). Motile bacteria are then released through host cell lysis (Fields et al. 2002). This cycle of infection and replication inside host cells is highly dependent on the Type-IVb (T4bSS) and Type-II (T2SS) secretion systems encoded by the *dot/icm* (defective in organelle trafficking/intracellular multiplication) and *lsp* (*Legionella* Type-II secretory pathway) genes respectively (reviewed in (Albert-Weissenberger et al. 2007; Vogel and Isberg 1999). Combined, T4bSS and T2SS translocate more than 300 effector proteins to the host cell cytoplasm. These effectors modify cell trafficking and ensure a safe replicative niche for the bacteria inside the LCV (Gomez-Valero et al. 2011; Heidtman et al. 2009).

It is believed that *L. pneumophila* displays a biphasic life style, switching between a replicative phase, during infection of host cells, and a transmissive phase, when the resources inside the cells are exhausted and infection of a new cell is needed (Molofsky and Swanson 2004). The replicative phase is characterized by exponential growth and repression of motility, and is mimicked by exponential growth in rich broth. On the other hand, bacteria in the transmissive phase become motile and express cytotoxicity-related traits, similar to what is seen when bacteria reach the postexponential phase in rich broth.

Apart from growing intracellularly in protozoans, *L. pneumophila* exists in three other states in the environment, namely the viable-but-non-culturable (VBNC) persistent state, the biofilm-associated sessile state, and the free-living planktonic state (Hussong et al. 1987; Rogers and Keevil 1992). All four states occur in natural and man-made water systems with optimal growth temperatures recorded between 25 and 37 °C (Ohno et al. 2003). Human infections are mainly a result of exposure to contaminated aerosols generated from man-made water reservoirs, such as air-conditioning systems, humidifiers, cooling towers, water heating systems, and even showers (Blatt et al. 1993; Hanrahan et al. 1987; Yiallouros et al. 2013). The widespread use of these systems greatly increases the chance of contracting Legionellosis. In order to survive and replicate in the water environment, *L. pneumophila* must adapt to various stresses, such as starvation and predation. Then, when it is aerosolized and transported to the human lungs, it must adapt to yet another, new set of conditions and stresses. Adaptation to these conditions requires changes in gene expression. In *L. pneumophila*, these changes are regulated by elements such as two-component systems (LetA/S, PmrA/B, and CpxR/A), sigma factors (RpoS), and RNA-binding proteins (CsrA) to name a few (Altman and Segal 2008; Gal-Mor and Segal 2003; Hovel-Miner et al. 2009; Molofsky and Swanson 2003b; Zusman et al. 2007). These regulators are critical for the switch between the replicative and transmissive phases. Small regulatory RNAs (sRNAs) are also major players of gene regulation in bacteria (Waters and Storz 2009). Moreover, this regulation by sRNA is likely faster and more cost-effective than regulation by polypeptides (Altuvia et al. 1997). In this review, we will discuss recent advances in the identification and characterization of sRNAs that were found in *L. pneumophila* and we will review a few examples of sRNAs involved in regulatory systems of other bacteria that might also be present in *L. pneumophila*.

2 Small Regulatory RNAs

Bacteria have an intrinsic ability to respond to changes in the environment, and regulatory networks encoded in the bacterial genome control such responses. Some important players in these networks are regulatory RNA molecules such as riboswitches and sRNAs (Waters and Storz 2009). Regulatory RNA molecules can control transcription, translation, and the stability of their target mRNAs and thus, influence genetic programs (Waters and Storz 2009). Regulatory RNA molecules were identified in bacteria prior to the discovery of the first microRNAs (miRNAs) and short interfering RNAs (siRNAs) in eukaryotes (Waters and Storz 2009).

Early work reported sRNA molecules associated with the control of replication of the ColE1 plasmid and transposition of Tn10 (Simons and Kleckner 1983; Stougaard et al. 1981). However, these early examples of sRNA were encoded on mobile genetic elements, and the first true bacterial sRNA encoded on the chromosome was only reported in 1984. This sRNA, MicF, was discovered in *Escherichia coli* and was shown to inhibit translation of the major outer membrane porin OmpF by a mechanism involving direct interaction with its target mRNA (Mizuno et al. 1984). Subsequent to these developments, several sRNAs were identified in the genome of *E. coli* by 2D PAGE (Wassarman et al. 1999). Shortly thereafter, the functions of some of these sRNAs were characterized and homologs were found in other bacterial species (Wassarman et al. 1999). Since then, predictive bioinformatic tools have been used to find putative regulatory RNAs. Bacterial genome sequencing, multilayered computational searches, deep sequencing, and tiled microarrays with full genome coverage have facilitated the discovery of more regulatory RNAs in several other bacteria and revealed a widespread and significant role for sRNAs in the regulation of gene expression (Waters and Storz 2009).

Small RNAs are short RNA molecules that are approximately 80 to 500 nucleotides in length (Waters and Storz 2009). They can be transcribed from their own promoter or can be produced through the processing of a larger transcript (Waters and Storz 2009). Most sRNAs do not encode proteins, but a few exceptions have been described, including SgrS, expressed by *E. coli*, and RNAIII, expressed by *Staphylococcus aureus* (Novick and Geisinger 2008; Wadler and Vanderpool 2007). sRNAs are involved in a series of regulatory functions in processes, such as stress response, virulence, and competence (Papenfort and Vogel 2010; Waters and Storz 2009; Yamamoto et al. 2011). The widespread use of these molecules in bacteria to regulate gene expression is probably due to their reduced metabolic cost and faster rate of regulation compared to protein regulators (Beisel and Storz 2010). Depending on the nature of their targets, sRNAs can be grouped into two functional categories: base-pairing and protein-binding sRNAs.

2.1 Base-Pairing Small RNAs

Base-pairing sRNAs are the most common type of sRNAs (Faucher and Shuman 2011). They are short but highly structured RNA molecules complementary to their target mRNA strands to a certain extent, and they contribute to the control of gene expression by acting as posttranscriptional regulators (Brantl 2007). They carry out this role through several mechanisms of action that require hybridization of the sRNA to its target mRNA. They can up-regulate the expression of a gene either by increasing the stability of the target or by inducing a modification in the secondary structure of the mRNA to expose the ribosomal binding site (RBS) (Waters and Storz 2009). Alternatively, they can down-regulate the expression of a gene either by inducing the decay of the target, or by inhibiting translation by blocking access of the ribosome either by binding directly to the RBS or by inducing a change in the secondary structure that efficiently hides the RBS (Lalaouna et al. 2013; Waters and Storz 2009). Moreover, this class of sRNAs represents a way to regulate products encoded on polycistronic mRNA independently from one another (Balasubramanian and Vanderpool 2013).

2.1.1 Cis-Encoded sRNAs

Base-pairing sRNAs can be encoded on the complementary strand of their target, in which case they are referred to as cis-encoded sRNAs (Brantl 2007). These share extensive sequence complementarity, but do not necessarily form long RNA duplexes with their target mRNA (Brantl 2007). Here, it is important to consider that an mRNA consists of 5' and 3' untranslated sequences (UTR), and a central coding region. Cis-encoded sRNAs antisense to either one of these regions could lead to posttranscriptional regulation of a target mRNA. However, additional experimental evidence is necessary to better understand the effects of the cis-encoded sRNA on its target gene(s) and the mechanisms involved. Jager et al. (2012) published evidence that cis-encoded sRNAs could mediate regulation of gene expression *in trans* and that the interaction of the sRNA with the cis-encoded and the trans-encoded targets occurs by means of two distinct domains.

2.1.2 Trans-Encoded sRNAs

In contrast to cis-encoded sRNAs, trans-encoded sRNAs are found outside the coding sequence of their target genes. Consequently, the level of homology between the sRNA and its target is usually low, and the hybridization of the sRNA to the mRNA occurs via short, imperfectly homologous sequences. Therefore, trans-encoded sRNAs often require the RNA-binding protein Hfq to interact with their target, contrary to cis-encoded sRNAs that usually do not require Hfq assistance (Waters and Storz 2009). It is also not uncommon for trans-encoded sRNAs to regulate the expression of many different mRNAs. In *E. coli*, the sRNA RyhB controls the expression of approximately 56 genes in response to changes in iron homeostasis (Massé et al. 2005).

2.1.3 Role of the RNA-Binding Protein Hfq

Hfq was first described as *E. coli* host factor I required for bacteriophage Qβ replication (Franze de Fernandez et al. 1968). It acts as a global regulator implicated in posttranscriptional regulation in many bacteria. In the last decade, elucidating the role of Hfq in sRNA-mediated gene regulation has been an area of intense research. Phenotypes due to Hfq deficiency have been studied in many bacteria, including pathogens such as *E. coli* (Tsui et al. 1994), *Salmonella typhimurium* (Sittka et al. 2007), and *Vibrio cholerae* (Ding et al. 2004). As a result, Hfq has been implicated in the regulation of growth, stress response, virulence, and biofilm formation (reviewed in Sobrero and Valverde 2012).

Hfq plays a central role in some bacterial sRNA pathways while being dispensable for others. For example, *Helicobacter pylori* expresses hundreds of sRNAs, but no Hfq homolog has been identified yet, suggesting that Hfq is dispensable for sRNA-mediated regulation in this bacterium (Sharma et al. 2010). However, in the case of *E. coli* and *V. cholerae*, some sRNAs require Hfq for effective regulation of their targets, while others carry out their respective functions without Hfq mediation (Lenz et al. 2004; Song et al. 2008; Zhang et al. 2003). According to the postulates of Jousselin et al. (2009), Hfq dependency is positively correlated with a high GC content in the bacterial genome, but is inversely proportional to the length of the interacting sequence between the sRNA and its target mRNA. Finally, the presence of a C-terminal extension in the Hfq sequence, which forms the mRNA interaction surface, is predictive of the involvement of Hfq in sRNA-mediated regulation. While the GC content in *L. pneumophila* is low (38 %), the C-terminal region of Hfq is similar to that of *V. cholera* where Hfq is required for some, but not all sRNA–mRNA interactions (Lenz et al. 2004; Song et al. 2008; Zhang et al. 2003). As discussed in detail in our previous publication (Faucher and Shuman 2011), the *L. pneumophila* Hfq protein might be required for precise sRNA–mRNA interactions in some regulatory pathways, including iron homeostasis (McNealy et al. 2005) described in Sect. 4.3. The following paragraphs will review what is known about Hfq in the model organism *E. coli* to help readers appreciate the known mechanism by which Hfq affects gene expression and to guide further analysis of its implication in *L. pneumophila* gene regulation.

Hfq acts as a posttranscriptional regulator by stabilizing the sRNA and facilitating its interaction with mRNA targets (Geissmann and Touati 2004; Gottesman 2004). This results in the efficient regulation of various genes (Gottesman et al. 2006; Gottesman and Storz 2010). Structural studies in *E. coli* revealed that Hfq belongs to a large family of Sm and Sm-like proteins characterized by an RNA-binding activity and formation of a homohexameric ring (Møller et al. 2002;

Schumacher et al. 2002; Zhang et al. 2002). Analysis of the RNA-binding site showed that mRNA sequences containing uridine residues bind to the proximal face of the Hfq protein, and that adenine-rich sequences bind to its distal face (Link et al. 2009). This structure suggests that Hfq could potentially bind two RNA molecules simultaneously, effectively enhancing molecular interaction by drawing complementary RNA sequences close together (Link et al. 2009).

Recently, Hwang et al. (2011) investigated how Hfq achieves both "annealing" and "unwinding" functions for two different RNA substrates. To gain insight into efficient annealing of an sRNA to the proximal binding site of Hfq, the sRNA DsrA, and its target mRNA *rpoS* were investigated. Hfq-dependent base pairing of DsrA with the leader sequence of *rpoS* enhances the translation of the latter by exposing the RBS and the start codon (Lease and Woodson 2004). In spite of the ring-shaped structure of Hfq hexamers that provide multiple binding surfaces both in distal and proximal positions, DsrA and *rpoS* compete for the same proximal sites. This competition makes the interaction between Hfq and RNA dynamic, and increases the binding efficiency by placing the two RNAs in close proximity (Soper et al. 2010), or by promoting structural remodeling of one of the RNA partners to match the structure of the other (Maki et al. 2010). What remains unclear is the order of binding, and if this particular mechanism is specific for sRNA–mRNA pairs. Further investigation will be of interest to see if these models can also be used to explain sRNA-mediated, negative regulation of gene expression. Moreover, whether Hfq directly affects ribosome binding to the mRNA, or whether it influences the action of ribonucleases subsequent to base pairing of the sRNA is still a matter of debate.

The example of DsrA discussed above serves as a model for positive regulation. A general model for negative, Hfq-mediated sRNA regulation was also described in which the sRNA binds the translation initiation region of its target mRNA (nucleotides −20 to +19 relative to the initiation codon) by pairing with the RBS sequence and/or the start codon, resulting in direct competition with the translation initiating 30S ribosomal subunit, and rapid degradation of the mRNA (Beyer et al. 1994; Huttenhofer and Noller 1994). One of the best characterized Hfq-associated sRNAs that fit this model is RyhB which is expressed under iron starvation (Massé and Gottesman 2002) and regulates many mRNAs encoding nonessential iron proteins (reviewed in Sect. 4.3). In recent years, many cases of sRNAs binding outside the translation initiation region of mRNAs have been reported with different binding mechanisms in each case. Desnoyers and Massé (2012) showed a second regulatory mechanism involving Hfq and the sRNA Spot42, which binds the 5′ UTR of its target mRNA *sdhC* and recruits Hfq to the translation initiation region. In this manner, the sRNA recruits Hfq to interfere with ribosome binding. This model of regulation was the first of its kind, proposing a novel mechanism where the sRNA is not the main effector, but rather an intermediary factor for the recruitment of Hfq, which directly modulates translation initiation.

2.2 *Protein-Binding Small RNA*

While base-pairing sRNAs modulate gene expression by interacting directly with mRNA molecules, protein-binding sRNAs influence gene expression by binding protein regulators of transcription (Waters and Storz 2009). To date, there are only two sRNAs known to regulate gene expression through interaction with proteins: CsrB and 6S RNA homologs. Both of these are encoded in intergenic regions.

2.2.1 CsrA–CsrB System

The CsrA regulation system was first discovered in *E. coli* where it controls the gene expression related to carbon metabolism (Romeo et al. 1993). This system includes three major components: CsrA, CsrB, and CsrC. CsrA is a protein with sequence homology to several other RNA-binding proteins (Romeo et al. 1993). It binds to mRNAs containing a CsrA binding site (a GGA motif) at, or close to the ribosomal binding site and prevents translation (Baker et al. 2002; Wang et al. 2005). In some cases, CsrA binds farther upstream from the RBS, in the 5′ untranslated region of its target, thus negatively affecting mRNA stability and/or translation rate (Babitzke and Romeo 2007; Baker et al. 2002; Romeo 1998; Wang et al. 2005).

The activation of genes repressed by CsrA occurs through a two-component system named BarA/UvrY (Lapouge et al. 2008; Lucchetti-Miganeh et al. 2008) (Kay et al. 2006; Valverde et al. 2003; Weilbacher et al. 2003). Upon activation, the response regulator UvrY binds to specific DNA elements located in the promoter regions of the sRNAs CsrB and CsrC, and strongly activates their expression. Both CsrB and CsrC contain multiple GGA motifs and when expressed, they bind to and sequester multiple CsrA molecules, and subsequently relieve the target mRNAs from CsrA regulation (reviewed in Lapouge et al. 2008; Lucchetti-Miganeh et al. 2008). CsrB-like sRNAs are widely distributed among bacterial species including *Legionella* (reviewed in Sect. 3.2), and it is common to find multiple homologs per organism, and in some cases, multiple CsrA-like proteins as well (reviewed in Sonnleitner and Haas 2011).

2.2.2 6S RNA

In 2000, Wassarman and colleagues reported, for the first time, the function of the 6S RNA identified almost 40 years earlier (Wassarman and Storz 2000). 6S RNA binds to the σ^{70} and the β/β' subunits of RNA polymerase (RNAP) and inhibits transcription from σ^{70}-dependent promoters containing a weak -35 element and an extended -10 element (Cavanagh et al. 2008; Klocko and Wassarman 2009; Wassarman 2007). By binding preferentially to the σ^{70}-RNAP holoenzyme, 6S RNA is able to affect the overall balance between different RNAP holoenzymes, and therefore adjust transcriptional programs on a global scale. Interestingly, this sRNA is widespread in bacteria, and so it is tempting to speculate that 6S RNA represents an essential bacterial element for efficient switching between different transcriptomic programs (Barrick 2005; Trotochaud and Wassarman 2005).

3 Small RNAs Expressed By *Legionella pneumophila*

3.1 Identification of sRNAs Expressed By L. pneumophila

Different approaches have been used to identify sRNAs expressed by *L. pneumophila*. First, an in silico approach was used to identify intergenic sRNAs by searching for the presence of Rho-independent terminators in intergenic regions (Faucher et al. 2010). Although 143 sRNAs were predicted in this study, microarray experiments showed that only a subset is actively transcribed. Out of the expressed sRNAs, six were confirmed by Northern blot and rapid amplification of cDNA ends (RACE), including a 6S RNA homolog, which will be discussed below. RNA sequencing was used to identify sRNAs expressed by the Philadelphia-1 strain during exponential (E) and postexponential (PE) growth in AYE (rich nutrient broth) and during infection of amoeba (Weissenmayer et al. 2011). A second group has used the same method to identify sRNAs expressed by the Paris strain during the E and PE phases in AYE (Sahr et al. 2012). Sahr et al. (2012) published an extensive analysis regarding the conservation of sRNAs between the Philadelphia-1 (Chien et al. 2004) and Paris (Cazalet et al. 2004) strains, and the reader is referred to this publication for a complete list of sRNAs identified in *L. pneumophila*. While the Philadelphia-1 report identifies a mere 33 cis-encoded sRNAs, the Paris study reports 622. This discrepancy is likely due to different methodologies used by the respective teams, which would have affected the signal, and therefore the number of transcripts that show a signal above the detection threshold, rather than a biological cause. It is beyond the scope of this review to list all the sRNAs identified to date, especially since most of them are still uncharacterized. More interestingly, transcription start sites (TSS) were identified in the Paris strain, which provide clues as to the architecture of the operons of *L. pneumophila*, and may help distinguish between sRNAs originating from primary transcripts and those that are produced through the cleavage of longer transcripts. To date, only a few sRNAs have been studied in detail and those will be discussed below (Table 1).

3.2 RsmX/Y/Z and CsrA

When *L. pneumophila* infects a host cell, switching off survival and transmission genes, and turning on genes important for intracellular replication become

Table 1 Phenotype-associated sRNAs expressed by *L. pneumophila*

Name	5' end[a]	3' end[a]	Size (nts)	Regulator	Target	Phenotype/Note	Reference
Protein-binding sRNAs							
RsmX	*3397553*	*3397653*	101	LetAS, RpoS	CsrA[b]	Δ*rsmX* shows a slight defect in ICM	(Sahr et al. 2012)
RsmY	7168	7059	110	LetAS, RpoS	CsrA	Δ*rsmYZ* is defective for ICM	(Faucher and Shuman 2011; Rasis and Segal 2009; Sahr et al. 2012)
RsmZ	1892720	1892592	132	LetAS, RpoS	CsrA		
6S RNA (*ssrS*)	951819	951673, 951638	147, 182		RNAP	Δ*ssrS* is unable to compete against the WT for ICM	(Faucher et al. 2010)
Cis-encoded sRNAs							
Lppnc0584	*2854265*	*2854119*	146	LetAS, RpoS	*phoA*[b]	Δ*lppnc0584* shows a slight defect in ICM	(Sahr et al. 2012)
Trans-encoded sRNAs							
Lppnc405	1911295	1911105	196			Δ*lppnc0405* shows a slight defect in ICM	(Sahr et al. 2012)
Lpr0035	1355695	1355444	251		*lpg1228*[b] *lpg1229*[b] other?	Δ*lpr0035* shows a reduction in host internalization and ICM	(Jayakumar et al. 2012)

[a] The borders of the sRNA refer to the Philadelphia-1 genome or the Paris genome (italicized) for sRNAs not conserved in Philadelphia-1

[b] This target is presumed, but its binding to the sRNA has not been investigated

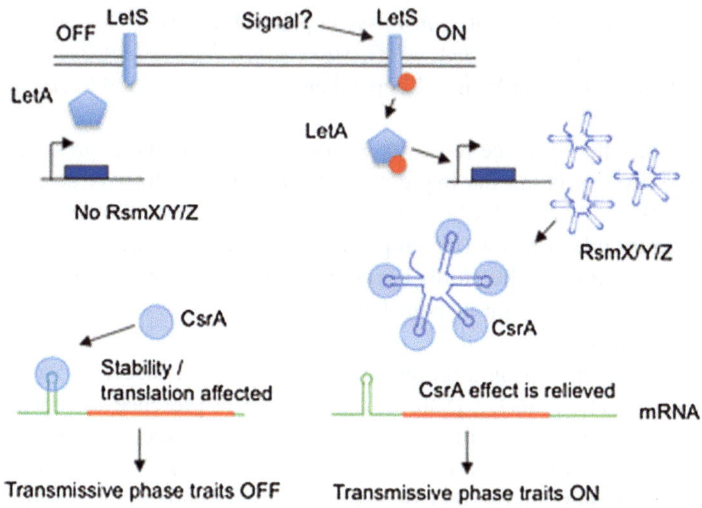

Fig. 1 Regulation of gene expression by RsmX/Y/Z. In the absence of RsmX/Y/Z, CsrA is able to bind to GGA motifs in hairpin loops in the 5' UTR region of its target mRNAs (*left*). This binding affects the stability and/or translation of the mRNA, therefore affecting gene expression. Upon sensing an, as yet unidentified, signal, the LetS sensor is autophosphorylated and activates the LetA regulator by phosphorylation (*right*). Then, LetA activates transcription of the sRNAs RsmX/Y/Z. These sRNAs possess multiple CsrA-binding domains, which bind and sequester the protein thereby reliving regulation of the mRNA targets by CsrA

essential. This is accomplished partly by a regulatory cascade composed of the two-component system LetA/S and CsrA, which relieve repression of several postexponential phase traits, as well as virulence factors necessary for intracellular growth in amoeba (Bachman and Swanson 2004; Gal-Mor and Segal 2003; Hammer et al. 2002; Lynch et al. 2003; Shi et al. 2006) and in macrophages (Byrne and Swanson 1998). LetA/S are homologs of the BarA/UvrY system that controls the expression of the sRNAs CsrB/C in *E. coli* (see Sect. 2.2.1). In 2006, CsrB homologs were predicted in *L. pneumophila* by the in silico approach that identified intergenic regions enriched for the GGA motif, characteristic of the *E. coli* CsrB (Kulkarni et al. 2006). The two homologs found in *L. pneumophila* were named RsmY/Z because of their small size and architectural resemblance to the RsmY/Z CsrB homologs expressed by *P. aeruginosa* (Lapouge et al. 2008). It was later confirmed that RsmY/Z were the missing regulatory determinants between the LetA/S two-component system and CsrA in *L. pneumophila* (Rasis and Segal 2009; Sahr et al. 2009). Importantly, they are crucial for the expression of some virulence determinants, and for the switch between the replicative and transmissive phases of the bacterium.

Upon detection of a yet unidentified signal, LetA/S activates the expression of *rsmY/Z*. LetA binds directly to a conserved consensus sequence upstream of *rsmY/Z* to induce their expression in the postexponential phase (Sahr et al. 2009). The two sRNAs then bind to CsrA and prevent it from interacting with target mRNAs leading to the expression of transmissive phase traits (Fig. 1). Some Icm/Dot effector genes (*ralF*, *sidC*, *sdeA*, *sdeC*, *sidF*, *sdhB*, *legC7*, *legC2*, and *vipA*) are regulated by LetA/S in an RsmY/Z-dependant manner (Rasis and Segal 2009; Sahr et al. 2009).

In addition, the virulence regulator RpoS also controls the expression of these two sRNAs by regulating the expression of *letS* (Hovel-Miner et al. 2009; Rasis and Segal 2009). Pitre et al. (2013) recently showed that RpoS induces expression of the integration host factor (*ihf*) in postexponential phase, which in turn induces expression of *rsmY/Z*. Therefore, RpoS influences the expression of *rsmY/Z* by controlling the transcription of both *letS* and *ihf*.

Deep RNA sequencing of the *L. pneumophila* Paris strain revealed the presence of another sRNA harboring a promoter containing a sequence very similar to the LetA binding site of *rsmY/Z* (Sahr et al. 2012). The presence of repeated GGA motifs in stem-loop regions of its predicted secondary structure suggested that it could be a third CsrB homolog, RsmX. Similar to *rsmY/Z*, expression analyses revealed high expression of *rsmX* in postexponential phase and expression studies in *letA* and *rpoS* deletion mutants demonstrated that it is indeed controlled by the same regulators as *rsmY/Z* (Sahr et al. 2012). Taken together, these observations indicate that this new sRNA may be part of the LetA/S-CsrA regulatory cascade. Furthermore, RsmX is necessary for full virulence of *L. pneumophila* (Sahr et al. 2012). While LetA/S positively regulates motility, through control of flagella expression (Fettes et al. 2001; Forsbach-Birk et al. 2004; Molofsky and Swanson 2003a) this regulation is surprisingly independent of RsmY/Z (Sahr et al. 2009) and RsmX (Sahr et al. 2012), which suggests a direct regulation of flagella formation by LetA/S.

RsmX is absent from the *L. pneumophila* Philadelphia-1 strain and *L. longbeachae*, but it is present and highly conserved in the Corby, Lens, and Lorraine strains. The presence of a third CsrB homolog is reminiscent of the system found in *P. fluorescence* where three homologs, RsmX/Y/Z, regulate CsrA activity, but only RsmY/Z are highly conserved among the different strains (Kay et al. 2005), while RsmX is somewhat dispensable. The evolutionary benefit of possessing a third homolog still needs to be addressed.

3.3 6S RNA

L. pneumophila 6S RNA was identified based on its expression pattern and predicted secondary structure which is similar to the *E. coli* 6S RNA, and by its capacity to bind to *L. pneumophila* RNAP (Fig. 2) (Faucher et al. 2010). The holoenzyme to which it binds preferentially is still unknown, but there is evidence that 6S RNA and RpoS regulate a distinct transcriptomic program in *L. pneumophila* (Faucher et al. 2010). A 6S RNA deficient strain was shown to have a defect in intracellular multiplication (ICM) in *A. castellanii* and cultured human macrophages. The mutant was also unable to compete against the wild-type strain

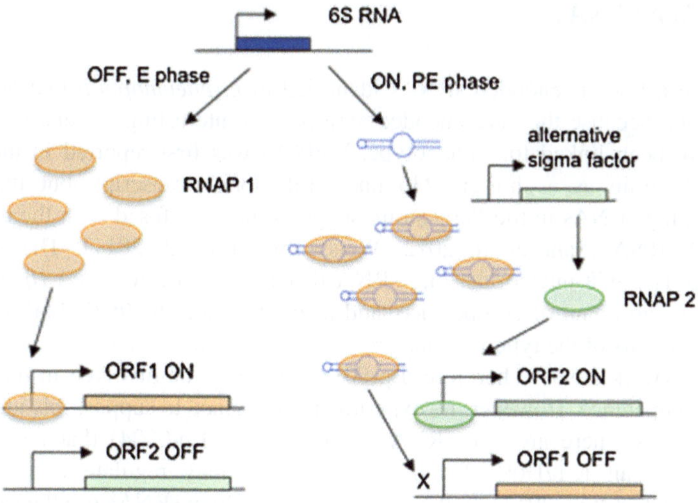

Fig. 2 Regulation of gene expression by 6S RNA. This model is based on the 6S RNA of *E. coli*. During exponential growth, a specific type of holoenzyme (RNAP 1, *orange*) is produced and is responsible for the bulk of transcription. Upon reaching postexponential phase, 6S RNA is expressed and binds to specific holoenzyme(s), in this case RNAP 1, which is then blocked from binding to promoters. In postexponential phase, alternative sigma factors are also expressed which lead to the formation of alternative holoenzymes (RNAP 2, *green*) insensitive to the action of 6S RNA, and therefore able to bind to their specific promoters. Many different RNAP holoenzymes, with different sigma subunits, can be produced, but the identity of the specific holoenzyme to which 6S RNA binds in *L. pneumophila* is currently unknown

during infection of macrophages and *A. castellanii*. Therefore, 6S RNA is likely indispensable for optimal intracellular growth in nonclinical environments, such as cooling towers, where many strains can be found, competing for the same, limited resources. By optimizing the switch between distinct transcriptional programs, 6S RNA improves the fitness and competitiveness of the strains that possess it.

Analysis of the transcriptomic impact of losing the 6S RNA in *L. pneumophila* revealed that it promotes expression of 127 genes in postexponential phase including Icm/Dot genes and those involved in the transport of small molecules, replication and DNA repair, and amino acids and fatty acids metabolic pathways (Faucher et al. 2010). This observation suggests that the 6S RNA mutant is nutritionally disadvantaged during ICM, which affects its growth rate and its ability to compete against the wild type for host cells (Faucher et al. 2010).

A second 6S RNA, named 6S2 RNA was identified in the Philadelphia-1 strain along with an antisense transcript (Weissenmayer et al. 2011) while only the sense transcript was detected in the Paris strain (Sahr et al. 2012). Binding of 6S2 RNA to RNAP was not investigated, and its function and identity as a 6S RNA homolog were based solely on similarity of its predicted structure to the 6S RNA of *E. coli* (Weissenmayer et al. 2011). Whether or not the putative 6S2 RNA truly operates as a homolog needs to be assessed experimentally.

3.4 Other sRNAs

There are a few cis-encoded sRNAs identified in *L. pneumophila* that are worth mentioning because they are encoded antisense to interesting targets, or because they have been linked to a phenotype. Lpr0018 was first reported in the Philadelphia-1 strain as a shorter (236 nucleotide long) transcript, but the report investigating sRNAs in the Paris strain suggests that it is fused with the upstream Lpr0019 sRNA (Sahr et al. 2012; Weissenmayer et al. 2011). The resulting transcript is a 978 nucleotide long sRNA that is antisense to *comA* (*lpg0626*), a putative determinant of competence, and to *lpg0627* and *lpg0628*, both of which encode subunits of the type IV pilus previously associated with competence (Stone and Abu Kwaik 1999). Therefore, Lpr0018 could also be involved in the regulation of competence. However, no experimental evidence to support this idea exists yet. Similarly, there are two sRNAs (Lpr0003 and Lpr0004) that are encoded antisense to an Icm/Dot effector, LegA10, and may regulate its expression (Weissenmayer et al. 2011). In the Paris strain, these two sRNAs use the same TSS and are probably expressed as a single sRNA (Sahr et al. 2012). The apparent discrepancy regarding separate or combined transcription of the sRNAs between the two studies for Lpr0003/Lpr0004 and Lpr0018/Lpr0019 is hard to reconcile. One possibility is that Weissenmayer et al. (2011) might have identified a cleavage product mediated by an endonuclease cut of the sRNA–mRNA duplexes that presumably could affect expression of the target gene.

Lppnc0584 is encoded antisense to *phoA* (*lpp2499*, *lpg2432*) and overlaps the gene by 37 nucleotides at the 3' end (Sahr et al. 2012). Its deletion resulted in a slight ICM defect in *A. castellanii*, but the underlying mechanism resulting in the observed defect was not investigated (Sahr et al. 2012). This sRNA is not conserved in the Philadelphia-1 strain (Sahr et al. 2012). Another cis-encoded sRNA (Lppnc0223) is regulated by RpoS and LetA, two important regulators of virulence in *L. pneumophila* (Sahr et al. 2012). However, deletion of Lppnc0223 did not result in any intracellular growth defect.

There are many trans-encoded sRNAs expressed by *L. pneumophila*, but only a few of them have been studied to some extent. The *lpr0035* gene overlaps the *attR* site of pLP45, a mobile genetic element encoding the *lvh/lvr* type IV secretion system (T4SS) (Segal et al. 1999). The mutant strain for *lpr0035* was defective for internalization by host cells, and also showed reduced ability to grow within macrophages (Jayakumar et al. 2012). Moreover, deletion of *lpr0035* also removed the direct repeat of the *attR* site and locked pLP45 in the chromosome. Complementation studies showed that the phenotypes observed were due to the deletion of *lpr0035*, and were unrelated to the resulting inhibition of pLP45 excision (Jayakumar et al. 2012). The inability to enter host cells was effectively complemented by overexpression of *lpg1228* or *lpg1229*, two genes encoded at the right end of pLP45, adjacent to *lpr0035* but transcribed divergently from the sRNA. Lpr0035 seems to regulate expression of these two genes, but the

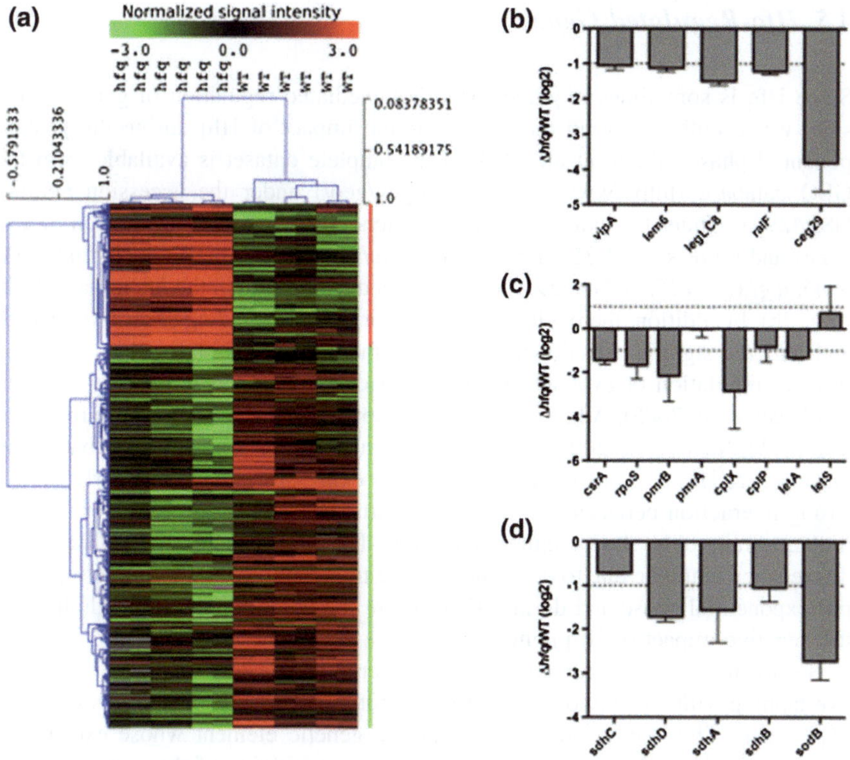

Fig. 3 Genes regulated by Hfq during postexponential phase in *L. pneumophila*. The data were retrieved from our previous publication (Trigui et al. 2013). The wild-type and *hfq* mutant strains were grown to postexponential phase and RNA was extracted, labeled, and hybridized to whole-genome microarrays. (**a**) Hierarchical clustering of genes showing a ratio to control value of $\pm$ 2 fold and a $P \leq 0.005$. The normalized signal intensity is shown. The cluster of genes induced in the *hfq* mutant is shown by a vertical red line. A vertical green line shows the cluster of genes that is repressed. The expression ratio of selected genes is shown: (**b**) Icm/Dot effectors (**c**) regulators and (**d**) genes regulated by RyhB in *E. coli*

mechanism by which it does so remains to be identified (Jayakumar et al. 2012). The intracellular growth defect, however, was not complemented by overexpression of *lpg1228, lpg1229*, or the *lvh/lvr* type IV secretion system. Therefore, the sRNA is likely a regulator of another virulence determinant that is yet to be identified. It is also unknown whether Hfq is required for Lpr0035 mediated regulation. Notably, the strong ICM defect of the *lpr0035* mutant is contrasting with the slight ICM defect of the *hfq* mutant (McNealy et al. 2005) and so, it is unlikely that Hfq is essential for the efficient regulation of target mRNA by Lpr0035.

Deletion of yet another trans-encoded sRNA, Lppnc405, resulted in a slight ICM defect (Sahr et al. 2012). However, its target was not identified in the study.

3.5 *Hfq-Regulated Genes*

Since Hfq is sometimes required for sRNA-mediated regulation of gene expression, we recently analyzed the transcriptomic impact of Hfq during the postexponential phase (Trigui et al. 2013). The complete dataset is available from the GEO database (http://www.ncbi.nlm.nih.gov/geo/) under the accession number GSE42905. Under this condition, Hfq deficiency resulted in the induction of 138 genes and repression of 357 genes. Some Icm/Dot effector genes, including *vipA*, *lem6*, *legLC8*, *ralF*, and *ceg29*, showed reduced expression in the *hfq* mutant strain (Fig. 3b). In addition, many virulence regulator genes, including *csrA*, *rpoS*, *pmrB*, and *letA* are significantly downregulated in the *hfq* mutant strain (Fig. 3c). This positive regulation of *csrA* and *rpoS* expression by Hfq was previously reported (McNealy et al. 2005). Additionally, it was shown that LetA is a positive regulator of Hfq (McNealy et al. 2005), and our results suggest that a positive feedback loop exists where Hfq also positively regulates *letA* expression. Our data clearly show a strong interaction between Hfq and major virulence regulators; however, the *hfq* mutant strain only shows minor defects in ICM (McNealy et al. 2005). This discrepancy could be due to differences in the transcriptomic program between the postexponential phase and during ICM, where the latter condition could alleviate the negative impact of Hfq deficiency.

Surprisingly, a large genomic region bordered by *lpg0973* and *lpg1070*, partly overlapping with the previously identified "efflux island" was overexpressed in the *hfq* mutant. This region represents a mobile genetic element whose excision is repressed by Hfq. We also showed that the overexpression of the genes encoded within it is genetically linked with excision of the element. Regulation of excision by Hfq could be due to a direct effect on the stability of a gene necessary for excision, such as the integrase (*lpg1070*) or an indirect effect through an sRNA (Trigui et al. 2013).

4 *Legionella pneumophila* Regulatory Systems That May Contain sRNAs

There are a number of regulatory systems in model bacterial species, such as *E. coli*, that are known for their use of sRNAs. Since these same regulatory pathways are found in *L. pneumophila*, it is highly possible that the corresponding sRNAs are also present. It is not trivial to find functionally equivalent sRNAs, since the sequence is often poorly conserved across species. For example, the presence of CsrB homologs in *L. pneumophila* was suspected long before their identification, but none were ever found by BLAST analysis. Similarly, both PrrF and RyhB are regulated by FUR and control expression of similar genes in *P. aeruginosa* and *E. coli* respectively; however, there is no sequence homology between them (Wilderman et al. 2004). This suggests that functionally equivalent sRNAs found in different species are not necessarily evolutionarily related, and that they may arise independently to fulfill a useful function (Waters and Storz 2009). Therefore, it should not be surprising that none of the sRNAs identified in *L. pneumophila* were readily identified as homologous to well-known sRNAs. In the following three sections, we will be reviewing a few regulatory pathways found in other bacteria that use sRNA-mediated regulation and for which an sRNA is likely to be found in *L. pneumophila* as well. These include the OxyR and Fur regulatory pathways, and the many sRNAs that regulate RpoS.

4.1 OxyR System and Oxidative Stress

OxyS is a 109 nucleotide sRNA regulated by the global oxidative stress response regulator OxyR (Altuvia et al. 1997; Kullik et al. 1995; Storz et al. 1990). Its regulon and mode of action are best characterized in *E. coli*. It is a trans encoded, posttranscriptional regulator of up to 40 genes (Altuvia et al. 1997). The expression of *oxyS* is increased in response to both exogenous and endogenous reactive oxygen species (ROS) and thus, expression is growth phase dependent under noninducing conditions (Gonzalez-Fecha and Demple 1999; Zheng et al. 2001). An antimutator effect observed in mutant studies was specifically attributed to OxyS-dependent regulation of DNA repair systems (Altuvia et al. 1997). The best understood genes under OxyS control in *E. coli* are *fhlA* and *rpoS*. Their regulation has been shown to require the chaperone protein Hfq (Zhang et al. 2002). OxyS negatively regulates both of these genes (Franch and Gerdes 2000) and does so slightly differently.

In *E. coli*, *fhlA* codes for a positive transcriptional regulator that controls the expression of the formatehydrogen lyase complex (Altuvia et al. 1997; Sankar et al. 1988; Schlensog and Böck 1990). This protein complex uses metal cofactors that are thought to promote extended cellular damage in the presence of H_2O_2. Therefore, when exogenous oxidative stress, producing additional H_2O_2, is present, OxyS inhibits the formation of the complex by repressing translation of the *fhlA* mRNA (Altuvia et al. 1998). At least two binding sites are used by OxyS in *fhlA* regulation: one blocking the RBS, and a second in the downstream coding region (Altuvia et al. 1998; Salim and Feig 2010). Earlier investigations confirmed that OxyS has 3' stem-loop structures denoted a, b, and c from 5' to 3' respectively, and that stem-loop c at the 3' end of OxyS was essential for binding the RBS sequence (Altuvia et al. 1998). Subsequently, Hfq involvement was discovered at the second OxyS-binding site, inside the coding sequence of *fhlA* (Salim and Feig 2010). This latter site may play a more significant role in Hfq-mediated stabilization of the mRNA-sRNA-Protein complex, while the 3' stem-loop c of OxyS may facilitate direct sRNA–mRNA interaction. The RNA duplex formation is initiated with a kissing complex of stem loops between the sRNA and its target, followed by more stable RNA–RNA pairing (Argaman and Altuvia 2000). The end result is a translational block of the *fhlA* mRNA.

The second OxyS target, *rpoS*, is a well-studied alternative sigma factor characteristically activated in stationary phase and in response to an assortment of stresses in *E. coli* (Reviewed in Battesti et al. 2011). Compared to *fhlA*, *rpoS* regulation is a much more complex process, involving mRNA binding competition with DsrA, RprA (see Sect. 4.2) and the newly identified ArcZ, three *rpoS*-activating sRNAs (Majdalani et al. 1998; Mandin and Gottesman 2010; Sledjeski et al. 1996). In addition to the requirement of stem-loop c, deletion experiments by Altuvia et al. (1998) demonstrated that the 27-nucleotide linker region between the stem-loops b and c is also essential for OxyS binding to *rpoS*. Moreover, ribonucleases seem to have a bearing on the sRNA's stability (Basineni et al. 2009). Massé et al. (2003) hypothesize that OxyS is degraded upon binding with its mRNA target (Altuvia et al. 1997). In this model, the RNA duplex between OxyS and its target initiates degradation of both RNAs. Hfq binding is thought to protect free OxyS from the RNase E-mediated degradasome until its interaction with the target mRNA (Massé et al. 2003). Since the chaperone protein is required by the *rpoS* upregulating sRNAs, Hfq sequestration by OxyS (depleting the chaperones available for *rpoS* activation) is also thought to be part of the mechanism by which *rpoS* is repressed. While this apparent protective role of Hfq, and OxyS turnover by RNase E has been confirmed, the implication of other ribonucleases in the OxyS regulatory pathway is yet to be fully elucidated (Basineni et al. 2009). And so, further studies are required to fully understand the elaborate and seemingly tightly regulated relationship of *rpoS* regulation by OxyS.

The *L. pneumophila* OxyR homolog was identified as a regulator of *ahpC2D*, an alkyl hydroperoxide reductase, and was shown to partially complement an *E. coli oxyR* mutation (Leblanc et al. 2008). While no OxyS homolog has been identified in *L. pneumophila* thus far, the sRNAs LprA and LprB are regulated by OxyR (Faucher et al. 2010). Like OxyS, LprA is upregulated in response to exogenous H_2O_2 during the exponential phase, but the size of LprA versus that of the *E. coli* OxyS are quite different (Faucher et al. 2010). Moreover, *lprA* was originally annotated as an ORF of unknown function in the Paris strain and the identification of the TSS by RNA-seq seems to indicate that it encodes a protein and thus, its current status as an sRNA is uncertain (Sahr et al. 2012). On the other hand, *lpr0042* is expressed antisense to the 5' end of *lprA* and could regulate expression of the ORF encoded within it (Weissenmayer et al. 2011). As is the case with *lprA*, expression of *lprB* is also induced by exogenous H_2O_2 during the exponential growth phase, but is repressed upon reaching the postexponential phase. But unlike the case of *lprA*, OxyR seems to act as a negative regulator of *lprB* during exponential growth in the absence of exogenous oxidative stress. These last two observations are contradictory to the regulation of *oxyS* in *E.* coli and argue against LprB being an OxyS homolog. Moreover, LprB is not highly conserved in the Paris strain. Nevertheless, it is conceivable that OxyS would have an equivalent in *L. pneumophila*, since it possesses an OxyR homolog (Leblanc et al. 2008).

4.2 RpoS-Regulating sRNA

RpoS, the transcriptional regulatory sigma factor that responds to stress and is essential in the stationary phase, is subject to complex regulations on multiple levels, including transcription, translation, and proteolysis. In *E. coli*, translation of *rpoS* is regulated by sRNAs that depend upon the 5' UTR of the *rpoS* mRNA. While the most frequent outcome of sRNA–mRNA pairings is negative regulation of the target mRNA, (Massé et al. 2003; Morita et al. 2005; Udekwu et al. 2005) the outcome of this particular pairing with *rpoS* results in a positive regulation. This translational upregulation is dependent upon the RNA chaperone Hfq and the two, main regulatory sRNAs, DsrA and RprA (Majdalani et al. 2001). DsrA and RprA stimulate the translation of *rpoS* by base pairing with the 5' UTR. Expression of these sRNAs increases both the accumulation and the half-life of their target mRNA (Updegrove and Wartell 2011).

DsrA is necessary for activation of *rpoS* translation in response to low temperature and osmotic shock (Majdalani et al. 2001), while RprA increases expression of *rpoS* both in response to osmotic shock (Majdalani et al. 2002) and cell envelope stress (Garcia-Calderon et al. 2005; Majdalani et al. 2002). Mu-Cullen et al. (2010) suggest that the major effect of DsrA on *rpoS* mRNA accumulation and efficient translation is in overcoming RNase E-dependent degradation of the latter. *rpoS* mRNA is also subject to degradation by an additional pathway mediated by RNase III, which, in contrast to the RNase E-mediated pathway, occurs both in the presence or absence of DsrA or RprA. Unlike RNaseE, the role of RNase III is to reduce the translation of *rpoS* even when the sRNAs are acting to stimulate translation (McCullen et al. 2010). The essential aspects of the efficient interaction between DsrA, *rpoS* and Hfq, and the mechanistic steps of both RNA annealing and unwinding in real time have been reported (Hwang et al. 2011). Annealing of DsrA to the *rpoS* mRNA occurs through three successive, distinct steps: (i) transient unproductive binding events, (ii) partial annealing, and (iii) full annealing. Although the competition of the two RNAs for the same binding site on Hfq resulted in many unproductive formations of transient ternary complexes, the annealing rate in the presence of Hfq was still higher than in its absence. Hwang et al. (2011) reported that when *rpoS* and DsrA fragments were engineered to use different RNA-binding sites on Hfq, the annealing efficiency was dramatically decreased suggesting that proximity of the sRNA and mRNA is key to producing an efficient, fruitful interaction between the two. In *L. pneumophila*, RpoS is involved in the regulation of virulence and stress response (Abu-Zant et al. 2006; Hales and Shuman 1999; Hovel-Miner et al. 2009). It is likely that some of the sRNAs that have already been identified are functional homologs of DsrA and RprA; however, a BLAST search did not readily identified homologs in the *L. pneumophila* genomes.

4.3 Fur and Iron Uptake

Fur is a central regulator of iron uptake and the synthesis of iron-containing proteins in bacteria (reviewed in Massé et al. 2007). In *E. coli*, the latter function is mediated, in part, through the negative regulation of the sRNA, RyhB. When the intracellular concentration of iron is high, Fur is activated and inhibits expression of the iron-uptake genes and *ryhB*. In contrast, inactivation of Fur by low concentrations of iron leads to the expression of iron-uptake genes and *ryhB*, which in turn inhibits the translation of iron-containing proteins, such as *sdhCDAB* and *sodB* (Massé and Gottesman 2002). The interaction of RyhB with its target mRNA is dependent upon Hfq (Massé et al. 2003). This posttranscriptional regulation allows the cell to preserve sufficient intracellular iron for essential cellular functions (Jacques et al. 2006; Massé et al. 2005; Salvail et al. 2010). In *L. pneumophila*, Fur was identified through complementation of an *E. coli fur* mutant (Hickey and Cianciotto 1994). Then, it was shown that Fur regulates expression of *frgA*, a gene with homology to the aerobactin synthetase of *E. coli* (Hickey and Cianciotto 1997). Another locus important for iron assimilation, encoding *iraAB*, was subsequently identified but its regulation by Fur has not been investigated (Viswanathan et al. 2000). Interestingly, both the *frgA* and the *iraAB* loci are necessary for intracellular multiplication. The presence of a Fur homolog, as well as an iron-uptake system and iron-containing proteins in *L. pneumophila* suggest that this bacterium most likely possesses a functional homolog of the sRNA RyhB. Unsurprisingly, no homolog of RyhB was found in *L. pneumophila* by BLAST but its presence is further supported by the fact that the *hfq* mutant of *L. pneumophila* is defective for growth in low-iron medium (McNealy et al. 2005). McNealy et al. (2005) also demonstrated Hfq-dependent regulation of *fur* expression. Our analysis of the transcriptome of the *hfq* mutant strain revealed that expression of the *E. coli* RyhB targets, *sdhCDAB* and *sodB*, in *L. pneumophila* are Hfq-dependent (Fig. 3d), which suggests a similar regulatory system for those genes in both bacteria (Trigui et al. 2013). Further experiments are required to identify the regulatory pathways involved in iron homeostasis in *L. pneumophila*.

5 Future Perspective

In the past 3 years, three studies have reported the identification of a large number of sRNAs expressed by *L. pneumophila*. Those that have been characterized the most, including 6S RNA and RsmX/Y/Z, are involved in the regulation of ICM-related genes, and their deletion results in strong ICM defects (Table 1). Nonetheless, there are still fundamental questions that have yet to be answered about the activation, binding interactions, and function of these sRNAs. For example, 6S RNA was shown to bind *L. pneumophila* RNAP, but it is still unclear if it has any preference for specific holoenzymes, and if so, to what extent. Cavanagh et al. (2008) showed that only some σ^{70}-dependent promoters are sensitive to the action of 6S RNA. The identification of TSS and consensus sequences for RpoD (σ^{70})- and RpoS-dependent promoters in the Paris strain could help understand the regulatory details of 6S RNA (Sahr et al. 2012). Moreover, the dissection of the regulons of these sRNAs has the potential to lead us to new genetic determinants of virulence and other processes such as competence and response to extreme environmental conditions, including oxidative stress. Now that we have identified a substantial amount of sRNAs expressed by *L. pneumophila*, the most interesting discoveries are yet to come. How these sRNAs work, how they regulate gene expression and influence the expression of virulence determinants and other traits in *L. pneumophila* are questions that need to be answered. The best strategy to reach this goal is to study the phenotype of mutant and overexpressor strains, since sRNAs can be either positive or negative regulators. Once a phenotype is established, it becomes easier to pinpoint the targets of the sRNAs. A variety of approaches can be used for this purpose, including in silico prediction, genome-wide transcriptomic approaches, 2D-gel analysis, and co-precipitation of the sRNA with the target (Faucher and Shuman 2011).

Acknowledgments Our research is supported by FRQNT *Établissement de nouveaux chercheurs* grant 2013-NC-164725 and NSERC Discovery grant 418289-2012. We would like to thank the reviewers for helpful suggestions.

References

Abu-Zant A, Asare R, Graham J, Abu Kwaik Y (2006) Role for RpoS but not RelA of *Legionella pneumophila* in modulation of phago-some biogenesis and adaptation to the phagosomal microenvironment. Infect Immun 74:3021–3026

Albert-Weissenberger C, Cazalet C, Buchrieser C (2007) *Legionella pneumophila*—a human pathogen that co-evolved with fresh water protozoa. Cell Mol Life Sci 64:432–448. doi:10.1007/s00018-006-6391-1

Altman E, Segal G (2008) The response regulator CpxR directly regulates expression of several *Legionella pneumophila icm/dot* components as well as new translocated substrates. J Bacteriol 190:1985–1996

Altuvia S, Weinstein-Fischer D, Zhang A, Postow L, Storz G (1997) A small, stable RNA induced by oxidative stress: role as a pleiotropic regulator and antimutator. Cell 90:43–53

Altuvia S, Zhang A, Argaman L, Tiwari A, Storz G (1998) The *Escherichia coli* OxyS regulatory RNA represses *fhlA* translation by blocking ribosome binding. EMBO J 17:6069–6075

Argaman L, Altuvia S (2000) *fhlA* repression by OxyS RNA: kissing complex formation at two sites results in a stable antisense-target RNA complex. J Mol Biol 300:1101–1112. doi:http://dx.doi.org/10.1006/jmbi.2000.3942

Babitzke P, Romeo T (2007) CsrB sRNA family: sequestration of RNA-binding regulatory proteins. Curr Opin Microbiol 10:156–163

Bachman M, Swanson M (2004) Genetic evidence that *Legionella pneumophila* RpoS modulates expression of the transmission phenotype in both the exponential phase and the stationary phase. Infect Immun 72:2468–2476

Baker C, Morozov I, Suzuki K, Romeo T, Babitzke P (2002) CsrA regulates glycogen biosynthesis by preventing translation of *glgC* in *Escherichia coli*. Mol Microbiol 44:1599–1610

Balasubramanian D, Vanderpool C (2013) New developments in post-transcriptional regulation of operons by small RNAs. RNA Biol 10:337–341
Barrick JE (2005) 6S RNA is a widespread regulator of eubacterial RNA polymerase that resembles an open promoter. RNA 11:774–784
Basineni SR, Madhugiri R, Kolmsee T, Hengge R, Klug G (2009) The influence of Hfq and ribonucleases on the stability of the small non-coding RNA OxyS and its target *rpoS* in *E. coli* is growth phase dependent. RNA Biol 6:584–594
Battesti A, Majdalani N, Gottesman S (2011) The RpoS-mediated general stress response in *Escherichia coli*. Annu Rev Microbiol 65:189–213. doi:10.1146/annurev-micro-090110-102946
Beisel C, Storz G (2010) Base pairing small RNAs and their roles in global regulatory networks. FEMS Microbiol Rev 34:866–882. doi:10.1111/j.1574-6976.2010.00241
Benin AL, Benson RF, Besser RE (2002) Trends in Legionnaires' disease, 1980–1998: declining mortality and new patterns of diagnosis. Clin Infect Dis 35:1039–1046. doi:10.1086/342903
Beyer D, Skripkin E, Wadzack J, Nierhaus K (1994) How the ribosome moves along the mRNA during protein synthesis. J Biol Chem 269:30713–30717
Blatt SP, Parkinson MD, Pace E, Hoffman P, Dolan D, Lauderdale P, Zajac RA, Melcher GP (1993) Nosocomial Legionnaires' disease: aspiration as a primary mode of disease acquisition. Am J Med 95:16–22. doi:10.1016/0002-9343(93)90227-G
Brantl S (2007) Regulatory mechanisms employed by cis-encoded antisense RNAs. Curr Opin Microbiol 10:102–109
Butler JC, Breiman RF (1998) Legionnaires' disease: clinical, epidemiological, and public health perspectives. Semin Respir Infect 13:84–89
Byrne B, Swanson M (1998) Expression of *Legionella pneumophila* virulence traits in response to growth conditions. Infect Immun 66:3029–3034
Cavanagh A, Klocko A, Liu X, Wassarman K (2008) Promoter specificity for 6S RNA regulation of transcription is determined by core promoter sequences and competition for region 4.2 of σ 70. Mol Microbiol 67:1242–1256
Cazalet C, Rusniol C, Brüggemann H, Zidane N, Magnier A, Ma L et al (2004) Evidence in the *Legionella pneumophila* genome for exploitation of host cell functions and high genome plasticity. Nat Genet 36:1165–1173
Chien M, Morozova I, Shi S, Sheng H, Chen J, Gomez S et al (2004) The genomic sequence of the accidental pathogen *Legionella pneumophila*. Science 305:1966–1968. doi:10.1126/science.1099776
Ding Y, Davis B, Waldor M (2004) Hfq is essential for *Vibrio cholerae* virulence and downregulates sigma expression. Mol Microbiol 53:345–354
Faucher SP, Shuman H (2011) Small regulatory RNA and *Legionella pneumophila*. Front Microbiol 2:98. doi:10.3389/fmicb.2011.00098
Faucher SP, Friedlander G, Livny J, Margalit H, Shuman H (2010) *Legionella pneumophila* 6S RNA optimizes intracellular multiplication. Proc Natl Acad Sci 107:7533–7538. doi:10.1073/pnas.0911764107
Fettes P, Forsbach-Birk V, Lynch D, Marre R (2001) Overexpresssion of a *Legionella pneumophila* homologue of the *E. coli* regulator *csrA* affects cell size, flagellation, and pigmentation. Int J Med Microbiol 291:353–360. doi:10.1078/1438-4221-00141
Fields BS (1996) The molecular ecology of *Legionellae*. Trends Microbiol 4:286–290. doi:10.1016/0966-842X(96)10041-X
Fields BS, Benson RF, Besser RE (2002) *Legionella* and Legionnaires' disease: 25 years of investigation. Clin Microbiol Rev 15:506–526
Forsbach-Birk V, McNealy T, Shi C, Lynch D, Marre R (2004) Reduced expression of the global regulator protein CsrA in *Legionella pneumophila* affects virulence-associated regulators and growth in *Acanthamoeba castellanii*. Int J Med Microbiol 294:15–25. doi:10.1016/j.ijmm.2003.12.003
Franch T, Gerdes K (2000) U-turns and regulatory RNAs. Curr Opin Microbiol 3:159–164

Franze de Fernandez M, Eoyang L, August J (1968) Factor fraction required for the synthesis of bacteriophage Qbeta-RNA. Nature 219:588–590

Fraser DW, Tsai TR, Orenstein W, Parkin WE, Beecham HJ, Sharrar RG, Harris J, Mallison GF, Martin SM, McDade JE, Shepard CC, Brachman PS (1977) Legionnaires' disease: description of an epidemic of pneumonia. N Engl J Med 297:1189–1197. doi:10.1056/NEJM197712012972201

Gal-Mor O, Segal G (2003) Identification of CpxR as a positive regulator of *icm* and *dot* virulence genes of *Legionella pneumophila*. J Bacteriol 185:4908–4919

Garcia-Calderon C, Garcia-Quintanilla M, Casadesus J, Ramos-Morales F (2005) Virulence attenuation in *Salmonella enterica rcsC* mutants with constitutive activation of the Rcs system. Microbiology 151:579–588

Geissmann T, Touati D (2004) Hfq, a new chaperoning role: Binding to messenger RNA determines access for small RNA regulator. EMBO J 23:396–405

Gomez-Valero L, Rusniok C, Cazalet C, Buchrieser C (2011) Comparative and functional genomics of *Legionella* identified eukaryotic like proteins as key players in host-pathogen interactions. Front Microbiol 2:208. doi:10.3389/fmicb.2011.00208

Gonzalez-Fecha B, Demple B (1999) Role for the *oxyS* gene in regulation of intracellular hydrogen peroxide in *Escherichia coli*. J Bacteriol 181:3833–3836

Gottesman S (2004) The small RNA regulators of *Escherichia coli*: roles and mechanisms. Annu Rev Microbiol 58:303–328

Gottesman S, Storz G (2010) Bacterial small RNA regulators: versatile roles and rapidly evolving variations. Cold Spring Harb Perspect Biol 3:a003798. doi:10.1101/cshperspect.a003798

Gottesman S, McCullen C, Guillier M, Vanderpool C, Majdalani N, Benhammou J, Thompson K, FitzGerald P, Sowa N, FitzGerald D (2006) Small RNA regulators and the bacterial response to stress. Cold Spring Harbor Symp Quant Biol 71:1–11

Hales L, Shuman H (1999) The *Legionella pneumophila rpoS* gene is required for growth within *Acanthamoeba castellanii*. J Bacteriol 181:4879–4889

Hammer B, Tateda E, Swanson M (2002) A two-component regulator induces the transmission phenotype of stationary-phase *Legionella pneumophila*. Mol Microbiol 44:107–118

Hanrahan JP, Morse DL, Scharf VB, Debbie JG, Schmid GP, McKinney RM, Shayegani M (1987) A community hospital outbreak of Legionellosis: transmission by potable hot water. Am J Epidemiol 125:639–649

Heidtman M, Chen EJ, Moy M-Y, Isberg RR (2009) Large scale identification of *Legionella pneumophila* Dot/Icm substrates that modulate host cell vesicle trafficking pathways. Cell Microbiol 11:230–248

Hickey E, Cianciotto N (1994) Cloning and sequencing of the *Legionella pneumophila fur* gene. Gene 143:117–121

Hickey E, Cianciotto N (1997) An iron- and fur-repressed *Legionella pneumophila* gene that promotes intracellular infection and encodes a protein with similarity to the *Escherichia coli* aerobactin synthetases. Infect Immun 65:133–143

Horwitz MA (1983a) Formation of a novel phagosome by the Legionnaires' disease bacterium (*Legionella pneumophila*) in human monocytes. J Exp Med 158:1319–1331. doi:10.1084/jem.158.6.2108

Horwitz MA (1983b) The Legionnaires' disease bacterium (*Legionella pneumophila*) inhibits phagosome-lysosome fusion in human monocytes. J Exp Med 158:2108–2126. doi:10.1084/jem.158.6.2108

Hovel-Miner G, Pampou S, Faucher S, Clarke M, Morozova I, Morozov P, et al (2009) SigmaS controls multiple pathways associated with intracellular multiplication of *Legionella pneumophila*. J Bacteriol 191:2461–2473. doi:http://dx.doi.org/10.1128/ JB.01578-08

Hussong D, Colwell R, O'Brien M, Weiss E, Pearson A, Weiner R, Burge W (1987) Viable *Legionella pneumophila* not detectable by culture on agar media. Bio-Technology 5:947–950

Huttenhofer A, Noller H (1994) Footprinting mRNA–ribosome complexes with chemical probes. EMBO J 13:3892–3901

Hwang W, Arluison V, Hohng S (2011) Dynamic competition of DsrA and *rpoS* fragments for the proximal binding site of Hfq as a means for efficient annealing. Nucleic Acids Res 39:5131–5139

Jacques J, Jang S, Prévost K, Desnoyers G, Desmarais M, Imlay J, Massé E (2006) RyhB small RNA modulates the free intracellular iron pool and is essential for normal growth during iron limitation in *Escherichia coli*. Mol Microbiol 62:1181–1190

Jarraud S, Descours G, Ginevra C, Lina G, Etienne J (2013) Identification of *Legionella* in clinical samples. Methods Mol Biol 954:27–56. doi:10.1007/978-1-62703-161-5_2

Jayakumar D, Early J, Steinman H (2012) Virulence phenotypes of *Legionella pneumophila* associated with non-coding RNA *lpr0035*. Infect Immun 80:4143–4153

Kay E, Dubuis C, Haas D (2005) Three small RNAs jointly ensure secondary metabolism and biocontrol in *Pseudomonas fluorescens* CHA0. Proc Natl Acad Sci 102:17136–17141

Kay E, Humair B, Denervaud V, Riedel K, Spahr S, Eberl L, Valverde C, Haas D (2006) Two GacA-dependent small RNAs modulate the quorum-sensing response in *Pseudomonas aeruginosa*. J Bacteriol 188:6026–6033

Klocko A, Wassarman K (2009) 6S RNA binding to Eσ 70requires a positively charged surface of σ 70 region 4.2. Mol Microbiol 73:152–164

Kulkarni P, Cui X, Williams J, Stevens A, Kulkarni R (2006) Prediction of CsrA-regulating small RNAs in bacteria and their experimental verification in *Vibrio fischeri*. Nucleic Acids Res 34:3361–3369

Kullik I, Toledano MB, Tartaglia LA, Storz G (1995) Mutational analysis of the redox-sensitive transcriptional regulator OxyR: Regions important for oxidation and transcriptional activation. J Bacteriol 177:1275–1284

Lalaouna D, Simoneau-Roy M, Lafontaine D, Massé E (2013) Regulatory RNAs and target mRNA decay in prokaryotes. Biochim Biophys Acta 1829:742–747. doi:10.1016/j.bbagrm.2013.02.013

Lapouge K, Schubert M, Allain F, Haas D (2008) Gac/Rsm signal transduction pathway of gamma-proteobacteria: from RNA recognition to regulation of social behaviour. Mol Microbiol 67:241–253

Lease R, Woodson S (2004) Cycling of the Sm-like protein Hfq on the DsrA small regulatory RNA. J Mol Biol 344:1211–1223

Leblanc JJ, Brassinga AKC, Ewann F, Davidson RJ, Hoffman PS (2008) An ortholog of OxyR in *Legionella pneumophila* is expressed postexponentially and negatively regulates the Alkyl hydroperoxide reductase (*ahpC2D*) operon. J Bacteriol 190:3444–3455

Lenz D, Mok K, Lilley B, Kulkarni R, Wingreen N, Bassler B (2004) The small RNA chaperone Hfq and multiple small RNAs control quorum sensing in *Vibrio harveyi* and *Vibrio cholerae*. Cell 118:69–82

Link T, Valentin-Hansen P, Brennan R (2009) Structure of *Escherichia coli* Hfq bound to polyriboadenylate RNA. Proc Natl Acad Sci 106:19292–19297

Lucchetti-Miganeh C, Burrowes E, Baysse C, Ermel G (2008) The post-transcriptional regulator CsrA plays a central role in the adaptation of bacterial pathogens to different stages of infection in animal hosts. Microbiology 154:16–29

Lynch D, Fieser N, Gloggler K, Forsbach-Birk V, Marre R (2003) The response regulator LetA regulates the stationary-phase stress response in *Legionella pneumophila* and is required for efficient infection of *Acanthamoeba castellanii*. FEMS Microbiol Lett 219:241–248

Majdalani N, Cunning C, Sledjeski D, Elliott T, Gottesman S (1998) DsrA RNA regulates translation of RpoS message by an anti-antisense mechanism, independent of its action as an antisilencer of transcription. Proc Natl Acad Sci 95:12462–12467. doi:10.1073/pnas.95.21.12462

Majdalani N, Chen S, Murrow J, St. John K, Gottesman S (2001) Regulation of RpoS by a novel small RNA: the characterization of RprA. Mol Microbiol 39:1382–1394

Majdalani N, Hernandez D, Gottesman S (2002) Regulation and mode of action of the second small RNA activator of RpoS translation, RprA. Mol Microbiol 46:813–826

Maki K, Morita T, Otaka H, Aiba H (2010) A minimal base-pairing region of a bacterial small RNA SgrS required for translational repression of ptsG mRNA. Mol Microbiol 76:782–792

Mandin P, Gottesman S (2010) Integrating anaerobic/aerobic sensing and the general stress response through the ArcZ small RNA. EMBO J 29:3094–3107

Marston BJ, Lipman HB, Breiman RF (1994) Surveillance for Legionnaires' disease: risk factors for morbidity and mortality. Arch Int Med 154:2417–2422. doi:10.1001/archinte.1994.00420210049006

Massé E, Gottesman S (2002) A small RNA regulates the expression of genes involved in iron metabolism in *Escherichia coli*. Proc Natl Acad Sci 99:4620–4625

Massé E, Escorcia F, Gottesman S (2003) Coupled degradation of a small regulatory RNA and its mRNA targets in *Escherichia coli*. Genes Dev 17:2374–2383

Massé E, Vanderpool CK, Gottesman S (2005) Effect of RyhB small RNA on global iron use in *Escherichia coli*. J Bacteriol 187:6962–6971

Massé E, Salvail H, Desnoyers G, Arguin M (2007) Small RNAs controlling iron metabolism. Curr Opin Microbiol 10:140–145

McCullen C, Benhammou J, Majdalani N, Gottesman S (2010) Mechanism of positive regulation by DsrA and RprA small noncoding RNAs: pairing increases translation and protects *rpoS* mRNA from degradation. J Bacteriol 192:5559–5571

McNealy T, Forsbach-Birk V, Shi C, Marre R (2005) The Hfq homolog in *Legionella pneumophila* demonstrates regulation by LetA and RpoS and inter-acts with the global regulator CsrA. J Bacteriol 187:1527–1532

Mizuno T, Chou M, Inouye M (1984) A unique mechanism regulating gene expression: translational inhibition by a complementary RNA transcript (micRNA). Proc Natl Acad Sci 81:1966–1970

Møller T, Franch T, Hojrup P, Keene D, Bachinger H, Brennan R, Valentin-Hansen P (2002) Hfq: a bacterial Sm-like protein that mediates RNA–RNA interaction. Mol Cell 9:23–30

Molofsky A, Swanson M (2003) *Legionella pneumophila* CsrA is a pivotal repressor of transmission traits and activator of replication. Mol Microbiol 50:445–461

Molofsky A, Swanson M (2004) Differentiate to thrive: lessons from the *Legionella pneumophila* life cycle. Mol Microbiol 53:29–40

Morita T, Maki K, Aiba H (2005) RNase E-based ribonucleoprotein complexes: mechanical basis of mRNA destabilization mediated by bacterial non-coding RNAs. Genes Dev 19:2176–2186

Novick RP, Geisinger E (2008) Quorum sensing in Staphylococci. Annu Rev Genet 42:541–564. doi:10.1146/annurev.genet.42.110807.091640

Ohno A, Kato N, Yamada K, Yamaguchi K (2003) Factors influencing survival of *Legionella pneumophila* serotype 1 in hot spring water and tap water. App Env Microbiol 69:2540–2547

Papenfort K, Vogel J (2010) Regulatory RNA in bacterial pathogens. Cell Host & Microbe 8:116–127. doi:10.1016/j.chom.2010.06.008

Pitre CAJ, Tanner JR, Patel P, Brassinga AKC (2013) Regulatory control of temporally expressed integration host factor (IHF) in *Legionella pneumophila*. Microbiology 159:475–492. doi:10.1099/mic.0.062117-0

Rasis M, Segal G (2009) The LetA-RsmYZ-CsrA regulatory cascade, together with RpoS and PmrA, post-transcriptionally regulates stationary phase activation of *Legionella pneumophila* Icm/Dot effectors. Mol Microbiol 72:995–1010. doi:10.1111/j.1365-2958.2009.06705.x

Reimer AR, Au S, Schindle S, Bernard KA (2010) *Legionella pneumophila* monoclonal antibody subgroups and DNA sequence types isolated in Canada between 1981 and 2009: Laboratory Component of National Surveillance. Eur J Clin Microbiol Infect Dis 29:191–205

Rogers J, Keevil C (1992) Immunogold and fluorescein immunolabelling of *Legionella pneumophila* within an aquatic biofilm visualized by using episcopic differential interference contrast microscopy. App Env Microbiol 58:2326–2330

Romeo T (1998) Global regulation by the small RNA-binding protein CsrA and the non-coding RNA molecule CsrB. Mol Microbiol 29:1321–1330

Romeo T, Gong M, Liu M, Brun-Zinkernagel A (1993) Identification and molecular characterization of *csrA*, a pleiotropic gene from *Escherichia coli* that affects glycogen biosynthesis, gluconeogenesis, cell size, and surface properties. J Bacteriol 15:4744–4755

Sahr T, Brüggemann H, Jules M, Lomma M, Albert-Weissenberger C, Cazalet C, Buchrieser C (2009) Two small ncRNAs jointly govern virulence and transmission in *Legionella pneumophila*. Mol Microbiol 72:741–762

Sahr T, Rusniok C, Dervins-Ravault D, Sismeiro O, Coppee J, Buchrieser C (2012) Deep sequencing defines the transcriptional map of *L. pneumophila* and identifies growth phase-dependent regulated ncRNAs implicated in virulence. RNA Biol 9:503–519

Salim NN, Feig AL (2010) An upstream Hfq binding site in the *fhlA* mRNA leader region facilitates the OxyS-*fhlA* interaction. PLoS One 5:e13028

Salvail H, Lanthier-Bourbonnais P, Sobota J, Caza M, Benjamin J, Mendieta M, Lepine F, Dozois C, Imlay J, Massé E (2010) A small RNA promotes siderophore production through transcriptional and metabolic remodeling. Proc Natl Acad Sci 107:15223–15228

Sankar P, Lee JH, Shanmugam KT (1988) Gene-product relationships of *fhlA* and *fdv* genes of *Escherichia coli*. J Bacteriol 195:5440–5445

Schlensog V, Böck A (1990) Identification and sequence analysis of the gene encoding the transcriptional activator of the formate hydrogenlyase system of *Escherichia coli*. Mol Microbiol 4:1319–1327. doi:10.1111/j.1365-2958.1990.tb00711.x

Schumacher M, Pearson R, Møller T, Valentin-Hansen P, Brennan R (2002) Structures of the pleiotropic translational regulator Hfq and an Hfq–RNA complex: a bacterial Sm-like protein. EMBO J 21:3546–3556

Segal G, Russo J, Shuman H (1999) Relationships between a new type IV secretion system and the *icm/dot* virulence system of *Legionella pneumophila*. Mol Microbiol 34:799–809

Sharma CM, Hoffmann S, Darfeuille F, Reignier J, Findeiss S, Sittka A, Chabas S, Reiche K, Hackermüller J, Reinhardt R, Stdler PF, Vogel J (2010) The primary transcriptome of the major human pathogen *Helicobacter pylori*. Nature 464:250–255. doi:10.1038/nature08756

Shi C, Forsbach-Birk V, Marre R, McNealy T (2006) The *Legionella pneumophila* global regulatory protein LetA affects DotA and Mip. Int J Med Microbiol 296:15–24

Simons RW, Kleckner N (1983) Translational control of IS10 transposition. Cell 34:683–691

Sittka A, Pfeiffer V, Tedin K, Vogel J (2007) The RNA chaperone Hfq is essential for the virulence of *Salmonella typhimurium*. Mol Microbiol 63:193–217

Sledjeski DD, Gupta A, Gottesman S (1996) The small RNA, DsrA, is essential for the low temperature expression of RpoS during exponential growth in *Escherichia coli*. EMBO J 15:3993–4000

Sobrero P, Valverde C (2012) The bacterial protein Hfq: much more than a mere RNA-binding factor. Crit Rev Microbiol 38:276–299. doi:10.3109/1040841X.2012.664540

Song T, Mika F, Lindmark B, Liu Z, Schild S, Bishop A, Zhu J, Camilli A, Johansson J, Vogel J, Wai S (2008) A new *Vibrio cholerae* sRNA modulates colonization and affects release of outer membrane vesicles. Mol Microbiol 70:100–111

Sonnleitner E, Haas D (2011) Small RNAs as regulators of primary and secondary metabolism in *Pseudomonas* species. Appl Microbiol Biotechnol 91:63–79

Soper T, Mandin P, Majdalani N, Gottesman S, Woodson S (2010) Positive regulation by small RNAs and the role of Hfq. Proc Natl Acad Sci 107:9602–9607

Stone B, Abu Kwaik Y (1999) Natural competence for DNA transformation by *Legionella pneumophila* and its association with expression of type IV pili. J Bacteriol 181:1395–1402

Storz G, Tartaglia L, Ames B (1990) The OxyR regulon. Antonie van Leeuwenhoek 58:157–161. doi:10.1007/BF00548927

Stougaard P, Molin S, Nordström K (1981) RNAs involved in copy-number control and incompatibility of plasmid R1. Proc Natl Acad Sci 78:6008–6012

Swanson MS, Hammer BK (2000) *Legionella pneumophila* pathogenesis: a fateful journey from amoebae to macrophages. Annu Rev Microbiol 54:567–613

Trigui H, Dudyk P, Sum J, Shuman H, Faucher S (2013) Analysis of the transcriptome of *Legionella pneumophila hfq* mutant reveals a new mobile genetic element. Microbiology. doi:10.1099.mic.0.067983.0

Trotochaud A, Wassarman K (2005) A highly conserved 6S RNA structure is required for regulation of transcription. Nat Struct Mol Biol 12:313–319

Tsui H, Leung H, Winkler M (1994) Characterization of broadly pleiotropic phenotypes caused by an *hfq* insertion mutation in *Escherichia coli* K-12. Mol Microbiol 13:35–49

Udekwu K, Darfeuille F, Vogel J, Reimegard J, Holmqvist E, Wagner E (2005) Hfq-dependent regulation of OmpA synthesis is mediated by an antisense RNA. Genes Dev 19:2355–2366

Updegrove T, Wartell R (2011) The influence of *Escherichia coli* Hfq mutations on RNA binding and sRNA•mRNA duplex formation in *rpoS* riboregulation. Biochim Biophys Acta 1809:532–540

Valverde C, Heeb S, Keel C, Haas D (2003) RsmY, a small regulatory RNA, is required in concert with RsmZ for GacA-dependent expression of biocontrol traits in *Pseudomonas fluorescens* CHA0. Mol Microbiol 50:1361–1379

Viswanathan V, Edelstein P, Pope C, Cianciotto N (2000) The *Legionella pneumophila iraAB* locus is required for iron assimilation, intracellular infection, and virulence. Infect Immun 68:1069–1079

Vogel JP, Isberg RR (1999) Cell biology of *Legionella pneumophila*. Curr Opin Microbiol 2:30–34. doi:10.1016/S1369-5274(99)80005-8

Wadler C, Vanderpool C (2007) A dual function for a bacterial small RNA: SgrS performs base pairing-dependent regulation and encodes a functional polypeptide. Proc Natl Acad Sci 104:20454–20459

Wang X, Dubey AK, Suzuki K, Baker CS, Babitzke P, Romeo T (2005) CsrA post-transcriptionally represses *pgaABCD*, responsible for synthesis of a biofilm polysaccharide adhesion of *Escherichia coli*. Mol Microbiol 56:1648–1663

Wassarman K (2007) 6S RNA: a regulator of transcription. Mol Microbiol 65:1425–1431

Wassarman K, Storz G (2000) 6S RNA regulates *E. coli* RNA polymerase activity. Cell 101:613–623

Wassarman KM, Zhang A, Storz G (1999) Small RNAs in *Escherichia coli*. Trends Microbiol 7:37–45

Waters LS, Storz G (2009) Regulatory RNAs in bacteria. Cell 136:615–628

Weilbacher T, Suzuki K, Dubey A, Wang X, Gudapaty S, Morozov I, Baker C, Georgellis D, Babitzke P, Romeo T (2003) A novel sRNA component of the carbon storage regulatory system of *Escherichia coli*. Mol Microbiol 48:657–670

Weissenmayer B, Prendergast J, Lohan A, Loftus B (2011) Sequencing illustrates the transcriptional response of *Legionella pneumophila* during infection and identifies seventy novel small non-coding RNAs. PLoS One 6:e17570

Wilderman PJ, Sowa NA, FitzGerald DJ, FitzGerald PC, Gottesman S, Ochsner UA, Vasil ML (2004) Identification of tandem duplicate regulatory small RNAs in pseudomonas aeruginosa involved in iron homeostasis. Proc Natl Acad Sci USA 101:9792–9797

Yamamoto S, Izumiya H, Mitobe J, Morita M, Arakawa E, Ohnishi M, Watanabe H (2011) Identification of a chitin-induced small RNA that regulates translation of the *tfoX* gene, encoding a positive regulator of natural competence in *Vibrio cholerae*. J Bacteriol 8:1953–1965

Yiallouros P, Papadouri T, Karaoli C, Papamichael E, Zeniou M, Pieridou-Bagatzouni D, Papageorgiou GT, Pissarides N, Harrison TG, Hadjidemetriou A (2013) First outbreak of nosocomial *Legionella* infection in term neonates caused by a cold mist ultrasonic humidifier. Clin Infec Dis 57:48–56. doi:10.103/cid/cit176

Yu VL, Plouffe JF, Pastoris MC, Stout JE, Schousboe M, Widmer A, Summersgill J, File T, Heath CM, Paterson DL, Chereshsky A (2002) Distribution of *Legionella* species and serogroups isolated by culture in patients with sporadic community-acquired Legionellosis: an international collaborative survey. J Infec Dis 186:127–128. doi:10.1086/341087

Zhang A, Wassarman KM, Ortega J, Steven AC, Storz G (2002) The Sm-like Hfq protein increases OxyS RNA interaction with target mRNAs. Molec Cell 9:11–22. doi:10.1016/S1097-2765(01)00437-3

Zhang A, Wassarman K, Rosenow C, Tjaden B, Storz G, Gottesman S (2003) Global analysis of small RNA and mRNA targets of Hfq. Mol Microbiol 50:1111–1124

Zheng M, Wang X, Templeton LJ, Smulski DR, LaRossa RA, Storz G (2001) DNA Microarray-mediated transcriptional profiling of the *Escherichia coli* response to hydrogen peroxide. J Bacteriol 183:4562–4570

Zusman T, Aloni G, Halperin E, Kotzer H, Degtyar E, Feldman M, Segal G (2007) The response regulator PmrA is a major regulator of the *icm/dot* type IV secretion system in *Legionella pneumophila* and *Coxiella burnetii*. Mol Microbiol 63:1508–1523

Type II Secretion and *Legionella* Virulence

Nicholas P. Cianciotto

Abstract Type II secretion (T2S) is one of six systems that can occur in Gram-negative bacteria for the purpose of secreting proteins into the extracellular milieu and/or into host cells. This chapter will describe the T2S system of *Legionella pneumophila*. Topics to be covered include the genetic basis of T2S in *L. pneumophila*, the numbers (>25), types, and novelties of *Legionella* proteins that are secreted via T2S, and the many ways in which T2S and its substrates promote *L. pneumophila* physiology, ecology, and virulence. Within the aquatic environment, T2S plays a major role in *L. pneumophila* intracellular infection of multiple types of (*Acanthamoeba*, *Hartmannella*, and *Naegleria*) amoebae. Within the mammalian host, T2S promotes bacterial persistence in lungs, intracellular infection of both macrophages and epithelial cells, and a dampening of the host innate immune response. In this context, T2S may represent a potential target for both industrial and biomedical application.

Contents

1 Introduction 82
2 T2S Genes in *L. pneumophila* 84
  2.1 T2S Genes 84
  2.2 Sec Genes 84
  2.3 Tat Genes 85
3 T2S Substrates of *L. pneumophila* 85
  3.1 Substrate Identification Based Upon Secreted Enzyme Activities 85
  3.2 Substrate Detection Via Proteomic Analyses 86
  3.3 The Connection Between T2S and Mip 87
  3.4 The Connection Between T2S and Surfactant-Based Sliding 88

N. P. Cianciotto (✉)
Department of Microbiology-Immunology, Northwestern University Medical School, 320 East Superior St, Chicago, IL 60611, USA
e-mail: n-cianciotto@northwestern.edu

Current Topics in Microbiology and Immunology (2013) 376: 81–102

DOI: 10.1007/82_2013_339

Published Online: 31 July 2013

3.5 Overall Assessment 88
4 The Role of T2S in *L. pneumophila* Environmental Persistence 90
4.1 The Importance of T2S in Extracellular Survival 90
4.2 The Importance of T2S in Intracellular Infection of Amoebae 90
4.3 The Importance of T2S in Establishing a Broad Host Range for *L. pneumophila* 92
5 The Role of T2S in *L. pneumophila* Infection of Mammalian Hosts 93
5.1 The Importance of T2S in Lung Models of Infection 93
5.2 The Importance of T2S in Intracellular Infection of Macrophages 93
5.3 The Importance of T2S in Intracellular Infection of Epithelial Cells 94
5.4 The Importance of T2S in Dampening the Innate Immune Response 95
6 T2S and Other *Legionella* Species 96
7 Concluding Remarks 96
References 97

1 Introduction

Legionella pneumophila secretes a very large number of factors that promote virulence and/or intracellular infection of host cells. These factors encompass both proteins and non-protein molecules, although it is the export of myriad proteins that most critically promotes infection. In Gram-negative bacteria, protein secretion is a complicated process that requires transit across the inner membrane, periplasm, and then the outer membrane. As a group, Gram-negatives have six, and perhaps eight, systems that facilitate secretion from within the bacterium to the extracellular milieu and/or into target host cells; i.e., type I, II, III, etc. (Desvaux et al. 2009). Much research has shown that type II and type IV protein secretion are essential for *L. pneumophila*. This chapter will focus on type II secretion (T2S), highlighting its mechanism, the types of secreted proteins, and the role of this secretion in *L. pneumophila*'s interactions with its hosts.

T2S systems are common, albeit not universal, among Gram-negative bacteria (Cianciotto 2005). T2S is a multi-step process (Korotkov et al. 2012) (Fig. 1). Proteins that are to be secreted are first translocated across the inner membrane. In most cases, unfolded substrates cross the inner membrane through the Sec pathway, however, in some instances, folded substrates cross via the twin-arginine translocon (Tat). Once in the periplasm, unfolded substrates assume a tertiary conformation and in certain instances oligomerize. In the last step, substrates are moved across the outer membrane by a complex of proteins that is dedicated to T2S, i.e., the T2S apparatus. Evolutionarily related to the bacterial type IV pilus, the T2S machinery consists of twelve "core" proteins—a cytosolic ATPase (T2S E), three inner membrane proteins that create a platform for T2S E (T2S F, L, M), multiple major and minor pseudopilins which form a pilus-like structure that spans the periplasm (T2S G, H, I, J, K), an inner membrane peptidase that clips pseudopilins prior to their integration into the apparatus (T2S O), an outer membrane

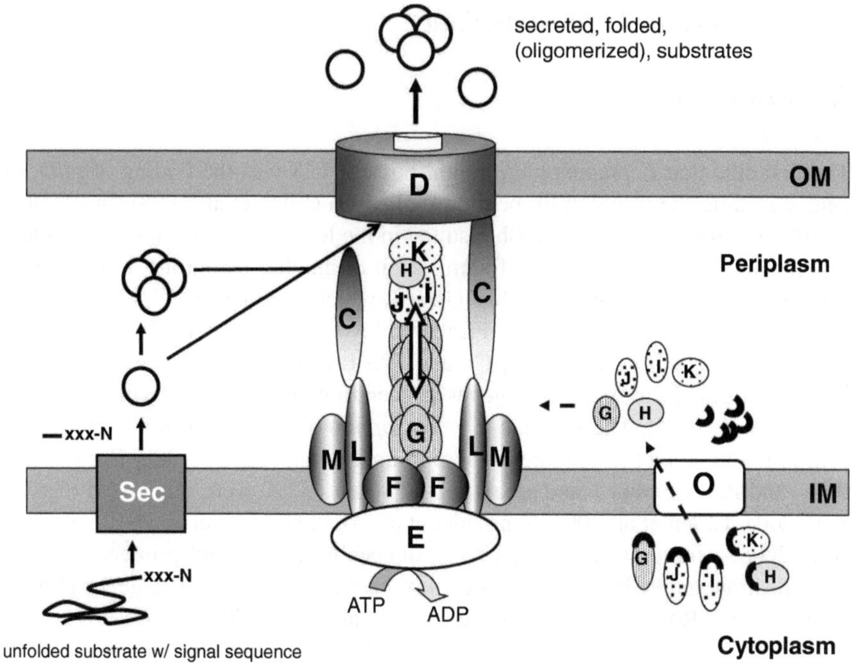

Fig. 1 Model of T2S in Gram-negative bacteria. The core components of T2S are indicated by single-letter designations, C–M. Following translocation across the inner membrane (IM) via Sec and folding (and in some cases, oligomerization) within the periplasm, protein substrates are recognized by the T2S apparatus. Using energy generated at the IM, a pilus-like structure that is made of major (G) and minor (H, I, J, K) pseudopilins acts to "push" substrates through the dodecameric secretin (D) in the outer membrane (OM). The pre-pilin peptidase (O) cleaves and N-methylates pseudopilins prior to their integration into the T2S apparatus. For substrates that are already folded within the cytoplasm, transport across the IM occurs via Tat (*not shown*)

protein (secretin) that oligomerizes to form the secretion pore (T2S D), and lastly a protein that links inner and outer membrane factors (T2S C). The current overall model is that substrates destined for secretion are recognized by the T2S apparatus and then, using energy generated at the inner membrane, the pseudopilus acts like a piston to push proteins through the secretin. The trait that defines a T2S substrate is still not clear but likely involves tertiary structure and initial interactions with T2S C and D. Studies document that T2S promotes the growth of environmental bacteria as well as the virulence of human, animal, and plant pathogens (Cianciotto 2005). In many ways, however, the examination of *L. pneumophila* has provided us the broadest understanding of the biological role of T2S (Cianciotto 2009).

2 T2S Genes in *L. pneumophila*

2.1 T2S Genes

The first clue that *L. pneumophila* might possess T2S was the finding of *pilD*, the gene encoding the pseudopilin peptidase (T2S O) (Liles et al. 1998). Inactivation of *pilD* in serogroup-1 strain 130b resulted in the loss of proteins in mutant culture supernatants (Liles et al. 1999). Examination of another serogroup-1 strain, Philadelphia-1, uncovered the locus *lspFGHIJK*, which encodes the T2S F, G, H, I, J, and K (Hales and Shuman 1999). Further study of 130b revealed genes encoding T2S D and E (*lspDE*), T2S C (*lspC*), and T2S L and M (*lspLM*), and mutation of *lspDE* confirmed the role of the genes in secretion (Rossier and Cianciotto 2001; Rossier et al. 2004). The fact that *L. pneumophila* has a full set of T2S genes was demonstrated when the genomes of serogroup-1 strains Alcoy, 130b, Corby, Lens, Paris, and Philadelphia 1 and serogroup-12 strain 43290 were sequenced (Cazalet et al. 2004; Chien et al. 2004; Schroeder et al. 2010; D'Auria et al. 2010; Glockner et al. 2008; Amaro et al. 2012). Using *lsp*-specific probes and primers, Southern hybridization and PCR analyses has confirmed the existence of T2S genes in many other strains (Rossier et al. 2004; Huang et al. 2006; Costa et al. 2011). T2S mutants of *L. pneumophila* grow normally in bacteriological media (e.g., buffered yeast extract [BYE] broth) at 37 °C (Rossier and Cianciotto 2001; Rossier et al. 2004). These mutants also display a typical efficiency of plating on buffered charcoal yeast extract agar at 37 °C. Thus, T2S seems not to be required for extracellular growth under standard laboratory conditions.

2.2 Sec Genes

The presence of T2S genes in *L. pneumophila* implied that there must also be a mechanism for moving proteins across the inner membrane prior to secretion via the T2S apparatus. Genome sequencing has now confirmed that *L. pneumophila* encodes a complete Sec system (Cazalet et al. 2004; Chien et al. 2004; Lammertyn and Anne 2004). *L. pneumophila* has the machinery for co-translational and post-translational translocation. In co-translational translocation, nascent polypeptides are delivered by the signal recognition particle, with the help of YidC, to a channel made of SecYEG (Robson and Collinson 2006; Lee and Schneewind 2001). In post-translational translocation, cytoplasmic SecA ATPase and SecB chaperone deliver nascent protein to SecYEG in a process that also involves SecDF and YajC (Robson and Collinson 2006; Lee and Schneewind 2001). *L. pneumophila* encodes the inner membrane type I signal peptidase (LepB) that clips N-termini from proteins as they emerge on the periplasmic face of the inner membrane (Lammertyn et al. 2004). *L. pneumophila* also has the type II signal peptidase that clips translocated lipoproteins (Geukens et al. 2006). In silico analysis indicates that there are >500 proteins that have a typical signal peptide and are potential substrates for Sec transport (DebRoy et al. 2006a).

2.3 Tat Genes

Genome sequencing has also confirmed that L. pneumophila carries the genes encoding Tat, (Cazalet et al. 2004; Chien et al. 2004; De Buck et al. 2004; Rossier and Cianciotto 2005). Tat consists of three integral membrane proteins, named TatA, TatB, and TatC. The signal peptides of Tat-dependent substrates differ from those of Sec-dependent signal peptides in a several ways. First, they have a twin Arg consensus motif (RRXΦΦ, where Φ is a hydrophobic residue), with that motif being followed by a region that is less hydrophobic than that of Sec signals. Second, basic residues are often found before the signal peptidase cleavage site. Finally, Tat signal peptides are often much longer than Sec signal sequences. In the Escherichia coli model, Tat is considered a three-step event (Sargent et al. 2006). First, the signal peptides are bound by an inner membrane complex consisting of TatB and TatC, with TatC likely being the main factor. Then, a complex consisting of TatA is recruited in a process that requires membrane electrochemical gradient. Finally, the folded exoprotein passes through a channel in the TatA complex. After tatA, tatB, and tatC were found in L. pneumophila, RT-PCR analysis ascertained that they are expressed during extra- and intracellular growth (De Buck et al. 2004). Since L. pneumophila tatB mutants grow normally in BYE broth, Tat, like T2S, is not required for extracellular growth (Rossier and Cianciotto 2005). However, tat mutants are impaired for growth in iron-limiting conditions and during infection of macrophages and amoebae, implying that Tat substrates are needed for survival under specialized conditions (Rossier and Cianciotto 2005; De Buck et al. 2005). In silico analysis revealed ca. 35 putative Tat substrates (De Buck et al. 2004; Rossier and Cianciotto 2005).

3 T2S Substrates of *L. pneumophila*

3.1 Substrate Identification Based Upon Secreted Enzyme Activities

Initially, 12 substrates were demonstrated to be dependent upon T2S (Cianciotto 2005; Rossier et al. 2004; Banerji et al. 2005). This conclusion was based upon the loss of activities from culture supernatants of *lspDE, lspF, lspG, lspGH,* or *pilD* mutants of strain 130b grown in BYE broth at 37 °C (Liles et al. 1999; Rossier and Cianciotto 2001; Rossier et al. 2004; Aragon et al. 2000). The activities defined were the tartrate-sensitive and -resistant acid phosphatases, phospholipases C, phospholipase A, lysophospholipase A, glycerophospholipid cholesterol acyltransferase (GCAT), mono-, di-, and triacylglycerol lipases, ribonuclease, and protease (Liles et al. 1999; Hales and Shuman 1999; Rossier and Cianciotto 2001; Rossier et al. 2004; Banerji et al. 2005; Aragon et al. 2000, 2001, 2002; Flieger et al. 2001, 2002; DebRoy et al. 2006b). Since mutants specifically lacking type IV pili are not impaired for these activities, the change in secretion of the *pilD* mutants is due to the loss of T2S (Rossier and Cianciotto 2001; Rossier et al. 2004). Later analysis of strain Paris identified a starch- and glycogen-degrading activity (Herrmann et al. 2011), and a similar result has now been obtained for strain 130b (Tyson et al. 2013). In some instances, the structural gene encoding the secreted activity has been defined; these encompass *map* (Map) for the tartrate-sensitive acid phosphatase (Aragon et al. 2001), *plcA* (PlcA) for phospholipase C activity (Rossier and Cianciotto 2005; Aragon et al. 2002), *plaA* (PlaA) for the lysophospholipase A (Flieger et al. 2002), *plaC* (PlaC) for GCAT (Banerji et al. 2005), *lipA* (LipA) and *lipB* (LipB) for mono- and triacylglycerol lipases (Aragon et al. 2002), *gamA* (GamA) for the starch hydrolase (Herrmann et al. 2011), and *proA/msp* (ProA/Msp) for a metalloprotease (Liles et al. 1999; Hales and Shuman 1999). Based upon gene sequences, most of these T2S substrates are predicted to be secreted via Sec. However, by virtue of Arg in its signal peptide mutations and the fact that *tatB* mutations diminish secreted phospholipase C activity, PlcA of strain 130b is predicted to be a Tat substrate (Rossier and Cianciotto 2005). Mutations in these various structural genes do not completely abolish the corresponding activity, suggesting that *L. pneumophila* has more than one secreted phosphatase, phospholipase C, lysophospholipase A, lipase, and protease and that T2S must mediate the secretion of >12 proteins. The analysis of supernatants from *proA* mutants indicates that some exoproteins (e.g., PlaC) are cleaved and activated by ProA (Banerji et al. 2005; Flieger et al. 2002; Lang et al. 2012). Finally, one of the first substrates to be described, the Map phosphatase, has striking sequence similarity to eukaryotic enzymes (Aragon et al. 2001), suggesting that *L. pneumophila* has learned the strategies of its host during its evolution as an intracellular parasite. More examples of *Legionella* eukaryotic-like proteins have emerged from the study of type IV secretion system as well as the further study of T2S (below).

3.2 Substrate Detection Via Proteomic Analyses

Next, strain 130b and an *lspF* mutant were grown in BYE broth at 37 °C, supernatants were compared by 2D PAGE, and then mass spectrometry was used to identify proteins that were present for wild type but absent for the mutant

(DebRoy et al. 2006a). Three of the identified proteins, i.e., ProA, Map, and PlaA, had been previously named as T2S substrates. A fourth, i.e., SrnA, proved to confer a previously found ribonuclease activity (Rossier et al. 2009). Several others were predicted to be enzymes that are akin to those in other bacteria, and subsequent mutational analyses showed that CelA is an endoglucanase/cellulase, ChiA a chitinase, and LapA and LapB two different aminopeptidases (DebRoy et al. 2006a; Rossier et al. 2008, 2009; Pearce and Cianciotto 2009). Others, reminiscent of Map, showed high similarity to eukaryotic proteins, with one (Lcl) having collagen-like repeats, and the other (LegP) similarity to astacin-like proteases (DebRoy et al. 2006a). Later work showed that Lcl has heparin-binding activity and promotes some forms of adherence (Duncan et al. 2011; Vandersmissen et al. 2010). LegP was later found to be translocated by Dot/Icm type IV secretion when legionellae were growing within macrophages (de Felipe et al. 2008). This raises an intriguing scenario whereby some proteins may be secreted or influenced by multiple pathways, with environmental conditions potentially dictating which secretion pathway(s) is most critical. Interestingly, other T2S substrates displayed little to no similarity to any known protein in the database (DebRoy et al. 2006a). The genes encoding two of these novel substrates, i.e., NttA and NttB, have been recently cloned and subjected to mutational analysis (Tyson et al. 2013). Proteomic data similar to these obtained from strain 130b were subsequently reported for strain Philadelphia-1 (Galka et al. 2008).

3.3 The Connection Between T2S and Mip

Mip, the surface-associated, peptidyl-proline *cis–trans*-isomerase of *L. pneumophila*, has been linked to the elaboration of secreted activity (DebRoy et al. 2006b). Mip mutants of strain 130b have a 40–70 % reduction in secreted phospholipase C activity. When culture supernatants were examined by chromatography, the activity linked to Mip was T2S-dependent but distinct from previously defined PlcA, suggesting that Mip promotes the elaboration of a "new" exoprotein. A recent study identified a gene (*plcB*) that is predicted to encode a protein that is highly similar to PlcA (McCoy-Simandle et al. 2011). Although yet-to-be-proven, the putative PlcB phospholipase C is a candidate for being the Mip-dependent enzyme. Mip might be involved in the release of an active phospholipase C, by acting directly on the exoprotein or a protein that forms part of the T2S pathway. Alternatively, Mip could be associating with newly secreted protein and cause changes that convert it from inactive to active. These data represent the first case of a surface peptidyl-proline isomerase being linked to the secretion or activation of proteins beyond the outer membrane (DebRoy et al. 2006b).

3.4 The Connection Between T2S and Surfactant-Based Sliding

L. pneumophila displays surface translocation when it is grown on media containing 0.5–1.0 % agar (Stewart et al. 2009). The growing bacteria appear in an amorphous, lobed pattern that is most manifest at 25–30 °C. *L. pneumophila* mutants lacking flagella and/or type IV pili behave as wild type does, indicating that the observed surface translocation is not swarming or twitching motility. A translucent film composed of a lipid-containing surfactant is visible atop the agar in front of the spreading legionellae; thus, *L. pneumophila* exhibits "sliding motility" (Stewart et al. 2009, 2011). Interestingly, *L. pneumophila lsp* mutants are defective for surface translocation and surfactant expression (Stewart et al. 2009). However, mutants lacking the outer membrane efflux pump, TolC, are also lacking surface translocation and surfactant (Stewart et al. 2011). When the *tolC* and *lspF* mutants are grown next to each other, the *lsp* mutant secretes surfactant, indicating that TolC and T2S conjoin to mediate surfactant secretion, with one (TolC) being the conduit for surfactant export and the other (T2S) the exporter of a molecule that is required for induction or maturation of surfactant synthesis or secretion (Stewart et al. 2011). None of the currently available exoprotein mutants lack surfactant, indicating the existence of yet another (unknown) T2S substrate. Overall, these data represent a novel observation linking bacterial sliding, surfactant, and T2S.

3.5 Overall Assessment

Proteomic results combined with the various characterizations of secreted activities signal that the number of proteins secreted by *L. pneumophila* T2S is ≥25. The T2S output would be greater than 25 is based upon several considerations; i.e., (i) low-level expression and/or degradation likely impaired detection of some proteins, (ii) the comparisons between wild type and mutant used cultures grown under one condition, (iii) as noted above, mutations inactivating structural genes did not always completely abolish activity in supernatants, and (iv) in silico analysis of genomes reveals ≥60 proteins that contain a signal sequence and are predicted to be extracellular by at least one program (DebRoy et al. 2006a). As is apparent from our discussion, *L. pneumophila* secretes many different factors via its T2S system (Table 1). Several sorts of enzymes identified, including proteases, aminopeptidases, lipolytic enzymes, chitinases, and phosphatases, are secreted by other T2S systems (Cianciotto 2005; Rossier et al. 2008). In some instances, the *L. pneumophila* substrates are related to the exoproteins of others; e.g., PlcA is highly similar to a *P. fluorescens* phospholipase C (Aragon et al. 2002). However, *L. pneumophila* T2S uniquely encompasses proteins that show great similarity to eukaryotic proteins; e.g., Map and LegP (DebRoy et al. 2006a; Aragon et al. 2001).

Table 1 T2S, its substrates, and their roles in intracellular infection

Protein	Protein activity or sequence novelty	Role in intracellular infection of[a]					Reference(s)
		Mac	Epi	Ac	Hv	Nl	
LspF[b]	T2S apparatus	+	+	+++	+++	+++	Rossier et al. (2004, 2008) and Söderberg et al. (2008)
CelA	Endoglucanase	–	–	–	–	–	Pearce and Cianciotto (2009)
ChiA	Chitinase	–	–	–	–	–	DebRoy et al. (2006)
GamA	Glucoamylase	–	–	–	–	–	Herrmann et al. (2011)
LapA	Leu/tyr aminopeptidase	–	–	–	–	–	Rossier et al. (2008)
LapB	Lys/arg aminopeptidase	–	–	–	–	–	Rossier et al. (2008)
LegP	Eukaryotic-like	–	nd	–	–	–	Tyson et al. (2013)
LipA	Monoacylglyercol lipase	–	–	–	–	–	Aragon et al. (2002)
LipB	Triacylglycerol lipase	–	–	–	–	–	Aragon et al. (2002)
Map	Acid phosphatase	–	–	–	–	–	Aragon et al. (2001)
NttA	Novel	–	nd	++	–	–	Tyson et al. (2013)
NttB	Novel	–	nd	–	–	–	Tyson et al. (2013)
PlaA	Lysophospholipase A	–	–	–	–	–	Flieger et al. (2002)
PlaC	Cholesterol acyltransferase	–	–	–	+	+	Banerji et al. (2005)
PlcA	Phospholipase C	–	–	–	–	–	Aragon et al. (2002)
PlcB	Phospholipase C[c]	–	–	–	–	–	McCoy-Simandle et al. (2011)
ProA	Metalloprotease	–	–	–	+	++	Liles et al. (1999) and Hales and Shuman (1999)
SrnA	T2 ribonuclease	–	–	–	+	+	Rossier et al. (2009)

[a] Based upon comparing the infectivity of wild type 130b to a mutant lacking the indicated protein; (–) no difference, (+) mutant impaired two- to tenfold, (++) mutant impaired 11- to 100-fold mutant defect, (+++) mutant impaired >100-fold, *nd* not determined. *Mac* macrophages, *Epi* epithelial cells, *Ac A. castellanii*, *Hv H. vermiformis*, *Nl N. lovaniensis*

[b] Similar results for mutants lacking LspD, LspE, or LspG (Rossier and Cianciotto 2001)

[c] Putative, based upon very high similarity to PlcA

L. pneumophila is also unique, at least so far, in secretion of an RNAse and the effects on surfactant (Rossier et al. 2009; Stewart et al. 2011). The types of enzymes most prevalent in the *Legionella* T2S repertoire are lipolytic enzymes as well as proteases/peptidases, a result that is consistent with the fact that amino acids are the main carbon and energy source for *L. pneumophila* (George et al. 1980). Yet, the existence of a chitinase, endoglucanase, and glucoamylase suggests that *L. pneumophila* is also adept at degrading complex carbohydrates (DebRoy et al. 2006a; Herrmann et al. 2011; Pearce and Cianciotto 2009). Perhaps, most notable, some *L. pneumophila* substrates do not bear any similarity whatsoever to known proteins raising the possibility of there being entirely new proteins secreted by T2S. In light of the number of proteins uncovered and the types of factors detected, the examination of *L. pneumophila* has illuminated more than ever the role that T2S has on secretion and function (Cianciotto 2009).

4 The Role of T2S in *L. pneumophila* Environmental Persistence

4.1 The Importance of T2S in Extracellular Survival

Although T2S mutants of *L. pneumophila* strain 130b replicate normally at 30–37 °C, they are defective for growth in media at 25, 17, and 12 °C (Söderberg et al. 2004). In trials that mimic aquatic habitats, the *lsp* mutants show reduced survival in tap water incubated at 25, 17, 12, and 4 °C (Söderberg et al. 2008). The T2S mutants grow better at the low temperatures when they are plated next to wild-type or wild-type culture supernatants, indicating that a secreted factor(s) can promote growth at low temperatures (Söderberg et al. 2004, 2008). Mutants lacking known T2S substrates grow normally at the low temperatures, indicating the existence of yet-to-be-defined substrates. Supporting this hypothesis, when wild-type *L. pneumophila* is grown at 17 or 12 °C, new proteins appear in supernatants, including a Sec-dependent protein that is predicted to be a peptidyl-proline isomerase (Söderberg and Cianciotto 2008). In another study, transcriptional profiling revealed that some T2S substrate genes are hyperexpressed when *L. pneumophila* is grown at 20 °C in a biofilm (Hindre et al. 2008). Finally, the T2S-dependent surfactant of strain 130b impedes, directly or indirectly, the growth of other species of *Legionella* (Stewart et al. 2011). Taken together, these data implicate T2S as an important factor in the planktonic persistence of *L. pneumophila* in the environment and therefore implicate T2S as a factor in disease transmission.

4.2 The Importance of T2S in Intracellular Infection of Amoebae

T2S mutants of *L. pneumophila* are greatly impaired for intracellular infection of freshwater amoebae, including *Acanthamoeba castellanii, Hartmannella vermiformis,* and *Naegleria lovaniensis* (Liles et al. 1999; Hales and Shuman 1999; Rossier and Cianciotto 2001; Rossier et al. 2004; Tyson et al. 2013; Polesky et al. 2001). Indeed, T2S mutants of strains 130b and Philadelphia-1 show very little sign of growth in the amoebae. The impaired infectivity of the mutants is complemented (i.e., reversed) when a copy of the T2S gene is reintroduced, confirming that T2S is required for infection. The initial assessments of infection were done at 35–37 °C, however, later studies have shown that the T2S mutant defect is also evident when the amoebae are cultured at 22–25 °C (Söderberg et al. 2008). Additional tests have determined that the *lsp* mutants are not impaired for entry into amoebae (Söderberg et al. 2008), indicating that T2S is facilitating bacterial resistance to intracellular killing and/or replication itself. Among the various T2S substrates, the PlaC GCAT, ProA protease, and SrnA ribonuclease are required for optimal infection of *H. vermiformis* and *N. lovaniensis* (Tyson et al. 2013; Rossier et al. 2008, 2009). Compatible with these data, an analysis of molecular evolution among *L. pneumophila* strains concluded that ProA and SrnA have been selected due to a role in virulence (Costa et al. 2011). A double mutant lacking both ProA and SrnA has a defect that is greater than the corresponding single mutants, indicating that the role of T2S in amoebal infection is due to the combined effect of multiple secreted proteins (Rossier et al. 2009). Based upon the analysis of a double mutant lacking both PlaC and ProA, the role of ProA in *H. vermiformis* was connected to its ability to activate PlaC, whereas in *N. lovaniensis*, ProA appeared to have multiple functions. In contrast to these results, the novel T2S substrate NttA proved to be necessary for optimal infection of *A. castellanii* (Tyson et al. 2013). Taken together, these data demonstrate that the reduced ability of T2S *lsp* mutants to infect *A. castellanii*, *H. vermiformis,* and *N. lovaniensis* is due to the loss of secreted effectors versus being simply due to any potential changes in the bacterial cell envelope. The relatively modest effect of each of the substrate mutations is compatible with a scenario in which the importance of T2S derives from an additive effect of multiple secreted proteins. The importance of NttA was revealed after testing only three of the novel T2S substrates (i.e., NttA, NttB, LegP), whereas a greater effort that had focused on proteins with similarity to known enzymes yielded only three required promoters of intracellular infection (PlaC, ProA, SrnA) suggests that the T2S substrates that are more unique in sequence (structure) and specific to *Legionella* are especially critical for *L. pneumophila* persistence (Tyson et al. 2013).

Given the novelty of NttA, it is difficult to predict what the protein might be doing in infected acanthamoebae. PlaC, on the other hand, is a glycerophospholipid: cholesterol acyltransferase that has phospholipase A and lysophospholipase activities (Banerji et al. 2005). Thus, one can posit that PlaC is altering sterol-containing membranes of *H. vermiformis* and *N. lovaniensis* and thereby influence processes such as the trafficking of the bacterial phagosome or the movement of nutrients and other factors into or out of the bacterial vacuole. Based upon the analysis of the *plaC proA* mutant, it appears that the key role of ProA in *H. vermiformis* infection is to activate PlaC. This does not appear to be the case during infection of *N. lovaniensis*; hence, the protease may, in certain hosts, help *L. pneumophila* obtain amino acid nutrients, degrade host factors that are designed to control growth, or cleave other secreted *Legionella* proteins (besides PlaC) that promote infection. In infected macrophages, at least, ProA exists in both the *L. pneumophila* phagosome and the host cytoplasm (Rechnitzer et al. 1992), supporting the possibility of it having multiple targets. As the last of the known T2S-dependent potentiators of intracellular infection, SrnA might be degrading *H. vermiformis* and *N. lovaniensis* RNA in order to obtain nutritional nucleotides and phosphate or as a means to alter host cell function. Given the role of protozoa in *L. pneumophila* survival in water, these data further establish T2S as a major factor in *Legionella* persistence in the environment. Because infected amoebae might be part of the infective dose that initiates lung infection (Brieland et al. 1996; Cirillo et al. 1999), these data also signal the relevance of T2S for disease.

4.3 The Importance of T2S in Establishing a Broad Host Range for L. pneumophila

The fact that a *L. pneumophila nttA* mutant was defective in infection of *A. castellanii* but not *H. vermiformis* or *N. lovaniensis* and that a *plaC* mutant, *proA* mutant, and *srnA* mutant were impaired in infection of *H. vermiformis* and *N. lovaniensis* but not in *A. castellanii* demonstrate that the importance of a substrate can be dependent on the amoebal host being infected and that the exoprotein repertoire has a significant role in shaping host range. It would seem that the different amoebae present different targets, intracellular environments, stresses, or trafficking pathways for the infecting legionellae to engage and/or overcome. Based on the behavior of the substrate mutants, infection of *H. vermiformis* and infection of *N. lovaniensis* seem to be more akin to each other than they are to infection of *A. castellanii*.

L. pneumophila strains can infect at least eighteen additional amoebae, including seven other species of *Acanthamoeba*, five other species of *Naegleria*, another species of *Hartmannella*, and species of *Balamuthia*, *Dictyostelium*, *Echinamoeba*, *Vahlkampfia*, and *Willaertia* (Rowbotham 1980; Tyndall and Domingue 1982; Anand et al. 1983; Holden et al. 1984; Newsome et al. 1985; Rowbotham 1986; Barbaree et al. 1986; Henke and Seidel 1986; Harf and Monteil 1988; Fields et al. 1989; Wadowsky et al. 1991; Fields 1996; Michel et al. 1998; Hagele et al. 2000; Solomon et al. 2000; Molmeret et al. 2001; Miyamoto et al. 2003; Shadrach et al. 2005; Dey et al. 2009; Harada et al. 2010), as well as three types of *Tetrahymena* ciliates (Barbaree et al. 1986; Fields 1996; Fields et al. 1984, 1986; Kikuhara et al. 1994). It has often been argued that when a *Legionella* mutant that lacks a secreted protein does not exhibit an infection defect it is the result of redundancy; i.e., loss of one protein is compensated for by expression of another protein(s). Given the current data concerning T2S, which directly follow from earlier assessments of ProA (Rossier et al. 2008), mutants should best be tested in multiple protozoa. In addition to being relevant to T2S, this strategy may be even more important for the investigation of *L. pneumophila* type IV secretion which mediates the secretion of >270 effectors (Newton et al. 2010; O'Connor et al. 2012).

5 The Role of T2S in *L. pneumophila* Infection of Mammalian Hosts

5.1 The Importance of T2S in Lung Models of Infection

T2S mutants of *L. pneumophila* are very impaired in a murine model of Legionnaires' disease (Rossier et al. 2004; McCoy-Simandle et al. 2011). Whereas parental strain 130b increases at least tenfold in the lungs of the A/J mice, a T2S (*lspF*) mutant exhibits no increase in CFU and is cleared much more rapidly. Examination of sera obtained from mice infected with wild type further revealed that T2S-dependent proteins are expressed in vivo (Rossier et al. 2004). Hence, T2S is a key contributor to *L. pneumophila* virulence. Among all of the T2S substrates tested so far, the ChiA chitinase stands out as being required for bacterial survival in the lungs (DebRoy et al. 2006a; Rossier et al. 2008, 2009). Mutants specifically lacking ChiA are impaired ca. fourfold when tested in the mouse model, and immunoblot analysis documents that ChiA is one of the T2S substrates that is expressed in vivo (DebRoy et al. 2006a). Because *chiA* mutants grow fine in macrophages in vitro and since their reduced survival in the lung was manifest in the later stages of infection, ChiA likely promotes persistence versus initial replication. Since mammals do not have chitin, these data lead to the novel hypothesis that there is a chitin-like factor in lungs whose degradation aids in *L. pneumophila* persistence. Alternately, ChiA might be a bifunctional enzyme that has another substrate. A protein with chitinase activity can promote the survival of a pathogen in a mammalian host had not been previously seen, and now recent work has shown a similar result for *L. monocytogenes* and its secreted chitinase (Chaudhuri et al. 2010). Thus, factors that are traditionally thought of as being important in the environment may have real relevance to disease. Although *proA* mutants do not clearly exhibit reduced growth or survival in the lungs of experimental animals, ProA is believed to contribute to disease by promoting the destruction of lung tissue (DebRoy et al. 2006a; Moffat et al. 1994; Baskerville et al. 1986; Conlan et al. 1986; Williams et al. 1987; Conlan et al. 1988; Blander et al. 1990). ProA can also degrade transferrin and thus may promote iron acquisition (James et al. 1997).

5.2 The Importance of T2S in Intracellular Infection of Macrophages

Early on, it was determined that *L. pneumophila* T2S (i.e., *lspDE, lspF, lspG, lspK,* or *lspO/pilD*) mutants display a reduced ability to infect U937 cells, a human macrophage-like cell line (Liles et al. 1999; Rossier and Cianciotto 2001; Rossier et al. 2004; Polesky et al. 2001). In the U937 cells, the T2S mutants exhibit tenfold reduced recovery at 48 h post-inoculation (Rossier and Cianciotto 2001; Rossier et al. 2004; Polesky et al. 2001). The reduced infectivity of the mutants is complemented by reintroduction of the corresponding *lsp* gene (Rossier et al. 2004). Since mutants specifically lacking type IV pili are not defective for macrophage infection, the infectivity defect of the *lspO*/*pilD* mutant is due to the loss of T2S (Rossier et al. 2004). Recently, the results obtained with U937 cells were validated when it was observed that T2S mutants (but not a complemented mutant) are similarly impaired for infection of the MH-S mouse alveolar macrophage line, A/J mouse bone-marrow-derived (BMD) macrophages, and explanted A/J mouse alveolar macrophages (McCoy-Simandle et al. 2011). The *lsp* mutants (but not a complement) are also impaired for infection of human peripheral blood monocytes and THP-1 cells, another human macrophage-like cell line (C. Mallama and N. P. Cianciotto, unpublished results). Recent microscopic examination of infected U937 cells and BMD A/J macrophages has confirmed that the numbers of intracellular mutant bacteria are significantly less than that of wild type at time points as early as 16 h post-inoculation (M. M. Pearce, R. C. White, and N. P. Cianciotto, unpublished results). T2S mutants are not impaired for the induction of apoptosis or a pore-forming activity linked to egress from the host cells (Molmeret et al. 2002; Zink et al. 2002). *L. pneumophila celA*, *chiA*, *gamA*, *lapA*, *lapB*, *legP*, *lipA*, *lipB*, *map*, *nttA*, *nttB*, *plaA*, *plaC*, *plcA*, *plcB*, *proA*, and *srnA* mutants have been tested for alterations in infection of macrophages. Given that some of the secreted effectors may exhibit functional redundancy; e.g., LapA and LapB, a variety of double mutants were also tested. However, all of the available substrate mutants grow normally (DebRoy et al. 2006a; Banerji et al. 2005; Aragon et al. 2001, 2002; Flieger et al. 2002; Herrmann et al. 2011; Tyson et al. 2013; Rossier et al. 2008, 2009; Pearce and Cianciotto 2009), indicating that the T2S system secretes a yet-to-be-defined factor(s) that is necessary for macrophage infection. In sum, these data clearly indicate that T2S promotes *L. pneumophila* intracellular growth within macrophages, but the identity of the key effector(s) remains unknown. Furthermore, they indicate that the reduced survival of T2S mutants within the lungs is partly due to impaired intracellular infection of resident macrophages.

5.3 *The Importance of T2S in Intracellular Infection of Epithelial Cells*

Various studies have identified alveolar epithelial cells as a potential niche for *L. pneumophila* growth during lung infection (Cianciotto et al. 1995; Opitz et al. 2006; Schmeck et al. 2007; N'Guessan et al. 2007; Newton et al. 2006; Molmeret et al. 2007; Edelstein et al. 2003). Therefore, T2S mutants were examined for their ability to infect alveolar epithelial cell lines, i.e., the human A549 type II epithelial line, the human WI-26 VA4 type I epithelial line, and the murine TC-1 line (McCoy-Simandle et al. 2011). Both the *lspF* mutant and the *lspDE* mutant displayed an impaired ability to grow in the epithelial monolayers, as evidenced by five- to tenfold reduced recoveries at 24, 48, and 72 h post-inoculation. A defect was not evident at $t = 0$, even after treating the infected monolayers with gentamicin in order to kill any residual extracellular bacteria, indicated that the mutants are not impaired for entry into these epithelial cells. A complemented *lspF* mutant infected the lung epithelial lines analogously to wild type, confirming that the reduced infectivity exhibited by the mutants was due to the loss of T2S rather than a second-site mutation (McCoy-Simandle et al. 2011). The *lspF* mutant is also impaired for infection of HeLa cells, a non-lung epithelial line that also supports *L. pneumophila* growth (McCoy-Simandle et al. 2011). As was the case for macrophage infection, none of the substrate mutants tested displayed an infection defect in epithelial cells (McCoy-Simandle et al. 2011), suggesting, once again, that a yet-to-be-defined exoprotein(s) is promoting infection of mammalian host cells. In sum, these data indicate that an intact T2S system is required for optimal infection of epithelial cells by *L. pneumophila*. Thus, *Legionella* T2S likely promotes pulmonary infection by facilitating intracellular growth in both alveolar macrophages and lung epithelia.

5.4 The Importance of T2S in Dampening the Innate Immune Response

Because T2S mutant numbers do not increase in the lungs, whereas they do, although not optimally, in macrophages and epithelial cells, it was hypothesized that T2S promotes other processes that are relevant to disease. Subsequent experiments determined that, following infection of U937 cell macrophages, T2S mutants, but not a complemented mutant, elicit significantly higher levels of cytokines and chemokines, including IL-6, IL-8, IL-10, IL-1β, TNF-α, and MCP-1 (McCoy-Simandle et al. 2011). Similar data were obtained when infected lung epithelial cell lines and the lungs of infected A/J mice were examined (McCoy-Simandle et al. 2011). Recent experiments have also confirmed this result in human macrophages derived from peripheral blood monocytes (C. Mallama and N. P. Cianciotto, unpublished results). Infection with a *proA* mutant specifically lacking the ProA protease (but not a complemented *proA* mutant) results in a partial elevation of cytokine levels (McCoy-Simandle et al. 2011), suggesting that the T2S system dampens the cytokine and chemokine output of infected host cells. None of the other substrate mutants trigger an altered chemokine/cytokine profile. Based on quantitative RT-PCR analysis of infected host cells, a T2S mutant, but not the *proA* mutant, generated higher levels of cytokine (e.g., IL-6, IL-8) transcripts, implying that some yet-to-be-defined T2S substrate(s) dampens signal transduction and transcription, whereas others, such as ProA, act at a post-transcriptional step in cytokine expression (McCoy-Simandle et al. 2011). Early studies determined that TNF-α can act, in autocrine fashion, to render macrophages less permissive for bacterial growth (Blanchard et al. 1989; Skerrett and Martin 1996; McHugh et al. 2000). Thus, it was determined whether the addition of antibodies directed against the TNF-α receptor could reverse the growth defect of the *lspF* mutant in U937 cells. Although the antibody treatment did not affect the recovery of wild-type bacteria, it did lead to increased outgrowth of the T2S mutant (McCoy-Simandle et al. 2011). In contrast to these results, antibodies against the IL-6 receptor did not have any effect on recovery of the mutant. These data imply that, at least in the case of TNF-α, modulation of the cytokine response by T2S can impact the ability of *L. pneumophila* to grow in host cells. In sum, the growth of *L. pneumophila* in the lungs is likely promoted by an ability of T2S also to dampen the output of chemokines and cytokines from macrophages and epithelia. The mechanisms for this enhanced replication might include the maintenance of permissive macrophages, limitations on killing by neutrophils, and subversion of complement, among other possibilities.

6 T2S and Other *Legionella* Species

The *Legionella* genus consists of 57 species, with 26 of those species being implicated in disease (Pearce et al. 2012). Based upon Southern hybridization analysis and whole-genome sequencing, all sixteen non-*pneumophila* species examined had *lsp* genes encoding the T2S apparatus (Rossier et al. 2004; Tyson et al. 2013). In light of this and given the prevalence of T2S among other Proteobacteria (Cianciotto 2005), it is quite likely that *lsp* genes exist throughout the *Legionella* genus. However, the output of the T2S system varies considerably between *Legionella* species. For example, Southern blot and BLAST analysis of *nttA*, *nttB*, and *legP* indicate that the presence of substrate genes can vary among species (Tyson et al. 2013). Also, some T2S-dependent phenotypes, including protease, phosphatase, and lipolytic activities, surfactant, and low-temperature growth, are lacking in non-*pneumophila* species (Stewart et al. 2009; Söderberg et al. 2008; Pearce et al. 2012). Thus, the variations in T2S output are most likely due to differences in substrate gene content and/or expression versus the presence or absence of the T2S apparatus. It will be important to more systematically assess the substrate genotypes among different species, as it may give insight into ecological, pathogenic, and evolutionary relationships.

7 Concluding Remarks

In summary, the T2S system of *L. pneumophila* is notable for the number, variety, and novelty of its secreted substrates. Table 1 presents a summation of the known T2S-dependent exoproteins that have now been examined for their role in intracellular infection. Based upon these results, it will be useful for future efforts to further investigate additional T2S substrates and activities (Cianciotto 2009) as this has the potential to reveal other exoproteins that contribute to infection. The role of T2S in *L. pneumophila* pathogenesis is the result at least seven factors; i.e., (i) extracellular survival in water samples, which are the source of human infection, (ii) growth in multiple types of amoebae, which are the main replicative niche for *L. pneumophila* in water and may also be part of the infective dose, (iii) intracellular infection of macrophages, which are the primary host cell in the lung, (iv) intracellular infection of lung epithelial cells, which are an alternative host cell in vivo, (v) dampening of the chemokine and cytokine output of infected macrophages and epithelial cells, which likely minimizes to some degree the inflammatory cell infiltrate into the lung, allowing for prolonged bacterial growth, (vi) the elaboration of ChiA, which appears to promote intra-pulmonary persistence independent of macrophage and epithelial cell infection, and (vii) the secretion of ProA, which degrades both host cytokines and lung tissue. Further research should be aimed at identifying yet additional substrates that are responsible for the various T2S-dependent processes as well as deciphering the mechanisms by which the secreted effectors act in order to promote environmental survival and disease progression. It is also possible that further work will uncover even more roles for this multi-dimensional secretion system.

References

Amaro F, Gilbert JA, Owens S, Trimble W, Shuman HA (2012) Whole-genome sequence of the human pathogen *Legionella pneumophila* serogroup 12 strain 570-CO-H. J Bacteriol 194:1613–1614

Anand CM, Skinner AR, Malic A, Kurtz JB (1983) Interaction of *L. pneumophila* and a free living amoeba (*Acanthamoeba palestinensis*). J Hyg 91:167–178

Aragon V, Kurtz S, Flieger A, Neumeister B, Cianciotto NP (2000) Secreted enzymatic activities of wild-type and *pilD*-deficient *Legionella pneumophila*. Infect Immun 68:1855–1863

Aragon V, Kurtz S, Cianciotto NP (2001) *Legionella pneumophila* major acid phosphatase and its role in intracellular infection. Infect Immun 69:177–185

Aragon V, Rossier O, Cianciotto NP (2002) *Legionella pneumophila* genes that encode lipase and phospholipase C activities. Microbiology 148:2223–2231

Banerji S, Bewersdorff M, Hermes B, Cianciotto NP, Flieger A (2005) Characterization of the major secreted zinc metalloprotease- dependent glycerophospholipid:cholesterol acyltransferase, PlaC, of *Legionella pneumophila*. Infect Immun 73:2899–2909

Barbaree JM, Fields BS, Feeley JC, Gorman GW, Martin WT (1986) Isolation of protozoa from water associated with a legionellosis outbreak and demonstration of intracellular multiplication of *Legionella pneumophila*. Appl Environ Microbiol 51:422–424

Baskerville A, Conlan JW, Ashworth LA, Dowsett AB (1986) Pulmonary damage caused by a protease from *Legionella pneumophila*. Br J Exp Pathol 67:527–536

Blanchard DK, Friedman H, Klein TW, Djeu JY (1989) Induction of interferon-gamma and tumor necrosis factor by *Legionella pneumophila*: augmentation of human neutrophil bactericidal activity. J Leukoc Biol 45:538–545

Blander SJ, Szeto L, Shuman HA, Horwitz MA (1990) An immunoprotective molecule, the major secretory protein of *Legionella pneumophila*, is not a virulence factor in a guinea pig model of Legionnaires' disease. J Clin Invest 86:817–824

Brieland J, McClain M, Heath L, Chrisp C, Huffnagle G, LeGendre M, Hurley M, Fantone J, Engleberg C (1996) Coinoculation with *Hartmannella vermiformis* enhances replicative *Legionella pneumophila* lung infection in a murine model of Legionnaires' disease. Infect Immun 64:2449–2456

Cazalet C, Rusniok C, Bruggemann H, Zidane N, Magnier A, Ma L, Tichit M, Jarraud S, Bouchier C, Vandenesch F, Kunst F, Etienne J, Glaser P, Buchrieser C (2004) Evidence in the *Legionella pneumophila* genome for exploitation of host cell functions and high genome plasticity. Nat Genet 36:1165–1173

Chaudhuri S, Bruno JC, Alonzo F 3rd, Xayarath B, Cianciotto NP, Freitag NE (2010) Contribution of chitinases to *Listeria monocytogenes* pathogenesis. Appl Environ Microbiol 76:7302–7305

Chien M, Morozova I, Shi S, Sheng H, Chen J, Gomez SM, Asamani G, Hill K, Nuara J, Feder M, Rineer J, Greenberg JJ, Steshenko V, Park SH, Zhao B, Teplitskaya E, Edwards JR, Pampou S, Georghiou A, Chou IC, Iannuccilli W, Ulz ME, Kim DH, Geringer-Sameth A, Goldsberry C, Morozov P, Fischer SG, Segal G, Qu X, Rzhetsky A, Zhang P, Cayanis E, De Jong PJ, Ju J, Kalachikov S, Shuman HA, Russo JJ (2004) The genomic sequence of the accidental pathogen *Legionella pneumophila*. Science 305:1966–1968

Cianciotto NP (2005) Type II secretion: a protein secretion system for all seasons. Trends Microbiol 13:581–588

Cianciotto NP (2009) Many substrates and functions of type II protein secretion: Lessons learned from *Legionella pneumophila*. Future Microbiol 4:797–805

Cianciotto NP, Kim Stamos J, Kamp DW (1995) Infectivity of *Legionella pneumophila mip* mutant for alveolar epithelial cells. Curr Microbiol 30:247–250

Cirillo JD, Cirillo SLG, Yan L, Bermudez LE, Falkow S, Tompkins LS (1999) Intracellular growth in *Acanthamoeba castellani* affects monocyte entry mechanism and enhances virulence of *Legionella pneumophila*. Infect Immun 67:4427–4434

Conlan JW, Baskerville A, Ashworth LAE (1986) Separation of *Legionella pneumophila* proteases and purification of a protease which produces lesions like those of Legionnaires' disease in guinea pig lung. J Gen Microbiol 132:1565–1574

Conlan JW, Williams A, Ashworth LA (1988) In vivo production of a tissue-destructive protease by *Legionella pneumophila* in the lungs of experimentally infected guinea-pigs. J Gen Microbiol 134:143–149

Costa J, d'Avo AF, da Costa MS, Verissimo A (2011) Molecular evolution of key genes for type II secretion in *Legionella pneumophila*. Environ Microbiol 14:2017–2033

D'Auria G, Jimenez-Hernandez N, Peris-Bondia F, Moya A, Latorre A (2010) *Legionella pneumophila* pangenome reveals strain-specific virulence factors. BMC Genomics 11:181

De Buck E, Lebeau I, Maes L, Geukens N, Meyen E, Van Mellaert L, Anne J, Lammertyn E (2004) A putative twin-arginine translocation pathway in *Legionella pneumophila*. Biochem Biophys Res Commun 317:654–661

De Buck E, Maes L, Meyen E, Van Mellaert L, Geukens N, Anne J, Lammertyn E (2005) *Legionella pneumophila* Philadelphia-1 *tatB* and *tatC* affect intracellular replication and biofilm formation. Biochem Biophys Res Commun 331:1413–1420

de Felipe KS, Glover RT, Charpentier X, Anderson OR, Reyes M, Pericone CD, Shuman HA (2008) *Legionella* eukaryotic-like type IV substrates interfere with organelle trafficking. PLoS Pathog 4:e1000117

DebRoy S, Dao J, Soderberg M, Rossier O, Cianciotto NP (2006a) *Legionella pneumophila* type II secretome reveals unique exoproteins and a chitinase that promotes bacterial persistence in the lung. PNAS 103:19146–19151

DebRoy S, Aragon V, Kurtz S, Cianciotto NP (2006b) *Legionella pneumophila* Mip, a surface-exposed peptidylproline *cis-trans*-isomerase, promotes the presence of phospholipase C-like activity in culture supernatants. Infect Immun 74:5152–5160

Desvaux M, Hebraud M, Talon R, Henderson IR (2009) Secretion and subcellular localizations of bacterial proteins: a semantic awareness issue. Trends Microbiol 17:139–145

Dey R, Bodennec J, Mameri MO, Pernin P (2009) Free-living freshwater amoebae differ in their susceptibility to the pathogenic bacterium *Legionella pneumophila*. FEMS Microbiol Lett 290:10–17

Duncan C, Prashar A, So J, Tang P, Low DE, Terebiznik M, Guyard C (2011) Lcl of *Legionella pneumophila* is an immunogenic GAG binding adhesin that promotes interactions with lung epithelial cells and plays a crucial role in biofilm formation. Infect Immun 79:2168–2181

Edelstein PH, Hu B, Higa F, Edelstein MA (2003) *lvgA*, a novel *Legionella pneumophila* virulence factor. Infect Immun 71:2394–2403

Fields BS (1996) The molecular ecology of legionellae. Trends Microbiol 4:286–290

Fields BS, Shotts EB Jr, Feeley JC, Gorman GW, Martin WT (1984) Proliferation of *Legionella pneumophila* as an intracellular parasite of the ciliated protozoan *Tetrahymena pyriformis*. Appl Environ Microbiol 47:467–471

Fields BS, Barbaree JM, Shotts EB Jr, Feeley JC, Morrill WE, Sanden GN, Dykstra MJ (1986) Comparison of guinea pig and protozoan models for determining virulence of *Legionella* species. Infect Immun 53:553–559

Fields BS, Sanden GN, Barbaree JM, Morrill WE, Wadowsky RM, White EH, Feeley JC (1989) Intracellular multiplication of *Legionella pneumophila* in amoebae isolated from hospital hot water tanks. Curr Microbiol 18:131–137

Flieger A, Gong S, Faigle M, Stevanovic S, Cianciotto NP, Neumeister B (2001) Novel lysophospholipase A secreted by *Legionella pneumophila*. J Bacteriol 183:2121–2124

Flieger A, Neumeister B, Cianciotto NP (2002) Characterization of the gene encoding the major secreted lysophospholipase A of *Legionella pneumophila* and its role in detoxification of lysophosphatidylcholine. Infect Immun 70:6094–6106

Galka F, Wai SN, Kusch H, Engelmann S, Hecker M, Schmeck B, Hippenstiel S, Uhlin BE, Steinert M (2008) Proteomic characterization of the whole secretome of *Legionella pneumophila* and functional analysis of outer membrane vesicles. Infect Immun 76:1825–1836

George JR, Pine L, Reeves MW, Harrell WK (1980) Amino acid requirements of *Legionella pneumophila*. J Clin Microbiol 11:286–291

Geukens N, De Buck E, Meyen E, Maes L, Vranckx L, Van Mellaert L, Anne J, Lammertyn E (2006) The type II signal peptidase of *Legionella pneumophila*. Res Microbiol 157:836–841

Glockner G, Albert-Weissenberger C, Weinmann E, Jacobi S, Schunder E, Steinert M, Hacker J, Heuner K (2008) Identification and characterization of a new conjugation/type IVA secretion system (trb/tra) of *Legionella pneumophila* Corby localized on two mobile genomic islands. Int J Med Microbiol 298:411–428

Hagele S, Kohler R, Merkert H, Schleicher M, Hacker J, Steinert M (2000) *Dictyostelium discoideum*: a new host model system for intracellular pathogens of the genus *Legionella*. Cell Microbiol 2:165–171

Hales LM, Shuman HA (1999) *Legionella pneumophila* contains a type II general secretion pathway required for growth in amoebae as well as for secretion of the Msp protease. Infect Immun 67:3662–3666

Harada E, Iida K, Shiota S, Nakayama H, Yoshida S (2010) Glucose metabolism in *Legionella pneumophila*: dependence on the Entner-Doudoroff pathway and connection with intracellular bacterial growth. J Bacteriol 192:2892–2899

Harf C, Monteil H (1988) Interactions between free-living amoebae and *Legionella* in the environment. Water Sci Tech 20:235–239

Henke M, Seidel KM (1986) Association between *Legionella pneumophila* and amoebae in water. Isr J Med Sci 22:690–695

Herrmann V, Eidner A, Rydzewski K, Bladel I, Jules M, Buchrieser C, Eisenreich W, Heuner K (2011) GamA is a eukaryotic-like glucoamylase responsible for glycogen- and starch-degrading activity of *Legionella pneumophila*. Int J Med Microbiol 301:133–139

Hindre T, Bruggemann H, Buchrieser C, Hechard Y (2008) Transcriptional profiling of *Legionella pneumophila* biofilm cells and the influence of iron on biofilm formation. Microbiology 154:30–41

Holden EP, Winkler HH, Wood DO, Leinbach ED (1984) Intracellular growth of *Legionella pneumophila* within *Acanthamoeba castellanii* Neff. Infect Immun 45:18–24

Huang B, Yuan Z, Heron BA, Gray BR, Eglezos S, Bates JR, Savill J (2006) Distribution of 19 major virulence genes in *Legionella pneumophila* serogroup 1 isolates from patients and water in Queensland, Australia. J Med Microbiol 55:993–997

James BW, Mauchline WS, Dennis PJ, Keevil CW (1997) A study of iron acquisition mechanisms of *Legionella pneumophila* grown in chemostat culture. Curr Microbiol 34:238–243

Kikuhara H, Ogawa M, Miyamoto H, Nikaido Y, Yoshida S (1994) Intracellular multiplication of *Legionella pneumophila* in *Tetrahymena thermophila*. Sangyo Ika Daigaku Zasshi 16:263–275

Korotkov KV, Sandkvist M, Hol WG (2012) The type II secretion system: biogenesis, molecular architecture and mechanism. Nat Rev Microbiol 10:336–351

Lammertyn E, Anne J (2004) Protein secretion in *Legionella pneumophila* and its relation to virulence. FEMS Microbiol Lett 238:273–279

Lammertyn E, Van Mellaert L, Meyen E, Lebeau I, De Buck E, Anne J, Geukens N (2004) Molecular and functional characterization of type I signal peptidase from *Legionella pneumophila*. Microbiology 150:1475–1483

Lang C, Rastew E, Hermes B, Siegbrecht E, Ahrends R, Banerji S, Flieger A (2012) Zinc metalloproteinase ProA directly activates *Legionella pneumophila* PlaC glycerophospholipid:cholesterol acyltransferase. J Biol Chem 287:23464–23478

Lee VT, Schneewind O (2001) Protein secretion and the pathogenesis of bacterial infections. Genes Dev 15:1725–1752

Liles MR, Viswanathan VK, Cianciotto NP (1998) Identification and temperature regulation of *Legionella pneumophila* genes involved in type IV pilus biogenesis and type II protein secretion. Infect Immun 66:1776–1782

Liles MR, Edelstein PH, Cianciotto NP (1999) The prepilin peptidase is required for protein secretion by and the virulence of the intracellular pathogen *Legionella pneumophila*. Mol Microbiol 31:959–970

McCoy-Simandle K, Stewart CR, Dao J, Debroy S, Rossier O, Bryce PJ, Cianciotto NP (2011) *Legionella pneumophila* type II secretion dampens the cytokine response of infected macrophages and epithelia. Infect Immun 79:1984–1997

McHugh SL, Newton CA, Yamamoto Y, Klein TW, Friedman H (2000) Tumor necrosis factor induces resistance of macrophages to *Legionella pneumophila* infection. Proc Soc Exp Biol Med 224:191–196

Michel R, Muller KD, Amann R, Schmid EN (1998) *Legionella*-like slender rods multiplying within a strain of *Acanthamoeba* sp. isolated from drinking water. Parasitol Res 84:84–88

Miyamoto H, Taniguchi H, Yoshida S (2003) A simple qualitative assay for intracellular growth of *Legionella pneumophila* within *Acanthamoeba culbertsoni*. Kansenshogaku Zasshi 77:343–345

Moffat JF, Edelstein PH, Regula DP Jr, Cirillo JD, Tompkins LS (1994) Effects of an isogenic Zn-metalloprotease-deficient mutant of *Legionella pneumophila* in a guinea-pig pneumonia model. Mol Microbiol 12:693–705

Molmeret M, Jarraud S, Mori JP, Pernin P, Forey F, Reyrolle M, Vandenesch F, Etienne J, Farge P (2001) Different growth rates in amoeba of genotypically related environmental and clinical *Legionella pneumophila* strains isolated from a thermal spa. Epidemiol Infect 126:231–239

Molmeret M, Alli OA, Zink S, Flieger A, Cianciotto NP, Kwaik YA (2002) *icmT* is essential for pore formation-mediated egress of *Legionella pneumophila* from mammalian and protozoan cells. Infect Immun 70:69–78

Molmeret M, Santic M, Asare R, Carabeo RA, Abu Kwaik Y (2007) Rapid escape of the *dot/icm* mutants of *Legionella pneumophila* into the cytosol of mammalian and protozoan cells. Infect Immun 75:3290–3304

Newsome AL, Baker RL, Miller RD, Arnold RR (1985) Interactions between *Naegleria fowleri* and *Legionella pneumophila*. Infect Immun 50:449–452

Newton HJ, Sansom FM, Bennett-Wood V, Hartland EL (2006) Identification of *Legionella pneumophila*-specific genes by genomic subtractive hybridization with *Legionella micdadei* and identification of *lpnE*, a gene required for efficient host cell entry. Infect Immun 74:1683–1691

Newton HJ, Ang DK, van Driel IR, Hartland EL (2010) Molecular pathogenesis of infections caused by *Legionella pneumophila*. Clin Microbiol Rev 23:274–298

N'Guessan PD, Etouem MO, Schmeck B, Hocke AC, Scharf S, Vardarova K, Opitz B, Flieger A, Suttorp N, Hippenstiel S (2007) *Legionella pneumophila*-induced PKCa-, MAPK-, and NF-kB-dependent COX-2 expression in human lung epithelium. Am J Physiol Lung Cell Mol Physiol 292:L267–L277

O'Connor TJ, Boyd D, Dorer MS, Isberg RR (2012) Aggravating genetic interactions allow a solution to redundancy in a bacterial pathogen. Science 338:1440–1444

Opitz B, Vinzing M, van Laak V, Schmeck B, Heine G, Gunther S, Preissner R, Slevogt H, N'Guessan PD, Eitel J, Goldmann T, Flieger A, Suttorp N, Hippenstiel S (2006) *Legionella pneumophila* induced IFN-b in lung epithelial cells via IPS-1 and IRF3, which also control bacterial replication. J Biol Chem 281:36173–36179

Pearce MM, Cianciotto NP (2009) *Legionella pneumophila* secretes an endoglucanase that belongs to the family-5 of glycosyl hydrolases and is dependent upon type II secretion. FEMS Microbiol Lett 300:256–264

Pearce MM, Theodoropoulos N, Mandel MJ, Brown E, Reed KD, Cianciotto NP (2012) *Legionella cardiaca* sp. nov., isolated from a case of native valve endocarditis in a human heart. Int J Syst Evol Microbiol 62:2946–2954

Polesky AH, Ross JT, Falkow S, Tompkins LS (2001) Identification of *Legionella pneumophila* genes important for infection of amoebas by signature-tagged mutagenesis. Infect Immun 69:977–987

Rechnitzer C, Williams A, Wright JB, Dowsett AB, Milman N, Fitzgeorge RB (1992) Demonstration of the intracellular production of tissue-destructive protease by *Legionella pneumophila* multiplying within guinea-pig and human alveolar macrophages. J Gen Microbiol 138:1671–1677

Robson A, Collinson I (2006) The structure of the Sec complex and the problem of protein translocation. EMBO Rep 7:1099–1103

Rossier O, Cianciotto NP (2001) Type II protein secretion is a subset of the PilD-dependent processes that facilitate intracellular infection by *Legionella pneumophila*. Infect Immun 69:2092–2098

Rossier O, Cianciotto NP (2005) The *Legionella pneumophila tatB* gene facilitates secretion of phospholipase C, growth under iron-limiting conditions, and intracellular infection. Infect Immun 73:2020–2032

Rossier O, Starkenburg S, Cianciotto NP (2004) *Legionella pneumophila* type II protein secretion promotes virulence in the A/J mouse model of Legionnaires' disease pneumonia. Infect Immun 72:310–321

Rossier O, Dao J, Cianciotto NP (2008) The type II secretion system of *Legionella pneumophila* elaborates two aminopeptidases as well as a metalloprotease that contributes to differential infection among protozoan hosts. Appl Environ Microbiol 74:753–761

Rossier O, Dao J, Cianciotto NP (2009) A type II-secreted ribonuclease of *Legionella pneumophila* facilitates optimal intracellular infection of *Hartmannella vermiformis*. Microbiology 155:882–890

Rowbotham TJ (1980) Preliminary report on the pathogenicity of *Legionella pneumophila* for freshwater and soil amoebae. J Clin Pathol 33:1179–1183

Rowbotham TJ (1986) Current views on the relationships between amoebae, legionellae and man. Isr J Med Sci 22:678–689

Sargent F, Berks BC, Palmer T (2006) Pathfinders and trailblazers: a prokaryotic targeting system for transport of folded proteins. FEMS Microbiol Lett 254:198–207

Schmeck B, N'Guessan PD, Ollomang M, Lorenz J, Zahlten J, Opitz B, Flieger A, Suttorp N, Hippenstiel S (2007) *Legionella pneumophila*-induced NF-kB- and MAPK-dependent cytokine release by lung epithelial cells. Eur Respir J 29:25–33

Schroeder GN, Petty NK, Mousnier A, Harding CR, Vogrin AJ, Wee B, Fry NK, Harrison TG, Newton HJ, Thomson NR, Beatson SA, Dougan G, Hartland E, Frankel G (2010) The genome of *Legionella pneumophila* strain 130b contains a unique combination of type IV secretion systems and encodes novel Dot/Icm secretion system effector proteins. J Bacteriol 192:6001–6016

Shadrach WS, Rydzewski K, Laube U, Holland G, Ozel M, Kiderlen AF, Flieger A (2005) *Balamuthia mandrillaris*, free-living ameba and opportunistic agent of encephalitis, is a potential host for *Legionella pneumophila* bacteria. Appl Environ Microbiol 71:2244–2249

Skerrett SJ, Martin TR (1996) Roles for tumor necrosis factor alpha and nitric oxide in resistance of rat alveolar macrophages to *Legionella pneumophila*. Infect Immun 64:3236–3243

Söderberg MA, Cianciotto NP (2008) A *Legionella pneumophila* peptidyl-prolyl *cis-trans* isomerase present in culture supernatants is necessary for optimal growth at low temperatures. Appl Environ Microbiol 74:1634–1638

Söderberg MA, Rossier O, Cianciotto NP (2004) The type II protein secretion system of *Legionella pneumophila* promotes growth at low temperatures. J Bacteriol 186:3712–3720

Söderberg MA, Dao J, Starkenburg S, Cianciotto NP (2008) Importance of type II secretion for *Legionella pneumophila* survival in tap water and amoebae at low temperature. Appl Environ Microbiol 74:5583–5588

Solomon JM, Rupper A, Cardelli JA, Isberg RR (2000) Intracellular growth of *Legionella pneumophila* in *Dictyostelium discoideum*, a system for genetic analysis of host-pathogen interactions. Infect Immun 68:2939–2947

Stewart CR, Rossier O, Cianciotto NP (2009) Surface translocation by *Legionella pneumophila*: A form of sliding motility that is dependent upon type II protein secretion. J Bacteriol 191:1537–1546

Stewart CR, Burnside DM, Cianciotto NP (2011) The surfactant of *Legionella pneumophila* is secreted in a TolC-dependent manner and is antagonistic toward other *Legionella* species. J Bacteriol 193:5971–5984

Tyndall RL, Domingue EL (1982) Cocultivation of *Legionella pneumophila* and free-living amoebae. Appl Environ Microbiol 44:954–959

Tyson JY, Pearce MM, Vargas P, Bagchi S, Mulhern BJ, Cianciotto NP (2013) Multiple *Legionella pneumophila* type II secretion substrates, including an entirely novel protein, contribute to differential infection of the amoebae *Acanthamoeba castellanii, Hartmannella vermiformis*, and *Naegleria lovaniensis*. Infect Immun 81:1399–1410

Vandersmissen L, De Buck E, Saels V, Coil DA, Anne J (2010) A *Legionella pneumophila* collagen-like protein encoded by a gene with a variable number of tandem repeats is involved in the adherence and invasion of host cells. FEMS Microbiol Lett 306:168–176

Wadowsky RM, Wilson TM, Kapp NJ, West AJ, Kuchta JM, States SJ, Dowling JN, Yee RB (1991) Multiplication of *Legionella* spp. in tap water containing *Hartmannella vermiformis*. Appl Environ Microbiol 57:1950–1955

Williams A, Baskerville A, Dowsett AB, Conlan JW (1987) Immunocytochemical demonstration of the association between *Legionella pneumophila*, its tissue-destructive protease, and pulmonary lesions in experimental Legionnaires' disease. J Pathol 153:257–264

Zink SD, Pedersen L, Cianciotto NP, Abu-Kwaik Y (2002) The Dot/Icm type IV secretion system of *Legionella pneumophila* is essential for the induction of apoptosis in human macrophages. Infect Immun 70:1657–1663

Effector Translocation by the *Legionella* Dot/Icm Type IV Secretion System

Jiazhang Qiu and Zhao-Qing Luo

Abstract *Legionella pneumophila* is an opportunistic pathogen responsible for Legionnaires' disease. This bacterium survives and replicates within phagocytes by bypassing their bactericidal activity. Intracellular replication of *L. pneumophila* requires the Dot/Icm type IV secretion system made of approximately 27 proteins that presumably traverses the bacterial and phagosomal membranes. The perturbation of the host killing ability largely is mediated by the collective functions of the protein substrates injected into host cells via the Dot/Icm transporter. Proper protein translocation by Dot/Icm is determined by a number of factors, including signals recognizable by the translocator, chaperones that may facilitate the proper folding of substrates and transcriptional regulation and protein stability that determine the abundance and temporal transfer of the substrates. Although a large number of Dot/Icm substrates have been identified, investigation to understand the translocation is ongoing. Here we summarized the recent advancements in our understanding of the factors that determine the protein translocation activity of the Dot/Icm transporter.

Abbreviations

Dot/Icm	Defective in organelle trafficking/intracellular multiplication)
Arf	ADP ribosylation factor
GEFs	Guanine nucleotide exchange factors

J. Qiu · Z.-Q. Luo (✉)
Department of Biological Sciences, Purdue University,
West Lafayette, IN 47907, USA
e-mail: luoz@purdue.edu

Current Topics in Microbiology and Immunology (2013) 376: 103–115
DOI: 10.1007/82_2013_345

Published Online: 6 August 2013

Contents

1 Introduction 104
2 Characteristics of the Dot/Icm Substrates 105
3 Structure of the Dot/Icm Secretion System 107
4 Signals Required for Dot/Icm-Mediated Protein Translocation 108
5 Engagement of the Substrates with the Transporter and the Roles of Chaperones 109
6 Temporal and Spatial Regulation of Substrate Translocation 111
7 Conclusion 112
References 113

1 Introduction

Specialized protein translocation systems are essential for successful host colonization by many bacterial pathogens. These systems deliver virulence factors into host cells to facilitate diverse aspects of the infection process, including the entry into the host cell, inhibition of phagocytosis and innate immunity, intracellular survival and replication, and the spread of the pathogen between host cells (Thanassi et al. 2012). The conjugation-adapted Dot/Icm type IV secretion system associated with the opportunistic bacterial pathogen *Legionella pneumophila* represents one such example, which translocates effectors into host cells. The activity of Dot/Icm is essential for the success of *L. pneumophila* intracellular replication in phagocytes, cells that are designed to digest internalized particles, including bacterial cells (Isberg et al. 2009). *L. pneumophila* is a relatively newly recognized bacterial pathogen that causes Legionnaires' disease, a name that was given after the first fully documented outbreak occurred among attendees of the 1976 American Legion Convention in Philadelphia (Fields et al. 2002; Fraser et al. 1977). Although at least 16 serogroups have been identified, larger than 90 % of the reported clinical Legionnaires' diseases were associated with serogroup 1 *L. pneumophila* (Diederen 2008; Fields et al. 2002).

Aquatic environments are the major reservoir for *L. pneumophila* where the bacterium persists as an intracellular parasite of fresh water protozoan (Diederen 2008; Fields et al. 2002). Aerosolized water contaminated with the bacteria from natural niches or man-made facilities is the primary source of human infection (Diederen 2008; Fields et al. 2002). Because human-to-human transmission of *L. pneumophila* infection has not been observed and the fact that replication in mammalian macrophages and in its amoebae hosts requires similar bacterial factors (Abu Kwaik et al. 1998; Segal and Shuman 1999), the protozoan hosts are believed to provide the primary evolutionary pressure for the acquisition and maintenance of its virulence factors. Internalized *L. pneumophila* resides in a phagosome that blazes a route that bypasses the default endocytic maturation pathway. Within minutes, the phagosomal membranes are quickly converted into membranes resembling those of the endoplasmic reticulum (ER) and are eventually decorated with ribosomes (Hubber and Roy 2010; Tilney et al. 2001; Xu and Luo 2013). It is within this niche that the bacterium begins to replicate until the nutrient is depleted, which triggers a differentiation switch and the potentiation of the pathogen for the second round of infection (Molofsky and Swanson 2004).

Genetic analysis of mutants failed to replicate in macrophages led to the identification of several loci called *dot* or *icm* on *L. pneumophila* chromosome that are critical for its intracellular replication (Berger and Isberg 1993; Brand et al. 1994; Segal et al. 1998; Segal and Shuman 1997; Swanson and Isberg 1996; Vogel et al. 1998). Three lines of evidence suggest that these genes code for proteins that assemble into a membrane associated protein complex involved in substrate translocation. First, many of these predicted proteins are membrane proteins; second, some of these proteins share detectable homologies with proteins involved in building the conductive apparatus for plasmid conjugation and most importantly, these proteins are required for conjugal transfer of mobilizable elements derived from the IncQ plasmid RSF1010 (Segal et al. 1998; Vogel et al. 1998). The notion that the Dot/Icm system was adapted from a plasmid conjugal transfer system became more apparent when almost identically organized genes from a ca. 60kbp DNA element essential for conjugation in the IncI plasmids *col*I-P9 and R64 were found to code for proteins highly similar to Dot/Icm components (Komano et al. 2000; Wilkins and Thomas 2000).

2 Characteristics of the Dot/Icm Substrates

The high level similarity between the Dot/Icm proteins and those involved in plasmid conjugal transfer immediately suggests a model in which these proteins assemble into a transporter that delivers virulence factors into host cells and intensive efforts were then invested into the identification of such translocated factors (Segal et al. 1998; Vogel et al. 1998; Vogel and Isberg 1999). The first success came from bioinformatics analysis of then yet unfinished genome sequence data of strain Philadelphia 1 in the identification of a protein bearing a Sec7 motif (Nagai et al. 2002). This motif is present in all known members of the guanine nucleotide exchange factors (GEFs) for the Arf small GTPases involved in diverse functions including vesicular trafficking (Nagai et al. 2002). Designated as RalF, this protein was subsequently shown to be a substrate of the Dot/Icm transporter capable of catalyzing GDT-GTP nucleotide exchange on Arf small GTPases and was required for the recruitment of Arf1 onto the bacterial phagosome (Nagai et al. 2002).

The number of experimentally verified Dot/Icm substrates quickly increased to hundreds after the success of several screening strategies by some elegantly designed methods for candidate identification and the use of sensitive protein translocation reporters (Altman and Segal 2008; Campodonico et al. 2005; Conover et al. 2003; Gal-Mor and Segal 2003a; Luo and Isberg 2004; Shohdy et al. 2005). Later, a larger protein database of experimentally confirmed substrates allowed the development of algorithms effective for computation-based substrate identification (Burstein et al. 2009; Huang et al. 2010). Finally, several large-scale, comprehensive undertakes led to almost saturated identification of proteins capable of being translocated by this transporter (Huang et al. 2010; Zhu et al. 2011). So far, at least 290 protein substrates have been experimentally confirmed (Gomez-Valero et al. 2011; Lifshitz et al. 2013; Zhu et al. 2011). The total number of Dot/Icm substrates is estimated to be around 300, which is about 10 % of the genes predicted to code for proteins in the several *L. pneumophila* strains of known genome sequences (Gomez-Valero et al. 2011). Comparing to other bacterial pathogens with specialized protein transfer systems involved in virulence, the number of substrates of the Dot/Icm system arguably is the largest known so far. In addition to the large repertoire, substrates of the Dot/Icm transporter have several other features. First, most of these proteins do not have homologs in other species and are unique to *L. pneumophila*. Second, whereas there appear to have a core cohort of proteins shared by all *L. pneumophila* strains, significant plasticity exists among different isolates (Gomez-Valero et al. 2011; Lifshitz et al. 2013; Ninio et al. 2009; Schroeder et al. 2010; Zusman et al. 2008). Third, contrast to the fact that mutations in component genes of the transporter completely abolished intracellular bacterial replication; elimination of a single substrate gene rarely leads to detectable defects in intracellular growth under standard laboratory conditions. This phenomenon has been attributed to functional redundancy among groups of substrates that target similar host processes (Luo 2011; Luo and Isberg 2004; O'Connor et al. 2011, 2012). Fourth, replication within taxonomically different hosts requires specific sets of substrates. Deletion of some clusters of substrates affects virulence in some protozoan hosts but not in permissive hosts such as mammalian macrophages, suggesting that the host range of *L. pneumophila* expands by acquiring Dot/Icm substrates on the basis of the core repertoire necessary for replication in permissive hosts (O'Connor et al. 2011).

One important aspect of the study of Dot/Icm substrates is the understanding of their activity and role in *L. pneumophila* infection. We are beginning to appreciate that these proteins modulate a wide range of host processes with sophisticated mechanisms of great precision (Hubber and Roy 2010; Xu and Luo 2013). Whereas some of the host processes such as membrane trafficking targeted by these proteins correlated well with the life style of the bacterium revealed by morphological, cell biological and molecular studies, the benefits of others such as the inhibition of host protein synthesis are less clear (Belyi et al. 2006, 2008; Fontana et al. 2011; Shen et al. 2009). Another important aspect lies in the study of the mechanisms involved in the recognition and translocation by the transporter.

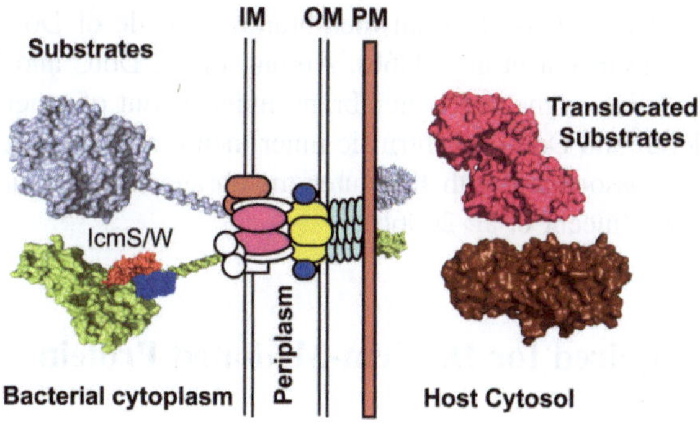

Fig. 1 A model for Dot/Icm-mediated protein translocation. The transporter was depicted as a multi-component complex that spans the bacterial inner membrane (*IM*), the periplasm, the outer membrane (*OM*) and the phagosomal membrane (*PM*). Two substrates that are actively being translocated in IcmS/W-dependent or independent manner were shown in the cytoplasm of the bacterium (*left* of the IM). The chaperone complex was shown to interact with the substrate of *green* color. The carboxyl termini of the two substrates being actively translocated were shown in the host cytosol (still connected to the PM). In addition, two substrates whose translocation has been completed were illustrated in *red* and *brown* color, respectively, in the host cytosol (*right* of the PM)

3 Structure of the Dot/Icm Secretion System

Based on gene organization, their importance in intracellular bacterial replication, homology with components of conjugation system or other specialized bacterial protein transporter, approximately 27 proteins are identified as essential components of the Dot/Icm complex (Segal et al. 1998; Sexton et al. 2004; VanRheenen et al. 2004; Vincent et al. 2006a; Vincent and Vogel 2006; Vogel et al. 1998; Zusman et al. 2004). Similar to classical conjugation machineries, the structure of the Dot/Icm transporter can be conceptually divided into three subcomplexes: the inner membrane receptor for engaging secreted substrates; the transmembrane core that bridges the inner and outer bacterial membranes and the out membrane core that extends from the bacterial surface, which presumably penetrates into the membranes of the recipient cell (Fig. 1). Considerable progress has been made in defining components of the inner and the transmembrane complexes, but virtually nothing is known about the outer membrane complex.

DotL is distantly related to VirD4, the coupling protein of the VirB/D4 type IV secretion system from *Agrobacterium tumefaciens* (Buscher et al. 2005). Members of this protein family is the primary component of the inner membrane complex directly involved in substrate recognition and in providing the energy that drives substrate secretion by its ATPase activity (Gomis-Ruth et al. 2004). In the Dot/Icm system, the inner substrate receptor complex appears to be formed by DotL, DotM and possibly DotN (Vincent et al. 2006a). DotM is an inner membrane protein that directly interacts with DotL, and DotN, a cysteine-rich protein important for stabilizing DotL and DotN (Vincent et al. 2006a). Current evidence suggests that the complex that links the two bacterial membranes is made of DotC, DotD, DotF, DotG and DotH (Vincent et al. 2006b). Among these, DotC and DotD are lipoproteins that localize in the outer membrane independent of other Dot/Icm components, and DotF and DotG are intrinsic inner membrane proteins (Vincent et al. 2006b). DotH is associated with the outer membrane in a manner that requires DotC and DotD (Vincent et al. 2006b).

4 Signals Required for Dot/Icm-Mediated Protein Translocation

The fact that only a small fraction of the approximately 3000 putative proteins in the cytoplasm of *L. pneumophila* (Cazalet et al. 2004; Chien et al. 2004) is translocated by the Dot/Icm system points to the existence of special features buried in the substrates. Such features fulfill the selectivity imposed by the transporter and thus are the signals recognized by the secretion system. By using a Cre/*lox*P-based reporter system, the signals for T4SS protein translocation were first mapped to the C-terminal portion of VirF and VirE2, two substrates of the *A. tumefaciens* VirB/D4 system (Vergunst et al. 2000). In *L. pneumophila,* Luo and Isberg first showed the C-terminal but not the N-terminal portion of RalF or LidA is capable of promoting the transfer of the Cre fusion between bacterial cells (Luo and Isberg 2004), suggesting that signals for substrate recognition resides in the carboxyl portion of the protein.

Our understanding of the residues shared among different Dot/Icm substrates evolves with the number of experimentally confirmed substrates. Based on the analysis of RalF and the Sid proteins, Nagai et al. (2005) revealed that the presence of three hydrophobic amino acids in a region close to the end of these proteins is important for translocation. Further study found that in addition to the hydrophobic residues, some small, polar, and charged amino acids (e.g., alanine, serine, threonine, glutamic acid) are found in a subset of substrates (Kubori et al. 2008). As the number of Dot/Icm substrates reached about 100, using information of the presence and absence of certain amino acids in the C-terminal portion of these proteins, mathematic algorithms were developed to effectively predict Dot/Icm substrate candidates (Burstein et al. 2009; Lifshitz et al. 2013). Later, glutamate clusters (E-block) were recognized as an important element for the translocation of many substrates (Huang et al. 2010). A recent study revealed that instead of specific amino acids at specific positions, it is amino acids with similar physicochemical properties in the last 35 residues that determine Dot/Icm-dependent protein translocation (Lifshitz et al. 2013). Among these, large glutamic acid stretches at positions -10 to -17 and several hydrophobic residues located at its C-terminal end are the most important (Lifshitz et al. 2013). In support of this notion, a stretch of composite amino acids synthesized based on these properties were shown to promote Dot/Icm-dependent protein translocation at high efficiencies (Lifshitz et al. 2013).

5 Engagement of the Substrates with the Transporter and the Roles of Chaperones

In contrast to the great progress in defining the nature of the amino acid composition of the signals important for Dot/Icm-dependent protein translocation, our understanding of the mechanisms underlying the recognition of the signals by the transporter is very limited. Direct interactions between the secretion signals and components of the transporter, the inner membrane receptor complex of the transporter in particular, presumably are necessary for substrate recognition (Fig. 1). Interestingly, earlier studies have indicated direct binding of many substrates to DotF (Luo and Isberg 2004), a component of the receptor (Vincent et al. 2006b). However, our unpublished results indicate that the domains involved in DotF binding in these substrates map to a central region but not the C-terminal end of the substrates.

In addition to the many membrane proteins essential for the activity of the Dot/Icm transporter, several small cytosolic acidic proteins such as IcmS, IcmW and LvgA, are known to be essential for the maximal activity of the transporter (Coers et al. 2000; Vincent and Vogel 2006). IcmS can directly interact with IcmW and LvgA to form the IcmS/IcmW and IcmS-LvgA complexes (Coers et al. 2000; Vincent and Vogel 2006). Accumulating evidence indicates that these complexes interact with a large subset of the substrates to facilitate their translocation by mechanisms akin to chaperones in type III secretion systems (Fattori et al. 2011). Deletion of any of these genes leads to severe attenuation but not abolishment of intracellular bacterial replication (Coers et al. 2000; Vincent and Vogel 2006). Furthermore, some substrates such as RalF and CegC3 can be translocated by the Δ*icmS/W* mutant at frequencies comparable to those by the wild type (Cambronne and Roy 2007; Lifshitz et al. 2013) (Fig. 1). How the binding of IcmS/W to the substrates facilitates their translocation is not well understood. It has been shown that for substrates such as SidG whose maximal translocation requires IcmS/W, interactions with the chaperones induce conformational changes but not the stability or solubility of this substrate (Cambronne and Roy 2007), which differs from the roles played by chaperones found in type III secretion systems whose primary function is to increase the solubility of the substrates (Fattori et al. 2011).

The direct consequence of the binding of IcmS/W may be the exposure of the signals to the transporter. The chaperones have been shown to bring the substrates into the proximity of the transporter by directly interacting with DotL, the coupling protein (Sutherland et al. 2012). DotB, a homolog of VirB11 of *A. tumefaciens* may also play a role in the selection of substrates by a yet uncharacterized mechanism (Sexton et al. 2005). It is likely that substrates whose translocation requires IcmS/W assume a conformation not optimal for recognition by the transporter prior to the binding of the chaperones.

After the initial substrate recognition, the next challenge in protein translocation is the conformation the substrate assumes when it is actively being transferred. Many of the Dot/Icm substrates are large proteins. For examples, 4 members of the SidE family are over 1,450 residues and SidH is a protein of 2,225 residues with a predicted size of more than 250 kDa (Luo and Isberg 2004). The fact that the translocation signals reside in the carboxyl terminal portion of the substrates suggests that translation of the proteins by the ribosome is complete and the polypeptides very likely have folded into their final conformations. In the current model of protein translocation, the substrates need to traverse through the channel formed by the coupling proteins, which, based on the structural analysis of TwrB, one such protein from the conjugative plasmid R388, has a size of 20 A (Gomis-Ruth et al. 2001). Although the *L. pneumophila* coupling protein DotL is considerably larger than TwrB (783 and 507 residues, respectively), the size of the channel it forms unlikely will be able to accommodate fully folded substrates, particularly these large ones. Thus, it is reasonable to predict that the substrates are translocated in a linearized form (Fig. 1). It is possible that the transporter induces the unfolding of the substrates after the initial engagement of the signals. Apparently, as evidenced by the fact that various reporter proteins have been successfully used to monitor Dot/Icm activity, such unfolding events can occur as long as signals recognizable by the transporter are available, with only limited, if any, requirements for the proteins they are fused to. This notion is consistent with the fact that the 100 C-terminal residues of many of these substrates alone are sufficient to promote high efficiency translocation of fused reporters (Huang et al. 2010; Lifshitz et al. 2013; Nagai et al. 2005). Despite these facts, the possibility that Dot/Icm substrates harbor intrinsic signals necessary for the unfolding during translocation cannot be ruled out. In support of this notion, it is worth noting that reporter proteins used for monitoring Dot/Icm-mediated substrate translocation so far, such as the Cre recombinase (Luo and Isberg 2004), the catalytic domain of the *Bordetella pertussis* adenylate cyclase (Cya) toxin (Bardill et al. 2005) and the β-lactamase (Charpentier et al. 2009) are all relatively small proteins with molecular weights less than 44 Kda. Further, both Cya and β-lactamase are secreted proteins, thus may be more amenable to the unfolding process. It is not clear whether large reporter proteins not derived from established Dot/Icm substrates or proteins of other irrelevant protein transporters can be similarly translocated. In our experiences, substrates fused to the green fluorescence protein (GFP) cannot be detectably translocated by the Dot/Icm system (unpublished results), further indicating that not all proteins fused to signals recognizable by the transporter can be transferred. No matter what the mechanism of the putative unfolding is, the chaperones IcmS/W may work together with the transporter to aid the putative unfolding of the substrate for maximal translocation.

6 Temporal and Spatial Regulation of Substrate Translocation

The life cycle of *L. pneumophila* can be divided into the transmissive and the replicative phases regulated by the stringent response signal ppGpp, which transduces cues such as nutrient availability (Hammer et al. 2002; Molofsky and Swanson 2004). Consistent with the observation that most of the *dot/icm* genes are expressed constitutively during different phases of bacterial growth (Gal-Mor and Segal 2003a, b; Gal-Mor et al. 2002), the transporter is required for most if not all stages of its intracellular life cycle (Liu et al. 2008). Thus, substrates are continuously translocated into host cells during infection, pointing to a potential hierarchical order of translocation.

In agreement with the two development phases exhibited by the bacterium, growth phase-dependent accumulation of Dot/Icm substrates in *L. pneumophila* is evident from early study of these proteins. Based on the phase of infection, the expression of substrate genes can occur at least four different time frames: prior to the infection, in the beginning of the infection, at the end of the infection and constitutive expression during the entire life cycle. Many substrate proteins such as RalF, and the Sids, accumulate to high levels in post-exponentially grown bacteria, a phase that is corresponding to the transmission stage (Bardill et al. 2005; Luo and Isberg 2004; Nagai et al. 2002). Consistent with the potential need of many of the substrates for a fast and effective "neutralization" of the relevant host defense processes important for the establishment replicative niche, the number of Dot/Icm substrates whose expression is induced at this stage appears large (Bardill et al. 2005; Jules and Buchrieser 2007; Luo and Isberg 2004; Nagai et al. 2002). The accumulation of some substrates does not become apparent until after the pathogen has entered a new replication cycle in the host cell. Examples of this category include LubX, the ubiquitin ligase responsible for the degradation of another effector SidH (Kubori et al. 2010) and SidK, a protein that inhibits v-ATPase activity (Xu et al. 2010). LepA and LepB, two substrates important for the exit of the bacteria from infected *Dictyostelium discoideum* presumably are induced at the end of its intracellular life cycle (Chen et al. 2004). Substrates whose expression is independent of bacterial growth phase include LidA (Conover et al. 2003), SidJ (Liu and Luo 2007) and Lem3 (Tan et al. 2011). These proteins may be important for each phase of the infection or their temporalities are regulated by different mechanisms.

The growth phase-dependent expression of Dot/Icm substrates mostly is regulated at the transcriptional level by sensing the environmental cues specific for each phase of its life cycle. Once the protein has been made, several factors can contribute to the order or the amount of proteins being translocated: First, the nature of the signals. Some substrates may harbor signals that are more readily recognized by the transporter; second, the affinity for the IcmS/W chaperones (for those that require), those with higher affinity may be recruited to the transporter earlier; third, the abundance of the protein. Substrates of higher quantity in the cytoplasm presumably will have a higher chance to engage the transporter. Apparently, for any given substrate, the timing and quantity of the protein delivered into the host cell will not be determined by one single but by a combination of multiple factors listed above.

Clearly, factors such as the level of protein in the bacterial cell, the strength of the translocation signals and the way the protein engages the transporter will greatly affect the timing of its translocation. However, it is important to note that temporal regulation of Dot/Icm substrate activity can be achieved by mechanisms such as the stability of the translocated protein and the function of other substrates. Several well-studied cases have been established for the second mechanism. For examples, the activity of SidH is down regulated by proteasome degradation mediated by the ubiquitin ligase LubX, whose expression does not become apparent 2 h after infection (Kubori et al. 2010). Similarly, the function of the multifunctional effector SidM is regulated by SidD, a deAMPylase that reverses its effects on Rab1 as infection has proceeded for over 4 h (Neunuebel et al. 2011; Tan and Luo 2011).

Comparing to the rich information on the temporal regulation of substrate translocation, spatial regulation of Dot/Icm activity is less understood. Immunostaining with specific antibody revealed polar distribution of some translocated substrates such as LidA and SidC in a portion of bacterial phagosomes (Conover et al. 2003; Luo and Isberg 2004). Whether such uneven distribution is a result of polar localization of the transporter with important functional implications or of the limitation of the detection method remains unknown. The VirB/D4 system of *A. tumefaciens* has been shown to specifically localize to the poles of the bacterial cells (Judd et al. 2005). It is not clear whether the Dot/Icm system assumes similar polar localization in the bacterial cell. The distribution of translocated substrates in infected cells conceivably should depend upon the site of their action but not the transporter itself. The sorting of these proteins into specific organelles such as the mitochondrion (Degtyar et al. 2009; Dolezal et al. 2012) and the nucleus (Rolando et al. 2013) clearly requires intrinsic targeting signals embedded in their structure.

7 Conclusion

Although the progress in the last decade has provided a clear picture of the inventory of proteins translocated by the Dot/Icm system as well as the features of the signals it recognizes, a number of important questions remain unresolved in this exciting research direction. First, very little is known about the state of the substrate actively being translocated (being trapped in the translocation apparatus). Future structural study of a Dot/Icm-substrate complex by technology such as cryo electron microscopy may reveal a "snake in a pipe" view of the transfer process. Second, virtually nothing is known about the mechanism underlying translocation signal recognition by the transporter. Third, when the protein of a pool of substrates is available, is there a hierarchical order of translocation? If so, how is such order of translocation achieved? Notably, current measurement of Dot/Icm-mediated protein transfer mostly is performed with reporter systems that significantly amplify the signals by specific enzymatic activity expressed from a multi-copy plasmid with a strong artificial promoter. These methods are useful in the identification of new substrates or in the comparison of the relative transfer efficiency, but will not be suitable for determining the order of transfer among different substrates. Thus, the development of new technology is required to address this question. Fourth, albeit the diversity of the translocation signals has been appreciated, how such signals were acquired by the substrates remains mysterious. This notion is particularly true for substrates such as LegS2 (de Felipe et al. 2005; Degtyar et al. 2009) that apparently were acquired by the bacterium from its hosts. The uniqueness of the carboxyl sequence of the hundreds of Dot/Icm substrates suggests limited roles of mechanism such as domain switch with existing substrates.

The Dot/Icm transporter is arguably the most important virulence determinant of *L. pneumophila* and *Coxiella burnetii* (Beare et al. 2011; Carey et al. 2011; Nagai and Kubori 2011). The study of the function of their substrates has generated great excitements in the study of the mechanisms of both infection and host signaling (Hubber and Roy 2010; Xu and Luo 2013). However, direct inhibition of the activity of individual substrate for therapeutic purpose less likely will be effective due to their dispensable role in infection. On the other hand, targeting the secretion system itself is a promising avenue in antibiotics research. Such anti-virulent antibiotics have several clear advantages over conventional ones, including the less likelihood of inducing resistance and the lower chance of killing nonpathogens and the beneficial microflora. More detailed characterization of the mechanism of substrate translocation by the Dot/Icm system and other transporter involved in bacterial virulence will be instrumental in the development of such agents.

Acknowledgments We thank Dr. Xiaodi Niu (Jilin University, Changchun, China) for help in the preparation of the figure and members of our laboratory for helpful discussion. Our work was supported by NIH-NIAID grants R01AI069344, K02AI085403 and R21AI092043.

References

Abu Kwaik Y, Gao LY, Stone BJ, Venkataraman C, Harb OS (1998) Appl Environ Microbiol 64:3127–3133

Altman E, Segal G (2008) J Bacteriol 190:1985–1996

Bardill JP, Miller JL, Vogel JP (2005) Mol Microbiol 56:90–103

Beare PA, Gilk SD, Larson CL, Hill J, Stead CM, Omsland A, Cockrell DC, Howe D, Voth DE, Heinzen RA (2011) MBio 2:e00175–11

Belyi Y, Niggeweg R, Opitz B, Vogelsgesang M, Hippenstiel S, Wilm M, Aktories K (2006) Proc Natl Acad Sci U S A 103:16953–16958

Belyi Y, Tabakova I, Stahl M, Aktories K (2008) J Bacteriol 190:3026–3035

Berger KH, Isberg RR (1993) Mol Microbiol 7:7–19

Brand BC, Sadosky AB, Shuman HA (1994) Mol Microbiol 14:797–808

Burstein D, Zusman T, Degtyar E, Viner R, Segal G, Pupko T (2009) PLoS Pathog 5:e1000508
Buscher BA, Conover GM, Miller JL, Vogel SA, Meyers SN, Isberg RR, Vogel JP (2005) J Bacteriol 187:2927–2938
Cambronne ED, Roy CR (2007) PLoS Pathog 3:e188
Campodonico EM, Chesnel L, Roy CR (2005) Mol Microbiol 56:918–933
Carey KL, Newton HJ, Luhrmann A, Roy CR (2011) PLoS Pathog 7:e1002056
Cazalet C, Rusniok C, Bruggemann H, Zidane N, Magnier A, Ma L, Tichit M, Jarraud S, Bouchier C, Vandenesch F, Kunst F, Etienne J, Glaser P, Buchrieser C (2004) Nat Genet 36:1165–1173
Charpentier X, Gabay JE, Reyes M, Zhu JW, Weiss A, Shuman HA (2009) PLoS Pathog 5:e1000501
Chen J, de Felipe KS, Clarke M, Lu H, Anderson OR, Segal G, Shuman HA (2004) Science 303:1358–1361
Chien M, Morozova I, Shi S, Sheng H, Chen J, Gomez SM, Asamani G, Hill K, Nuara J, Feder M, Rineer J, Greenberg JJ, Steshenko V, Park SH, Zhao B, Teplitskaya E, Edwards JR, Pampou S, Georghiou A, Chou IC, Iannuccilli W, Ulz ME, Kim DH, Geringer-Sameth A, Goldsberry C, Morozov P, Fischer SG, Segal G, Qu X, Rzhetsky A, Zhang P, Cayanis E, De Jong PJ, Ju J, Kalachikov S, Shuman HA, Russo JJ (2004) Science 305:1966–1968
Coers J, Kagan JC, Matthews M, Nagai H, Zuckman DM, Roy CR (2000) Mol Microbiol 38:719–736
Conover GM, Derre I, Vogel JP, Isberg RR (2003) Mol Microbiol 48:305–321
de Felipe KS, Pampou S, Jovanovic OS, Pericone CD, Ye SF, Kalachikov S, Shuman HA (2005) J Bacteriol 187:7716–7726
Degtyar E, Zusman T, Ehrlich M, Segal G (2009) Cell Microbiol 11:1219–1235
Diederen BM (2008) J Infect 56:1–12
Dolezal P, Aili M, Tong J, Jiang JH, Marobbio CM, Lee SF, Schuelein R, Belluzzo S, Binova E, Mousnier A, Frankel G, Giannuzzi G, Palmieri F, Gabriel K, Naderer T, Hartland EL, Lithgow T (2012) PLoS Pathog 8:e1002459
Fattori J, Prando A, Martins AM, Rodrigues FH, Tasic L (2011) Protein Pept Lett 18:158–166
Fields BS, Benson RF, Besser RE (2002) Clin Microbiol Rev 15:506–526
Fontana MF, Banga S, Barry KC, Shen X, Tan Y, Luo ZQ, Vance RE (2011) PLoS Pathog 7:e1001289
Fraser DW, Tsai TR, Orenstein W, Parkin WE, Beecham HJ, Sharrar RG, Harris J, Mallison GF, Martin SM, McDade JE, Shepard CC, Brachman PS (1977) N Engl J Med 297:1189–1197
Gal-Mor O, Segal G (2003a) J Bacteriol 185:4908–4919
Gal-Mor O, Segal G (2003b) Microb Pathog 34:187–194
Gal-Mor O, Zusman T, Segal G (2002) J Bacteriol 184:3823–3833
Gomez-Valero L, Rusniok C, Cazalet C, Buchrieser C (2011) Front Microbiol 2:208
Gomis-Ruth FX, Moncalian G, Perez-Luque R, Gonzalez A, Cabezon E, de la Cruz F, Coll M (2001) Nature 409:637–641
Gomis-Ruth FX, Sola M, de la Cruz F, Coll M (2004) Curr Pharm Des 10:1551–1565
Hammer BK, Tateda ES, Swanson MS (2002) Mol Microbiol 44:107–118
Huang L, Boyd D, Amyot WM, Hempstead AD, Luo ZQ, O'Connor TJ, Chen C, Machner M, Montminy T, Isberg RR (2010) Cell Microbiol 13:227–245
Hubber A, Roy CR (2010) Annu Rev Cell Dev Biol 26:261–283
Isberg RR, O'Connor TJ, Heidtman M (2009) Nat Rev Microbiol 7:13–24
Judd PK, Kumar RB, Das A (2005) Proc Natl Acad Sci U S A 102:11498–11503
Jules M, Buchrieser C (2007) FEBS Lett 581:2829–2838
Komano T, Yoshida T, Narahara K, Furuya N (2000) Mol Microbiol 35:1348–1359
Kubori T, Hyakutake A, Nagai H (2008) Mol Microbiol 67:1307–1319
Kubori T, Shinzawa N, Kanuka H, Nagai H (2010) PLoS Pathog 6:e1001216
Lifshitz Z, Burstein D, Peeri M, Zusman T, Schwartz K, Shuman HA, Pupko T, Segal G (2013) Proc Natl Acad Sci U S A 110:E707–E715
Liu Y, Gao P, Banga S, Luo ZQ (2008) Proc Natl Acad Sci U S A 105:9385–9390

Liu Y, Luo ZQ (2007) Infect Immun 75:592–603
Luo ZQ (2011) Front Microbiol 2:31
Luo ZQ, Isberg RR (2004) Proc Natl Acad Sci U S A 101:841–846
Molofsky AB, Swanson MS (2004) Mol Microbiol 53:29–40
Nagai H, Cambronne ED, Kagan JC, Amor JC, Kahn RA, Roy CR (2005) Proc Natl Acad Sci U S A 102:826–831
Nagai H, Kagan JC, Zhu X, Kahn RA, Roy CR (2002) Science 295:679–682
Nagai H, Kubori T (2011) Front Microbiol 2:136
Neunuebel MR, Chen Y, Gaspar AH, Backlund PS Jr, Yergey A, Machner MP (2011) Science 333:453–456
Ninio S, Celli J, Roy CR (2009) PLoS Pathog 5:e1000278
O'Connor TJ, Adepoju Y, Boyd D, Isberg RR (2011) Proc Natl Acad Sci U S A 108:14733–14740
O'Connor TJ, Boyd D, Dorer MS, Isberg RR (2012) Science 338:1440–1444
Rolando M, Sanulli S, Rusniok C, Gomez-Valero L, Bertholet C, Sahr T, Margueron R, Buchrieser C (2013) Cell Host Microbe 13:395–405
Schroeder GN, Petty NK, Mousnier A, Harding CR, Vogrin AJ, Wee B, Fry NK, Harrison TG, Newton HJ, Thomson NR, Beatson SA, Dougan G, Hartland EL, Frankel G (2010) J Bacteriol 192:6001–6016
Segal G, Purcell M, Shuman HA (1998) Proc Natl Acad Sci U S A 95:1669–1674
Segal G, Shuman HA (1997) Infect Immun 65:5057–5066
Segal G, Shuman HA (1999) Infect Immun 67:2117–2124
Sexton JA, Miller JL, Yoneda A, Kehl-Fie TE, Vogel JP (2004) Infect Immun 72:5983–5992
Sexton JA, Yeo HJ, Vogel JP (2005) Mol Microbiol 57:70–84
Shen X, Banga S, Liu Y, Xu L, Gao P, Shamovsky I, Nudler E, Luo ZQ (2009) Cell Microbiol 11:911–926
Shohdy N, Efe JA, Emr SD, Shuman HA (2005) Proc Natl Acad Sci U S A 102:4866–4871
Sutherland MC, Nguyen TL, Tseng V, Vogel JP (2012) PLoS Pathog 8:e1002910
Swanson MS, Isberg RR (1996) Infect Immun 64:2585–2594
Tan Y, Arnold RJ, Luo ZQ (2011) Proc Natl Acad Sci U S A 108:21212–21217
Tan Y, Luo ZQ (2011) Nature 475:506–509
Thanassi DG, Bliska JB, Christie PJ (2012) FEMS Microbiol Rev 36:1046–1082
Tilney LG, Harb OS, Connelly PS, Robinson CG, Roy CR (2001) J Cell Sci 114:4637–4650
VanRheenen SM, Dumenil G, Isberg RR (2004) Infect Immun 72:5972–5982
Vergunst AC, Schrammeijer B, den Dulk-Ras A, de Vlaam CM, Regensburg-Tuink TJ, Hooykaas PJ (2000) Science 290:979–982
Vincent CD, Buscher BA, Friedman JR, Williams LA, Bardill P, Vogel JP (2006a) J Bacteriol 188:8231–8243
Vincent CD, Friedman JR, Jeong KC, Buford EC, Miller JL, Vogel JP (2006b) Mol Microbiol 62:1278–1291
Vincent CD, Vogel JP (2006) Mol Microbiol 61:596–613
Vogel JP, Andrews HL, Wong SK, Isberg RR (1998) Science 279:873–876
Vogel JP, Isberg RR (1999) Curr Opin Microbiol 2:30–34
Wilkins BM, Thomas AT (2000) Mol Microbiol 38:650–657
Xu L, Luo ZQ (2013) Microbes Infect 15:157–167
Xu L, Shen X, Bryan A, Banga S, Swanson MS, Luo ZQ (2010) PLoS Pathog 6:e1000822
Zhu W, Banga S, Tan Y, Zheng C, Stephenson R, Gately J, Luo ZQ (2011) PLoS ONE 6:e17638
Zusman T, Degtyar E, Segal G (2008) Infect Immun 76:4581–4591
Zusman T, Feldman M, Halperin E, Segal G (2004) Infect Immun 72:3398–3409

Modulation of Small GTPases by *Legionella*

Roger S. Goody and Aymelt Itzen

Abstract The pathogenic bacterium *Legionella pneumophila* interacts intimately with signaling molecules during the infection of eukaryotic host cells. Among a diverse set of regulatory molecules, host small GTPases appear to be prominent and significant targets. Small GTPases are molecular switches that regulate cellular signaling via their respective nucleotide-bound states: When bound to GDP, they are inactive, but become activated upon binding to GTP. *Legionella* secretes specific bacterial proteins into the cytosol of the host cell that most prominently modulate the activities of small GTPases involved in vesicular trafficking, but probably also other G-proteins. The master regulators of vesicular trafficking, i.e., Rab and Arf proteins, are majorly targeted G-proteins of *Legionella* proteins, and among these, Rab1 experiences the most diverse modifications. Generally, the activities of small GTPases are modulated by GDP/GTP exchange (activation), GTP hydrolysis (deactivation), membrane recruitment, post-translational modifications (phosphocholination, adenylylation), and tight and competitive binding. Here, we discuss the consequences and molecular details of the modulation of small GTPases for the infection by *Legionella*, with a special but not exclusive focus on Rab and Arf proteins.

R. S. Goody · A. Itzen (✉)
Max-Planck-Institute of Molecular Physiology, Department of Physical Biochemistry,
Otto-Hahn Str. 11, 44227 Dortmund, Germany
e-mail: aymelt.itzen@tum.de

R. S. Goody
e-mail: roger.goody@mpi-dortmund.mpg.de

A. Itzen
Chemistry Department, Center for Integrated Protein Science Munich,
Technische Universität München, Lichtenbergstr. 4,
85747 Garching, Germany

Current Topics in Microbiology and Immunology (2013) 376: 117–133
DOI: 10.1007/82_2013_340

Published Online: 6 August 2013

Contents

1 Introduction 118
2 Function and Regulation of Small GTPases 118
3 *Legionella* Infection and Small GTPases 119
3.1 Modulation of Arf Proteins 120
3.2 Modulation of Rab1 Proteins 122
3.3 Modulation of Other Rab Proteins 127
3.4 Speculative Modulation of Further Small G-Proteins 129
4 Conclusions 129
References 130

1 Introduction

Legionnaires' disease is caused by the infection of human lung macrophages with the bacterial pathogen *Legionella pneumophila* (reviewed in (Isberg et al. 2009)). The establishment of this disease is a fascinating molecular process during which the bacterium needs to enter a cell, evade the destructive arsenal of the macrophage, and create the appropriate environment for intracellular replication. Since eukaryotic cells—and particularly professional phagocytes—are equipped with fine-tuned cellular defense mechanisms to encounter and destroy potential pathogens, *Legionella* have evolved ingenious strategies to evade the intracellular machineries of host attack. In order to efficiently interfere with potentially defensive host pathways, a plethora of secreted *Legionella* proteins appears to affect key components of eukaryotic signaling networks (e.g., apoptosis, phosphatidyl inositides, ubiquitination). Consequently, a variety of *Legionella* proteins specifically targets small GTPases (also known as small G-proteins) because these molecules serve as master regulators of intracellular signaling. We will review the diverse modulatory activities of secreted *Legionella* proteins with respect to small GTPases.

2 Function and Regulation of Small GTPases

Members of the family of small GTPases are guanosine diphosphate (GDP)/guanosine triphosphate (GTP)-binding proteins that function as molecular switches and thereby regulate and integrate cellular signaling (Cherfils and Zeghouf 2013). The switch-like behavior of these GTPases is established by regulated transitions between the bound nucleotides, since the GDP form is the inactive state whereas the GTP form defines the active state. The activity states of small GTPases are communicated to interaction partners via two regulatory loop regions called switch I and switch II that change their conformations dramatically between the active and inactive forms. The switch regions are conformationally flexible in the inactive GDP state, but become highly ordered and structurally restrained in the active GTP form. These structural changes control the interaction and specificity of G-proteins through upstream and downstream acting molecules.

The intracellular interconversion between the GDP and GTP bound forms requires the action of regulatory proteins: GDP/GTP exchange factors (GEFs) specifically catalyze nucleotide exchange and utilize the high intracellular GTP:GDP ratio to promote GTP loading and thus activation of small G-proteins. The functional adversaries of GEFs are GTPase activating proteins (GAPs) that terminate signaling by stimulating the intrinsic GTP hydrolysis activity of G-proteins, thereby returning it to the off-state. Generally, forward signaling by G-proteins is only possible from the active form and is mediated by the interactions with so-called GTPase effector proteins that have a much higher affinity for the GTP state compared to the GDP state.

These functional principles are common to all members of small GTPases. The protein class is subdivided into the Ras, Rho, Ran, Arf/Sar, and families (reviewed in (Cherfils and Zeghouf 2013; Novick and Hutagalung 2011)). Each family controls different routes of intracellular signaling: Ras controls cell differentiation, Rho organizes cytoskeleton rearrangements, Ran is involved in nuclear protein import, Arf/Sar controls membrane vesicle formation, and Rab coordinates vesicular trafficking. With the exception of Ran proteins, all small GTPases cycle between membrane bound and cytosolic forms, in which in most cases only the GTP state is associated with membranes, whereas the GDP state generally represents the cytosolic form. The interaction with the membrane is mainly mediated by virtue of post-translationally attached lipid moieties at the C-terminus (Ras, Rho, Rab) or N-terminus in some Arf/Sar proteins. Rho and Rab proteins are linked to one or two geranylgeranyl residues, respectively. The highly hydrophobic character of these 20 carbon unit lipids is able to confer a very stable membrane association of the protein. GDP dissociation inhibitors (GDI) act as chaperones for the membrane solubilization of the inactive proteins by binding the geranylgeranyl residues. RhoGDI and RabGDI are functionally similar, but are unrelated in amino acid sequence and 3-D atomic structure. In contrast, in some cases Arf/Sar proteins are lipidated at the N-terminus with a myristoyl moiety. The myristoyl group contains only 14 hydrophobic carbons and thus confers only moderate membrane affinity. Chaperoning of the lipid is achieved intramolecularly by the GTPase: In the GDP state, myristoyl is buried in a hydrophobic pocket, but is displaced and becomes exposed in the GTP state.

3 *Legionella* Infection and Small GTPases

Legionella are taken up by professional macrophages via phagocytosis. Shortly after uptake, the evasion of the destructive defense mechanisms of the host cell requires the coordinated manipulation of standard host cell signaling pathways. Instead of converting into a microbicidal and degradative phagolysosome that would eliminate the bacterial intruder, the phagosome is transformed into a replicative organelle—the *Legionella*-containing vacuole (LCV). The transformation of the phagosome into the LCV is directed by a large set of secreted bacterial proteins that manipulate and modulate specific signaling pathways, thereby ensuring pathogen survival. About 270 different bacterial proteins are transmitted via a Type IV secretion system into the cytosol of the host cell (often also called *Legionella* effectors; to avoid confusion with GTPase effectors, i.e., proteins that are regulated by active GTPases, the term *Legionella* proteins instead of *Legionella* effectors will be used throughout this article). The functions of these *Legionella* proteins are diverse, but one important targeted protein family appears to be small GTPases. Strikingly, although many other bacterial pathogens targeting small G-proteins mainly affect the Rho family (Aktories 2011; Sarantis and Grinstein 2012), proteins from *Legionella* manipulate members of the Arf and Rab families. The concentration of *Legionella* proteins on Arfs and Rabs as master regulators of vesicular trafficking emphasizes the significance of the interference with intracellular transport routes for the establishment of the LCV and the survival of the pathogen.

In the following paragraphs, we review the physiological role of Arfs and Rabs and their manipulation by *Legionella* (a summary of discussed *Legionella* proteins, their targets, and functions is listed in Table 1). Also, we hypothesize on the potential involvement of other as yet non-investigated GTPase families in the process of *Legionella* infection.

3.1 Modulation of Arf Proteins

ADP ribosylation factor (Arf) proteins regulate the intracellular transport of membranes and proteins and are involved in vesicle coat formation. Six different Arfs can be divided into three functional classes: Arf1, Arf2, and Arf3 constitute class I, Arf4 and Arf5 class II, and Arf6 class III.

Arf1 is localized to the cytosolic face of the LCV shortly after bacterial uptake into macrophages (Kagan and Roy 2002). The deposition of Arf1 is dependent on the *Legionella* protein RalF that is secreted by the Dot/Icm Type IV transporter system (Nagai et al. 2002). Arf1 is important for LCV formation as demonstrated with dominant negative Arf1 or by RNA interference-mediated knockdown of Arf1 (Dorer et al. 2006; Kagan and Roy 2002; Kagan et al. 2004). However, the exact physiological mechanism by which Arf1 supports LCV formation has not yet been identified. One of the characteristics of LCV formation is the change in membrane phospholipid identity that is in part mediated through phosphatidylinositol-4-phosphate kinase IIIβ (PI4 K IIIβ) (Brombacher et al. 2009; Godi et al. 1999). Since PI4 K IIIβ is activated by Arf1 in healthy mammalian cells, Arf1 recruitment and activation at the LCV could potentially lead to activation of PI4 K IIIβ at this membrane and thereby initiate phospholipid remodeling. Nevertheless, in the absence of the Arf1 activator RalF, which is an ArfGEF from

Table 1 *Legionella* proteins affecting small GTPases

Modulator[a]	GTPase target	Additional targets[b]	Activity	Biochemical consequences
DrrA	Rab1		GEF	Activation
DrrA	Rab1	Rab3, Rab4, Rab6, Rab8, Rab11, Rab13, Rab14, Rab37	ATase	Inhibition of binding to GAPs, effectors, RabGDI, but not LidA
SidD	Rab1		Deadenylylase	Reverses $DrrA_{ATase}$ effects
LepB	Rab1	Rab8, Rab13, Rab35	GAP	Deactivation
LidA	Rab1, Rab6, Rab8		Rab effector	Extremely tight binding, blocking of other interaction partners
AnkX	Rab1, Rab35		Phosphocholinase	Inhibition of binding to GAPs, effectors, RabGDI, but not LidA
Lem3	Rab1		Dephosphocholinase	Reverses AnkX effects
VipD	Rab5, Rab22		Rab effector	Blocks binding of human Rab effectors
RalF	Arf1	Arf6	GEF	Activation
?	Rab7, Rab8, Rab14		?	

[a] from *Legionella pneumophila*
[b] only biochemical data, not proven in vivo

Legionella, the level of phosphatidylinositol-4-phosphate (PI4P, the product of PI4 K IIIβ) is unaffected, as indicated by unimpaired recruitment of the *Legionella* PI4P-binding protein SidC (Brombacher et al. 2009). Therefore, changes in phospholipid levels and identity are probably not mediated through Arf1 and PI4 K IIIβ.

The *Legionella* protein RalF modulates Arf1 activity. It contains an N-terminal Sec7-domain that recruits to and activates Arf1 at the LCV via its GEF activity (Nagai et al. 2002). The GEF activity of RalF appears to be specific for Arf1, although some residual activity toward Arf6 is also observed in vitro (Alix et al. 2012). Interestingly, the GEF activity of RalF is autoinhibited by the binding of a C-terminal capping domain that entirely blocks RalF-stimulated nucleotide exchange (Alix et al. 2012). The capping domain targets RalF to the LCV by an unknown mechanism that is independent of phospholipids (Alix et al. 2012; Brombacher et al. 2009), leading to efficient recruitment of Arf1 to the same location. Also, the mode of release of autoinhibition is as yet unknown but (in contrast to mammalian Sec-domain ArfGEFs of the cytohesin family) appears to be independent of the interaction with phospholipids (Alix et al. 2012; Brombacher et al. 2009; DiNitto et al. 2007; Stalder et al. 2011).

The activity of GEFs is physiologically counteracted by GAPs that terminate the GTPase's active state by GTP hydrolysis. In the case of the Rab, Rho and Arf proteins, the GDP bound, inactive (in contrast to the GTP bound) G-protein is subsequently removed from the membrane either by solubilization with GDI proteins or by intramolecular binding of the lipid moiety. The latter scenario applies to Arf proteins. Interestingly, Arf1 accumulates on the LCV at early stages of phagosome formation but vanishes after approximately 10 h post-infection (Kagan and Roy 2002). The disappearance of Arf1 from aged LCVs could potentially indicate the recruitment of endogenous ArfGAPs or the presence of a *Legionella* ArfGAP that acts as an adversary to RalF. A similar scenario is observed for the Rab protein Rab1 that is under the control of the *Legionella* GEF DrrA/SidM and GAP LepB (discussed in detail below) (Ingmundson et al. 2007; Machner and Isberg 2006, 2007; Murata et al. 2006). However, a passive removal of Arf1 by diffusion or vesicular transport is also conceivable. In summary, the *Legionella* protein RalF controls Arf1 recruitment to and activation at the LCV at the beginning of LCV formation, but the molecular and physiological mechanisms downstream of Arf1activation remain elusive.

3.2 Modulation of Rab1 Proteins

With more than 60 members in humans, Rab proteins constitute the largest sub-family of the class of small GTPases (reviewed in (Novick and Hutagalung 2011)). Rabs are master regulators of vesicular trafficking and control the specificity, direction, timing, and coordination of eukaryotic intracellular vesicle transport (reviewed in (Cherfils and Zeghouf 2013; Novick and Hutagalung 2011)).

Each Rab cycles between a specific localization at a distinct compartment in the active state and the cytosol in complex with RabGDI in the GDP state. Active Rabs at a membrane serve as molecular beacons that attract a diverse set of Rab effectors (e.g., tethering factors, molecular motors, phospholipid modulators, etc.) and thereby determine vesicular trafficking at that membrane. Thus, to take control over the localization and activation of specific Rab proteins would allow pathogens to modulate specific vesicle traffic in order to block microbicidal actions, to camouflage themselves inside the host, to acquire nutrients, and to inhibit unprofitable trafficking events. It is therefore surprising that, currently, only a few intracellularly surviving bacteria are known to subvert intracellular trafficking via the modulation of Rab proteins, examples being *Shigella flexneri* and enteropathogenic *Escherichia coli* that provide the Rab1GAPs VirA and EspG (Dong et al. 2012), *Brucella abortus* that secretes a Rab2 binding protein (de Barsy et al. 2011), *Salmonella typhi* that releases the Rab29/Rab32/Rab38 protease GtgE (Spano et al. 2011), and *Chlamydia pneumoniae* Cpn0585 that interacts with active Rab1, Rab10, and Rab11 (Cortes et al. 2007).

In this respect, *Legionella* is an interesting exception, since multiple and surprising mechanisms have been identified by which bacterial proteins modulate the activity of Rab proteins during infection (discussed in detail below). The Rab1 subfamily is especially targeted by several *Legionella* proteins and is both activated and inhibited at different stages of LCV formation. The Rab1 subfamily comprises Rab1A, Rab1B, and Rab35 (also annotated as Rab1C). Rab1A and Rab1B have identical function and control ER-to-Golgi vesicular trafficking (Allan et al. 2000; Moyer et al. 2001), whereas Rab35 regulates the sorting of cargo from early endosomes and localizes to clathrin-coated vesicles (Allaire et al. 2010; Kouranti et al. 2006). The interference of *Legionella* proteins with early secretory processes was first observed with the recruitment of ER-derived vesicles—normally transported from the ER to the ER-Golgi intermediate compartment (ERGIC) and Golgi—to the LCV at early time points of phagosome formation (Tilney et al. 2001). The recruitment of ER-derived vesicles was then shown to be dependent on the previous deposition of Rab1 at the LCV (Kagan et al. 2004). We discuss below the process of Rab1 recruitment, activation, and modulation (summarized in Fig. 1).

3.2.1 Modes of Activation by Legionella Proteins

The recruitment of Rab1 to the LCV is dependent on the secreted *Legionella* protein DrrA/SidM (Machner and Isberg 2006; Murata et al. 2006). DrrA contains three distinct functional domains: An N-terminal adenylyl transferase domain (ATase) (Müller et al. 2010), a central Rab1-GEF domain (Machner and Isberg 2006; Murata et al. 2006), and a C-terminal phosphatidylinositol-4-phosphate binding domain (P4M) (Brombacher et al. 2009). The unusually high affinity of the P4M localizes DrrA efficiently to the LCV (Brombacher et al. 2009; Machner and Isberg 2006; Schoebel et al. 2010) and probably restricts the GEF and ATase activities to this compartment. It could eventually be shown that the GEF activity of DrrA is

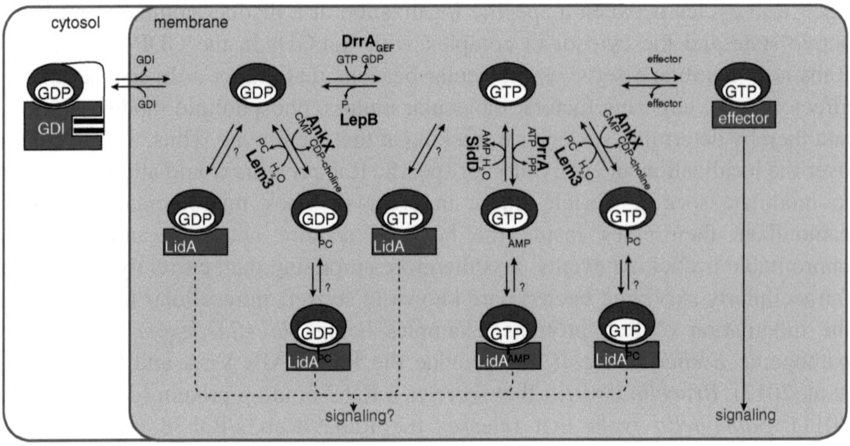

Fig. 1 Biochemical modulations of small GTPases exemplified for Rab1. Rab proteins exist in a soluble form in the cytosol in complex with GDI. GDP-GTP exchange by GEFs (e.g., *Legionella* DrrA) stabilizes Rabs in on membrane. This distribution can be reversed by GAPs (e.g., *Legionella* LepB) that deactivate Rabs by GTP hydrolysis. In addition to cyclic GEF and GAP reactions, the activity of Rabs appears to be modulated by strong binding (e.g., LidA) in both active and inactive states. Furthermore, reversible post-translational modifications (e.g., phosphocholination and adenylylation) additionally influence signaling from the Rab proteins. Adenylylation only occurs on active Rab1, whereas phosphocholination can target both active and inactive Rab1. In vitro, LidA will bind to modified and unmodified as well as to active and inactive Rabs

necessary and sufficient to specifically recruit and activate Rab1 from cytosolic Rab1:GDI complexes by exchanging GDP for GTP (Ingmundson et al. 2007; Machner and Isberg 2007; Schoebel et al. 2009). This unexpected observation of the significance of the bacterial GEF activity in localizing Rab1 has initiated systematic reinvestigation of the recruitment of Rab proteins and has ultimately demonstrated that RabGEFs are generally major determinants for Rab membrane targeting (Blümer et al. 2013). Although the principles of DrrA GEF activity are comparable to GEFs of other small GTPases, the amino acid sequence and protein structure of the central GEF domain are entirely unrelated to any other protein (Schoebel et al. 2009; Suh et al. 2010; Zhu et al. 2010). Although the Rab1GEF activity of DrrA and its significance for Rab1-mediated mistargeting of ER-derived vesicles to the LCV is evident, the molecular details of the recruitment of these vesicles and signaling events downstream of Rab1 remain elusive. Generally, Rab1 is involved in tethering of vesicles by binding to membrane-localized tethering factors, such as MICAL, GM130, or p115 (Allan et al. 2000; Moyer et al. 2001; Weide et al. 2003). This mechanism could, in principle, apply to LCV-localized Rab1 as well but the observation that adenylylation and/or phosphocholination of Rab1 by the *Legionella* proteins DrrA and/or AnkX inhibits effector interaction speaks against such a scenario (the effect of these post-translational modifications will be discussed in detail below) (Goody et al. 2012; Müller et al. 2010).

Despite the lack of data explaining the necessity for Rab1, one possible mechanism could be the interaction with the *Legionella* Rab effector LidA. LidA is secreted at early time points of and throughout infection (Conover et al. 2003; Derre and Isberg 2005) and localizes to intracellular membranes via binding to phosphatidylinositol-3-phosphates (PtdIns(3)P) (Brombacher et al. 2009; Neunuebel et al. 2012). It is an effector of Rab1, but also interacts with Rab6 and Rab8 (Machner and Isberg 2006). Similar to other Rab effectors, LidA preferably binds to active Rabs. However, whereas other effectors bind active Rabs with dissociation constants (K_D) in the micromolar to nanomolar range, the affinity between LidA and its interacting Rab proteins is several orders of magnitude higher (nanomolar to picomolar K_Ds) (Schoebel et al. 2011). Unusually, inactive Rabs also are bound with exceptionally high affinity. This extremely tight binding can be explained by the atomic structure of the Rab8:LidA complex, in which an unusually extended Rab:effector interface is apparent (Cheng et al. 2012; Schoebel et al. 2011). The high affinity of the complex also allows the binding to adenylylated and phosphocholinated Rab1 in vitro even though other effectors would be unable to bind (Fig. 1). Whether LidA has evolved with high affinity toward Rab proteins to overcome the negative effect of these post-translational modifications or whether it is necessary to efficiently compete with endogenous regulatory proteins (e.g. GAPs) is not clear yet (Neunuebel et al. 2012). Rab proteins are removed from the LCV at later stages of infection by a process that probably requires solubilization by RabGDI. Since RabGDI and LidA share the same binding site on Rabs, the question arises as to how LidA can be displaced from the high affinity Rab:LidA complex (the Rab:LidA complex is likely to have a half-life of several days in vitro in the absence of stimulatory factors or effects (Schoebel et al. 2011)). Also, whether LidA interacts with Rab proteins in cis (on the LCV) or in trans (Rab on the LCV, LidA on the vesicle) and thereby tethers the ER-derived vesicle remains elusive (Neunuebel et al. 2012).

3.2.2 Modes of Inhibition by Legionella Proteins

Legionella has not only evolved proteins that activate Rab1, but also with different mechanisms, to terminate or inhibit signaling from Rab1. During the course of infection, *Legionella* achieves Rab1 inhibition by either providing the Rab1GAP LepB or by introducing reversible post-translational modifications (i.e. adenylylation and phosphocholination).

LepB is a 1294 amino acid protein that is secreted by *Legionella* at later stages of infection and accumulates in the host cell (Ingmundson et al. 2007). LepB has been characterized as a Rab1GAP (Ingmundson et al. 2007), but also has GAP activities with lower efficiency toward Rab8, Rab13, and Rab35 in vitro (Gazdag et al. 2013). The GAP domain of LepB (amino acids 317-618) is unrelated in sequence and structure to any other protein. All known eukaryotic RabGAPs contain a TBC-domain (Tre-2, Bub2, and Cdc16) (Richardson and Zon 1995) that catalyzes GTP hydrolysis on Rab proteins by a dual finger mechanism. A catalytic arginine plus a specific glutamine are provided in trans by GAP to stimulate GTP to GDP conversion (Pan et al. 2006). Surprisingly, LepB operates by a GAP mechanism that is reminiscent of GAPs of non-Rab small GTPases. Here, only an arginine is provided in trans, but the catalytic glutamine from the G3 motif of Rab1 is used in cis (Gazdag et al. 2013). Thus, *Legionella* provides both a Rab1GEF (DrrA) and a Rab1GAP (LepB) that appear to act at different stages of infection (Ingmundson et al. 2007).

In addition to regulation of the nucleotide states by DrrA and LepB, the activity of Rab1 is also regulated by covalent modifications. At the N-terminus, DrrA possesses adenylyltransferase (ATase) activity that covalently modifies Rab1 with an adenosine monophosphate (AMP) moiety at Tyr77 (according to Rab1B numbering) (Müller et al. 2010). Although $DrrA_{ATase}$ is not entirely specific for Rab1 per se, it preferentially modifies the GTP bound rather than the GDP bound form in vitro (Müller et al. 2010, 2012) and therefore probably succeeds Rab1 activation by $DrrA_{GEF}$ in vivo. Adenylylation occurs in the regulatorily important switch II region, thereby blocking the access of the human Rab1GAP TBC1D20 and the *Legionella* GAP LepB to Rab1 and keeping Rab1 in the GTP bound state (Gazdag et al. 2013; Haas et al. 2007; Müller et al. 2010; Pan et al. 2006). Also, binding of GDI and therefore Rab solubilization from membranes is blocked by the presence of the AMP group (Oesterlin et al. 2012). Although adenylylation keeps Rab1 in the GTP state, this form cannot be unambiguously regarded as the active form, since the binding of effectors such as MICAL III (with the exception of *Legionella* LidA (Schoebel et al. 2011)) is not possible anymore (Müller et al. 2010). It therefore remains unclear why $DrrA_{GEF}$ activates Rab1 via GTP loading, but immediately afterwards essentially inactivates the protein by adenylylation through $DrrA_{ATase}$. However, the *Legionella* deadenylylase SidD reverses adenylylation via hydrolytic removal of the AMP group and hence demonstrates that this modification must have a regulatory function for Rab1 activity (Neunuebel et al. 2011; Tan and Luo 2011). The gene encoding for SidD is located directly adjacent to DrrA in the *Legionella* genome (Neunuebel et al. 2011). SidD is a catalytically efficient enzyme and does not discriminate between GDP or GTP loaded Rab1-AMP, raising the possibility that spontaneous or GAP-mediated GTP hydrolysis precedes deadenylylation (Müller et al. 2012).

In addition to adenylylation, the Rab1 family (including Rab35) is also a substrate of the *Legionella* phosphocholine transferase AnkX. The enzyme AnkX utilizes cytidine diphosphate choline (CDP-choline) to covalently transfer a phosphocholine group to $Ser76_{Rab1B}$ or $Thr76_{Rab35}$ with the concomitant release of cytidine monophosphate (CMP) (Mukherjee et al. 2011). Similar to adenylylation, the phosphocholination of Rab1 (but not Rab35) is hydrolytically reversible and the enzyme responsible for cleaving the modification is the *Legionella* dephosphocholinase Lem3 (Goody et al. 2012; Tan et al. 2011). The Lem3 gene (Lpg0696) is (similar to the situation with DrrA and SidD) located adjacent to the gene encoding for AnkX (Goody et al. 2012). The biochemical effects of Rab phosphocholination are similar to the alterations seen for adenylylated Rabs. Thus, the binding of GAPs (TBC1D20, LepB), RabGDI, $DrrA_{GEF}$, and mammalian effector molecules is severely impaired in phosphocholinated Rab1 as is the binding of the GEF Connecdenn/DENND1A in phosphocholinated Rab35 (Goody et al. 2012; Mukherjee et al. 2011; Müller et al. 2012). However, like adenylylation, the phosphocholination of Rab1 is still compatible with binding to LidA (Goody et al. 2012).

Despite these functional similarities between the adenylylation and phosphocholination of Rab proteins, the enzymatic characteristics of AnkX show a significant difference to $DrrA_{ATase}$. AnkX does not discriminate between active and inactive Rab1 and modifies both activity states with comparable catalytic efficiency (Goody et al. 2012), whereas $DrrA_{ATase}$ preferentially modifies the GTP bound form of Rab1. Consequently, AnkX (but not $DrrA_{ATase}$) efficiently modifies Rab1 even in the Rab1:GDP:RabGDI-complex, leading to effective displacement of RabGDI (Goody et al. 2012; Oesterlin et al. 2012). It is conceivable that not only GDP to GTP exchange enables Rab membrane recruitment but also post-translational modifications such as phosphocholination of Rab1 (Blümer et al. 2013; Oesterlin et al. 2012; Schoebel et al. 2009). Since AnkX is presumably membrane localized (Pan et al. 2008), its phosphocholination could recruit Rab1 and/or Rab35 to the same compartment. The significance of such a mechanism is still unclear, but it could theoretically create an enriched inactive pool of membrane-localized Rab1 that is readily activated by Lem3 and GEFs.

Obviously, *Legionella* has evolved with an astounding protein toolbox (DrrA-GEF, LepB, LidA, DrrAATase, SidD, AnkX, Lem3) (Fig. 1) that specifically addresses the activity of the Rab1 family. Although the effects of these proteins on Rab1 activities have been analyzed in great biochemical and structural detail, many of their physiological consequences and cellular implications for LCV formation and *Legionella* survival remain mysterious.

3.3 Modulation of Other Rab Proteins

Proteomic analyses have revealed that not only Arf1 and Rab1 proteins are localized to the LCV in an Icm/Dot dependent fashion. The secretory GTPase Rab8 and the endosomal GTPases Rab7 and Rab14 were identified by proteomic studies on LCV purified from Dictyostelium (Shevchuk et al. 2009; Urwyler et al. 2009b). The presence of these additional G-proteins on the LCV depends on a functional Icm/Dot secretion system (Urwyler et al. 2009b) and therefore indicates the potential existence of secreted *Legionella* proteins with Rab7, Rab8, and/or Rab14 recruitment activity. Since RabGEFs have been shown to act as major determinants for Rab membrane targeting (Blümer et al. 2013), either potential *Legionella* GEFs could be involved in recruiting these Rabs similar to the situation with $DrrA_{GEF}$ (Schoebel et al. 2009) and Rab1, or endogenous host RabGEFs may be mislocalized to the LCV by unknown mechanisms and thereby act as recruiting factors. This hypothesis is, however, entirely speculative. Alternatively, these Rabs may be acquired by direct fusion of the LCV with the respective compartments or by interaction with transport vesicles containing Rab7, Rab8, and/or Rab14 (Urwyler et al. 2009a).

Another possibility, at least for the recruitment of Rab8 to the LCV, is via LidA-mediated displacement of GDI from cytosolic Rab8:GDP:RabGDI complexes: Due to the enormous affinity of LidA for Rab8, LidA can efficiently compete with RabGDI for the binding of Rab8 as observed with Rab1 (Oesterlin et al. 2012). LidA forms a stable stoichiometric complex with Rabs (Schoebel et al. 2011) and the amount of Rab recruitment would be limited by the total LidA availability (in contrast to catalytically acting GEFs that can recruit many Rabs per enzyme (Blümer et al. 2013)). However, the observation that the additional LidA binding partner Rab6/6' is not localized to the LCV despite a LidA affinity comparable to Rab8 speaks against a direct contribution of LidA in Rab8 membrane recruitment (Chen and Machner 2013; Kagan et al. 2004; Schoebel et al. 2011).

Biochemical experiments have not only revealed the enzymatic details of the GAP reaction of LepB toward Rab1 but have also shown that LepB can stimulate GTP hydrolysis on Rab8, Rab13, and Rab35, albeit with 10–100 times decreased catalytic efficiency (Gazdag et al. 2013). Although they are worse substrates than Rab1, the high catalytic activity of LepB still makes Rab8, and Rab13 reasonable substrates, thus raising the possibility that their activity is also modulated by LepB during *Legionella* infection. As mentioned above, Rab8 has been detected on LCVs and is therefore a potential substrate, whereas there are no data suggesting the presence of Rab13 at the phagosome (Urwyler et al. 2009b).

The adenylylation by DrrA is not only restricted to the Rab1 subfamily, but also other Rab proteins have been identified as substrates: Rab1B, Rab3A, Rab4B, Rab6A, Rab8A, Rab11A, Rab13, Rab14, and Rab37 served as $DrrA_{ATase}$ substrates in vitro whereas Rab5A, Rab7A, Rab9A, Rab22A, Rab23, Rab27A, Rab31, Rab32, and Rab38 did not (Müller et al. 2010). The adenylylated Tyr77 of Rab1B is conserved among the other substrates along with adjacent sequences and therefore potentially explains this observation in vitro. Interestingly, the LCV-localized proteins Rab8 and Rab14 are substrates of $DrrA_{ATase}$ and may thus be recruited by its activity (Urwyler et al. 2009b). The observation that Rab6A is not detected on the LCV despite being modified by $DrrA_{ATase,}$ however, argues against such a simple explanation (Chen and Machner 2013; Kagan et al. 2004).

Currently, there is one example demonstrating that *Legionella* proteins in principle also modulate the activities of small GTPases distant from the LCV membrane: The secreted bacterial factor VipD is a bifunctional protein possessing an N-terminal phospholipase A2 activity and a C-terminal Rab5/Rab22 effector domain (Ku et al. 2012; Shohdy et al. 2005). Like Rab5 and Rab22, VipD localizes to early endosomes (Kauppi et al. 2002; Zhu et al. 2009) but is recruited independently of these Rabs (Ku et al. 2012). The C-terminal Rab-binding domain of VipD does not influence the nucleotide state of Rab5 or Rab22 but preferentially interacts with the active GTP bound form with moderate affinity in vitro ($K_{D,\text{active Rab5}} = 254$ nM, $K_{D,\text{active Rab22}} = 132$ nM) (Ku et al. 2012). It appears that VipD can efficiently compete with the binding of Rab5 and Rab22 effectors ($K_D > 0.9$ μM (Eathiraj et al. 2005; Mishra et al. 2010)) to the activated forms of these Rab proteins and thereby block downstream signaling (Ku et al. 2012). Thus, VipD interferes with the transition of early endosomes to late endosomes and inhibits their entry into the lysosomal pathway (Zhu et al. 2009). One possible result of this blockage could be the facilitation of survival of intracellular *Legionella* by preventing the fusion of phagocytic vesicles with lysosomes.

3.4 Speculative Modulation of Further Small G-Proteins

Currently, it appears as if *Legionella* has selectively placed major emphasis on the manipulation of small GTPases involved in the regulation of vesicular trafficking via modulation of Arf and Rab proteins. This is in contrast to many other bacterial pathogens since the literature mostly refers to manipulation of Rho/Rac/Cdc42 GTPases that regulate cytoskeleton rearrangements (reviewed in (Aktories 2011)). Such bacterial factors usually either promote bacterial uptake by controlling the cytoskeleton at an early stage of phagocytosis or induce cytotoxicity by globally inhibiting the Rho GTPase family. As such, it is surprising that currently no *Legionella* effectors have been reported to interfere with Rho GTPase signaling, neither in a stimulatory nor in an inhibitory manner. Whether RhoA manipulation could interfere with *Legionella* uptake dynamics, phagosome formation, intracellular LCV mobility, or *Legionella* release remains to be evaluated.

Curiously, a few other small GTPases (RasG, Ran) have been detected on purified LCVs in proteomic analyses (Urwyler et al. 2009b). Both Ran and its binding partner RanGAP have been identified in these experiments and therefore strengthen the significance of Ran localization to LCVs. The main function of Ran is to regulate nuclear import/export, but regulation of non-centrosomal microtubules has also been reported (Schulze et al. 2008). However, Ran has not been implicated in processes linking phagocytosis or bacterial evasion of host cell defense.

4 Conclusions

Legionella is a rich source for proteins that modulate different aspects of signaling and regulation by small GTPases. These secreted *Legionella* proteins utilize diverse mechanism to either stimulate or inhibit signaling, such as nucleotide exchange (GEF), RabGDI-displacement, GTP hydrolysis (GAP), exceptionally tight binding (LidA), or reversible and irreversible post-translational modifications (adenylylation, phosphocholination). Although much of the enzymology, biochemistry, and structural consequences of these modulatory mechanisms have been understood at the molecular level, the physiological significance for the survival and multiplication of *Legionella* in host cells remains obscure in many aspects. The research focus has been on events implicated in LCV formation and remodeling in the past.

Nevertheless, *Legionella* proteins may also interfere with GTPase signaling at other compartments during infection and modulate their activities.

Rab1 is the most manifoldly targeted protein by *Legionella* among the family of small GTPases. However, the physiological function of most of the 270 secreted *Legionella* proteins have not been identified yet and it is thus conceivable (though highly speculative) that also other small GTPases previously detected on LCVs (e.g., Rab7, Rab8, Rab14, Ran) are specifically modulated by bacterial proteins by as yet unknown strategies as well. It will require the combined efforts of microbiology, biochemistry, and proteomics to elucidate the full scope of modulation of small GTPases by *Legionella* and its secreted proteins.

References

Aktories K (2011) Bacterial protein toxins that modify host regulatory GTPases. Nat Rev Microbiol 9:487–498. doi: 10.1038/nrmicro2592

Alix E, Chesnel L, Bowzard BJ, Tucker AM, Delprato A, Cherfils J, Wood DO, Kahn RA, Roy CR (2012) The Capping domain in Ralf regulates effector functions. PLOS Pathog 8. doi: 10.1371/journal.ppat.1003012

Allaire PD, Marat AL, Dall'Armi C, Di Paolo G, McPherson PS, Ritter B (2010) The connecdenn DENN domain: a GEF for Rab35 mediating cargo-specific exit from early endosomes. Mol Cell 37:370–382. doi: 10.1016/j.molcel.2009.12.037

Allan BB, Moyer BD, Balch WE (2000) Rab1 recruitment of p115 into a cis-SNARE complex: Programming budding COPII vesicles for fusion. Science 289:444–448. doi:10.1126/science.289.5478.444

Blümer J, Rey J, Dehmelt L, Mazel T, Wu YW, Bastiaens P, Goody RS, Itzen A (2013) RabGEFs are a major determinant for specific Rab membrane targeting. J Cell Biol 200:287–300. doi: 10.1083/jcb.201209113

Brombacher E, Urwyler S, Ragaz C, Weber SS, Kami K, Overduin M, Hilbi H (2009) Rab1 guanine nucleotide exchange factor SidM is a major phosphatidylinositol 4-phosphate-binding effector protein of *Legionella pneumophila*. J Biol Chem 284:4846–4856. doi: 10.1074/jbc.M807505200

Chen Y, Machner MP (2013) Targeting of the small GTPase Rab6A' by the *Legionella pneumophila* effector LidA. Infect Immun. doi: 10.1128/IAI.00157-13

Cheng W, Yin K, Lu DF, Li BQ, Zhu DY, Chen YZ, Zhang H, Xu SJ, Chai JJ, Gu LC (2012) Structural insights into a Unique *Legionella pneumophila* effector LidA recognizing Both GDP and GTP Bound Rab1 in their active state. Plos Pathog 8. doi: 10.1371/journal.ppat.1002528.

Cherfils J, Zeghouf M (2013) Regulation of small GTPases by GEFs, GAPs, and GDIs. Physiol Rev 93:269–309. doi: 10.1152/physrev.00003.2012.

Conover GM, Derre I, Vogel JP, Isberg RR (2003) The *Legionella pneumophila* LidA protein: a translocated substrate of the Dot/Icm system associated with maintenance of bacterial integrity. Mol Microbiol 48:305–321. doi: 10.1046/j.1365-2958.2003.03400.x

Cortes C, Rzomp KA, Tvinnereim A, Scidmore MA, Wizel B (2007) *Chlamydia pneumoniae* inclusion membrane protein Cpn0585 interacts with multiple Rab GTPases. infection and immunity 75:5586–5596. doi: 10.1128/IAI.01020-07.

de Barsy M, Jamet A, Filopon D, Nicolas C, Laloux G, Rual JF, Muller A, Twizere JC, Nkengfac B, Vandenhaute J, Hill DE, Salcedo SP, Gorvel JP, Letesson JJ, De Bolle X (2011) Identification of a Brucella spp. secreted effector specifically interacting with human small GTPase Rab2. Cell Microbiol 13:1044–1058. doi: 10.1111/j.1462-5822.2011.01601.x

Derre I, Isberg RR (2005) LidA, a translocated substrate of the *Legionella pneumophila* type IV secretion system, interferes with the early secretory pathway. Infect Immun 73:4370–4380. doi: 10.1128/IAI.73.7.4370-4380.2005

DiNitto JP, Delprato A, Gabe Lee MT, Cronin TC, Huang S, Guilherme A, Czech MP, Lambright DG (2007) Structural basis and mechanism of autoregulation in 3-phosphoinositide-dependent Grp1 family Arf GTPase exchange factors. Mol Cell 28:569–583. doi: 10.1016/j.molcel.2007.09.017

Dong N, Zhu Y, Lu Q, Hu L, Zheng Y, Shao F (2012) Structurally distinct bacterial TBC-like GAPs Link Arf GTPase to Rab1 inactivation to counteract host defenses. Cell 150:1029–1041. doi: 10.1016/j.cell.2012.06.050

Dorer MS, Kirton D, Bader JS, Isberg RR (2006) RNA interference analysis of Legionella in Drosophila cells: exploitation of early secretory apparatus dynamics. PLoS Pathog 2:e34. doi: 10.1371/journal.ppat.0020034

Eathiraj S, Pan X, Ritacco C, Lambright DG (2005) Structural basis of family-wide Rab GTPase recognition by rabenosyn-5. Nature 436:415–419. doi: 10.1038/nature03798

Gazdag EM, Streller A, Haneburger I, Hilbi H, Vetter IR, Goody RS, Itzen A (2013) Mechanism of Rab1b deactivation by the *Legionella pneumophila* GAP LepB. EMBO Reports 14:199–205. doi: 10.1038/embor.2012.211

Godi A, Pertile P, Meyers R, Marra P, Di Tullio G, Iurisci C, Luini A, Corda D, De Matteis MA (1999) ARF mediates recruitment of PtdIns-4-OH kinase-beta and stimulates synthesis of PtdIns (4,5)P2 on the Golgi complex. Nat Cell Biol 1:280–287. doi: 10.1038/12993

Goody PR, Heller K, Oesterlin LK, Muller MP, Itzen A, Goody RS (2012) Reversible phosphocholination of Rab proteins by *Legionella pneumophila* effector proteins. EMBO J 31:1774–1784. doi: 10.1038/emboj.2012.16

Haas AK, Yoshimura S, Stephens DJ, Preisinger C, Fuchs E, Barr FA (2007) Analysis of GTPase-activating proteins: Rab1 and Rab43 are key Rabs required to maintain a functional Golgi complex in human cells. J Cell Sci 120:2997–3010. doi: 10.1242/jcs.014225

Ingmundson A, Delprato A, Lambright DG, Roy CR (2007) *Legionella pneumophila* proteins that regulate Rab1 membrane cycling. Nature 450:365–369. doi: 10.1038/nature06336

Isberg RR, O'Connor TJ, Heidtman M (2009) The *Legionella pneumophila* replication vacuole: making a cosy niche inside host cells. Nat Rev Microbiol 7:13–24. doi: 10.1038/nrmicro1967

Kagan JC, Roy CR (2002) *Legionella* phagosomes intercept vesicular traffic from endoplasmic reticulum exit sites. Nat Cell Biol 4:945–954. doi: 10.1038/ncb883

Kagan JC, Stein MP, Pypaert M, Roy CR (2004) *Legionella* subvert the functions of Rab1 and Sec22b to create a replicative organelle. J Exp Med 199:1201–1211. doi: 10.1084/jem.20031706

Kauppi M, Simonsen A, Bremnes B, Vieira A, Callaghan J, Stenmark H, Olkkonen VM (2002) The small GTPase Rab22 interacts with EEA1 and controls endosomal membrane trafficking. J Cell Sci 115:899–911

Kouranti I, Sachse M, Arouche N, Goud B, Echard A (2006) Rab35 regulates an endocytic recycling pathway essential for the terminal steps of cytokinesis. Curr Biol 16:1719–1725. doi: 10.1016/j.cub.2006.07.020

Ku B, Lee KH, Park WS, Yang CS, Ge J, Lee SG, Cha SS, Shao F, Heo WD, Jung JU, Oh BH (2012) VipD of *Legionella pneumophila* Targets Activated Rab5 and Rab22 to Interfere with Endosomal Trafficking in Macrophages. Plos Pathog 8. doi: 10.1371/journal.ppat.1003082

Machner MP, Isberg RR (2006) Targeting of host Rab GTPase function by the intravacuolar pathogen *Legionella pneumophila*. Develop Cell 11:47–56. doi: 10.1016/j.devcel.2006.05.013

Machner MP, Isberg RR (2007) A bifunctional bacterial protein links GDI displacement to Rab1 activation. Science 318:974–977. doi: 10.1126/science.1149121

Mishra A, Eathiraj S, Corvera S, Lambright DG (2010) Structural basis for Rab GTPase recognition and endosome tethering by the C2H2 zinc finger of early endosomal autoantigen 1 (EEA1). Proc Natl Acad Sci U S A 107:10866–10871. doi: 10.1073/pnas.1000843107

Moyer BD, Allan BB, Balch WE (2001) Rab1 interaction with a GM130 effector complex regulates COPII vesicle cis-Golgi tethering. Traffic 2:268–276. doi: 10.1034/j.1600-0854.2001.10007.x

Mukherjee S, Liu XY, Arasaki K, McDonough J, Galan JE, Roy CR (2011) Modulation of Rab GTPase function by a protein phosphocholine transferase. Nature 477:103–106. doi: 10.1038/Nature10335

Müller MP, Peters H, Blumer J, Blankenfeldt W, Goody RS, Itzen A (2010) The *Legionella* effector protein DrrA AMPylates the membrane traffic regulator Rab1b. Science 329:946–949. doi: 10.1126/science.1192276

Müller MP, Shkumatov AV, Oesterlin LK, Schoebel S, Goody PR, Goody RS, Itzen A (2012) Characterization of Enzymes from *Legionella pneumophila* involved in reversible adenylylation of Rab1 protein. J Biol Chem 287:35036–35046. doi: 10.1074/jbc.M112.396861

Murata T, Delprato A, Ingmundson A, Toomre DK, Lambright DG, Roy CR (2006) The *Legionella pneumophila* effector protein DrrA is a Rab1 guanine nucleotide-exchange factor. Nat Cell Biol 8:971–9777. doi: 10.1038/ncb1463

Nagai H, Kagan JC, Zhu X, Kahn RA, Roy CR (2002) A bacterial guanine nucleotide exchange factor activates ARF on *Legionella* phagosomes. Science 295:679–682. doi: 10.1126/science.1067025

Neunuebel MR, Chen Y, Gaspar AH, Backlund PS, Yergey A, Machner MP (2011) De-AMPylation of the small GTPase Rab1 by the pathogen *Legionella pneumophila*. Science 333:453–456. doi: 10.1126/science.1207193

Neunuebel MR, Mohammadi S, Jarnik M, Machner MP (2012) *Legionella pneumophila* LidA affects nucleotide binding and activity of the host GTPase Rab1. J Bacteriol 194:1389–1400. doi: 10.1128/JB.06306-11

Novick PJ, Hutagalung AH (2011) Role of Rab GTPases in membrane traffic and cell physiology. Physiol Rev 91:119–149. doi: 10.1152/physrev.00059.2009

Oesterlin LK, Goody RS, Itzen A (2012) Posttranslational modifications of Rab proteins cause effective displacement of GDP dissociation inhibitor. Proc Natl Acad Sci U S A 109:5621–5621. doi: 10.1073/pnas.1121161109

Pan X, Eathiraj S, Munson M, Lambright DG (2006) TBC-domain GAPs for Rab GTPases accelerate GTP hydrolysis by a dual-finger mechanism. Nature 442:303–306. doi: 10.1038/nature04847

Pan XX, Lührmann A, Satoh A, Laskowski-Arce MA, Roy CR (2008) Ankyrin repeat proteins comprise a diverse family of bacterial type IV effectors. Science 320:1651–1654. doi: 10.1126/science.1158160

Richardson PM, Zon LI (1995) Molecular cloning of a cDNA with a novel domain present in the tre-2 oncogene and the yeast cell cycle regulators BUB2 and cdc16. Oncogene 11:1139–1148

Sarantis H, Grinstein S (2012) Subversion of phagocytosis for pathogen survival. Cell Host Microbe 12:419–431. doi: 10.1016/j.chom.2012.09.001

Schoebel S, Blankenfeldt W, Goody RS, Itzen A (2010) High-affinity binding of phosphatidylinositol 4-phosphate by *Legionella pneumophila* DrrA. EMBO Reports 11:598–604. doi: 10.1038/embor.2010.97

Schoebel S, Cichy AL, Goody RS, Itzen A (2011) Protein LidA from Legionella is a Rab GTPase supereffector. Proc Natl Acad Sci U S A 108: 17945–17950. doi: 10.1073/pnas.1113133108

Schoebel S, Oesterlin LK, Blankenfeldt W, Goody RS, Itzen A (2009) RabGDI displacement by DrrA from *Legionella* is a consequence of its guanine nucleotide exchange activity. Mol Cell 36:1060–1072. doi: 10.1016/j.molcel.2009.11.014

Schulze H, Dose M, Korpal M, Meyer I, Italiano JE, Shivdasani RA (2008) RanBP10 is a cytoplasmic guanine nucleotide exchange factor that modulates noncentrosomal microtubules. J Biol Chem 283:14109–14119. doi: 10.1074/jbc.M709397200

Shevchuk O, Batzilla C, Hagele S, Kusch H, Engelmann S, Hecker M, Haas A, Heuner K, Glockner G, Steinert M (2009) Proteomic analysis of Legionella-containing phagosomes isolated from Dictyostelium. Int J Med Microbiol 299:489–508. doi: 10.1016/j.ijmm.2009.03.006

Shohdy N, Efe JA, Emr SD, Shuman HA (2005) Pathogen effector protein screening in yeast identifies Legionella factors that interfere with membrane trafficking. Proc Natl Acad Sci U S A 102: 4866–4871. doi: 10.1073/pnas.0501315102

Spano S, Liu XY, Galan JE (2011) Proteolytic targeting of Rab29 by an effector protein distinguishes the intracellular compartments of human-adapted and broad-host Salmonella. Proc Natl Acad Sci U S A 108:18418–18423. doi: 10.1073/pnas.1111959108

Stalder D, Barelli H, Gautier R, Macia E, Jackson CL, Antonny B (2011) Kinetic studies of the Arf activator Arno on model membranes in the presence of Arf effectors suggest control by a positive feedback loop. J Biol Chem 286:3873–3883. doi: 10.1074/jbc.M110.145532

Suh HY, Lee DW, Lee KH, Ku B, Choi SJ, Woo JS, Kim YG, Oh BH (2010) Structural insights into the dual nucleotide exchange and GDI displacement activity of SidM/DrrA. EMBO J 29:496–504. doi: 10.1038/emboj.2009.347

Tan YH, Arnold RJ, Luo ZQ (2011) *Legionella pneumophila* regulates the small GTPase Rab1 activity by reversible phosphorylcholination. Proc Natl Acad Sci U S A 108:21212–21217. doi: 10.1073/pnas.1114023109

Tan YH, Luo ZQ (2011) *Legionella pneumophila* SidD is a deAMPylase that modifies Rab1. Nature 475:506–509. doi: 10.1038/nature10307

Tilney LG, Harb OS, Connelly PS, Robinson CG, Roy CR (2001) How the parasitic bacterium *Legionella pneumophila* modifies its phagosome and transforms it into rough ER: implications for conversion of plasma membrane to the ER membrane. J Cell Sci 114: 4637–4650

Urwyler S, Brombacher E, Hilbi H (2009a) Endosomal and secretory markers of the *Legionella*-containing vacuole. Commun Integr Biol 2:107–109. doi: 10.1111/j.1600-0854

Urwyler S, Nyfeler Y, Ragaz C, Lee H, Mueller LN, Aebersold R, Hilbi H (2009b) Proteome analysis of *Legionella* vacuoles purified by magnetic immunoseparation reveals secretory and endosomal GTPases. Traffic 10:76–87. doi: 10.1111/j.1600-0854.2008.00851.x

Weide T, Teuber J, Bayer M, Barnekow A (2003) MICAL-1 isoforms, novel rab1 interacting proteins. Biochem Biophys Res Commun 306:79–86. doi: 10.1016/S0006-2961x(03)00918-5

Zhu HP, Liang ZM, Li GP (2009) Rabex-5 Is a Rab22 Effector and Mediates a Rab22-Rab5 signaling cascade in endocytosis. Mol Biol Cell 20:4720–4729. doi: 10.1091/mbc.E09-06-0453

Zhu Y, Hu L, Zhou Y, Yao Q, Liu L, Shao F (2010) Structural mechanism of host Rab1 activation by the bifunctional *Legionella* type IV effector SidM/DrrA. Proc Natl Acad Sci U S A 107:4699–4704. doi: 10.1073/pnas.0914231107

Host Lipidation: A Mechanism for Spatial Regulation of *Legionella* Effectors

Stanimir S. Ivanov and Craig Roy

Abstract Bacterial pathogens have evolved the capacity to translocate proteins into the cytosol of infected cells to manipulate host processes. How do pathogens regulate spatially these bacterial effector proteins once they are released into the host cell? One mechanism, which is used by *Legionella* and other bacterial pathogens, is to encode effectors that mimic the substrates of eukaryotic lipid transferases. In this review we discuss three membrane-targeting pathways in eukaryotes that are exploited by *Legionella* and other pathogens—prenylation, palmitoylation, and myristoylation. Lipidation of bacterial substrates primes the effectors for coincidence detection-mediated targeting onto membrane-bound organelles by increasing membrane affinity. Intracellular membrane-targeting strategies that exploit protein fatty acylation and prenylation direct bacterial effectors to compartments where their target substrates reside and thus are critical for effector function.

Abbreviations

CaaX motif	Cysteine-aliphatic-aliphatic-X amino acid motif
cAMP	cyclic AMP
CMPs	CaaX motif-containing proteins
Dot/Icm	Defect in organelle trafficking/intracellular multiplication
DHHC PATs	Asp-His-His-Cys palmitoyl acyltransferases
FTase	Farnesyltransferase
GEFs	Guanine nucleotide exchange factors
GGTase	Geranylgeranyltransferase

S. S. Ivanov · C. Roy (✉)
Department of Microbial Pathogenesis, Yale University School of Medicine,
Boyer Center for Molecular Medicine, 295 Congress Ave,
New Haven, CT 06519, USA
e-mail: craig.roy@yale.edu

S. S. Ivanov
e-mail: stanimir.ivanov@yale.edu

Current Topics in Microbiology and Immunology (2013) 376: 135–154
DOI: 10.1007/82_2013_344

Published Online: 6 August 2013

HR	Hypersensitivity response
LCV	*Legionella*-containing vacuole
NMT	N-myristoyltransferase
PATs	Palmitoyl acyltransferases
SCV	*Salmonella*-containing vacuole
T3SS	Type III secretion system
T4SS	Type IV secretion system

Contents

1 Introduction ... 136
2 Spatial Regulation of Lipidated Polypeptides in Eukaryotic Cells ... 137
3 Brief Overview of Protein Lipidation in Eukaryotes ... 138
3.1 Protein Fatty Acylation ... 139
3.2 Prenylation of CaaX Motif Proteins ... 140
3.3 Membrane Affinity of Lipidated Proteins ... 140
3.4 Mechanisms of Dynamic Membrane Association of Lipidated Proteins ... 141
4 Host Prenyltransferases Facilitate Membrane Localization of *Legionella pneumophila* Effectors ... 142
4.1 Identification of Legionella Effectors Containing Prenylation Domains ... 142
4.2 The CaaX Motifs Encoded by the Legionella CMPs are Membrane-Targeting Domains that Control Protein Localization ... 143
4.3 Does Prenylation Regulate the Legionella CMPs Function? ... 145
5 Mimicry of Eukaryotic Lipidation Substrates by Bacterial Effectors is a Conserved Strategy for Spatial Organization ... 146
5.1 Regulation of Salmonella enterica Effectors by Lipidation ... 147
5.2 Membrane Tethering of Pseudomonas syringae Effectors by Dual-Acylation ... 148
6 Conclusions and Perspectives ... 150
References ... 151

1 Introduction

Homeostasis is maintained in eukaryotic cells by tightly controlling protein functions. Maintaining the localization of proteins at specific subcellular compartments is important for regulating protein function. There are multiple mechanisms to target eukaryotic proteins to different membrane-bound compartments (Cho and Stahelin 2005; Resh 2006; Kutateladze 2010). Bacterial pathogens that translocate proteins into the cytosol of host cells can utilize spatial control systems of eukaryotes to target effector proteins to specific subcellular organelles. This review describes lipidation pathways that are used for targeting protein to membranes in eukaryotes that are subverted by bacterial pathogens. These include prenylation, palmitoylation and myristoylation. All three pathways conjugate different lipid groups onto specific residues within defined domains of their protein substrates. Depending on the enzymatic reaction employed, these lipid modifications are either transient or permanent. Most of the proteins from the eukaryote Ras superfamily of GTPases, which function at the cytosolic leaflet of the lipid bilayer to control vesicular trafficking, cellular proliferation and differentiation, generate membrane affinity by lipidation (Konstantinopoulos et al. 2007; Mizuno-Yamasaki et al. 2012). In general, single lipidation motifs are poor membrane anchors, however tandem lipidation motifs or lipid motifs in cooperation with additional targeting determinants generate strong membrane affinity. Thus, protein lipidation is a part of a complex molecular code that spatially and temporally regulates membrane localization.

Translocation of bacterial proteins into the cytosol of eukaryotic cells by dedicated secretion systems is a highly conserved mechanism in diverse groups of bacterial pathogens and symbionts that allows bacteria to manipulate the host (Galan and Wolf-Watz 2006; Voth et al. 2012). Recently, proteins translocated into the host cells by different bacterial pathogens, such as *Legionella*, *Salmonella* and *Pseudomonas*, were found to be substrates for host lipid transferases (Nimchuk et al. 2000; Reinicke et al. 2005; Ivanov et al. 2010; Hicks et al. 2011). As a result of lipidation, the modified bacterial effectors are imbedded into the cytosolic leaflet of the lipid bilayer of distinct membrane-bound organelles. This review highlights what is currently known about the lipidation of *Legionella* effector proteins and describes effectors from other bacterial pathogens that exploit eukaryotic lipid transferases to localize effector proteins in host cells.

2 Spatial Regulation of Lipidated Polypeptides in Eukaryotic Cells

Evolution has shaped the bacterial proteins functioning inside the cytosol of infected cells to exploit the spatial organization of eukaryotic cells and the mechanisms that regulate it. In eukaryotic cells, the combination of resident proteins and lipids on each cellular organelle generates a powerful molecular code that provides a distinct organelle identity (Krauss and Haucke 2007). The identity code ensures that only a specific set of proteins accumulate on distinct organelle membranes by requiring multiple spatial determinants (Carlton and Cullen 2005; Di Paolo and De Camilli 2006; Jean and Kiger 2012), which involves at least two distinct molecular interactions for stable organelle association. Numerous molecular determinants that encode combinations of protein–protein, protein-lipid and lipid–lipid interactions facilitate such spatial control in eukaryotes. One of the most important targeting mechanisms that prime proteins for association with membrane-bound organelles is lipidation.

Upon lipidation, a protein gains a membrane-targeting domain by acquiring a hydrophobic moiety that can embed in the lipid bilayer. However, the membrane affinity of proteins with a single lipid modification is generally weak with a high

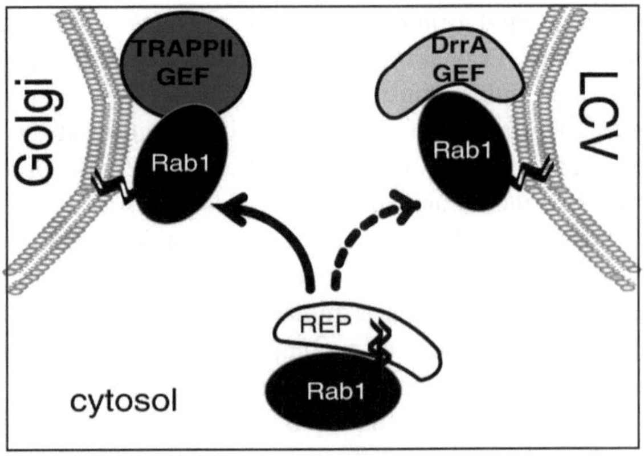

Fig. 1 Localization of lipidated Rab1 in Legionella infected cells

"off" rate, which allows lipidated proteins to sample membranes until a second event generates stable association (Resh 2006; Jean and Kiger 2012). Appending an additional lipid moiety, electrostatic interactions, protein–protein, and protein–lipid interactions represent examples of a second signal. Because lipidation primes proteins for membrane association often a single additional signal is sufficient to localize lipidated proteins to a distinct organelle (Jean and Kiger 2012). One example is the spatial compartmentalization of the lipidated small GTPase Rab1 (Hutagalung and Novick 2011). Rab1 is post-translationally lipidated and sequestered in the cytosol through an interaction with Rab escort protein (REP) (Alexandrov et al. 1994). Specific interactions with Rab1 activating proteins that function as guanine nucleotide exchange factors (GEFs) provides a second signal that directs the targeting of Rab proteins to specific organelles (Yamasaki et al. 2009; Blumer et al. 2013). Interestingly, a distinct pool of membrane-localized Rab1 is found on the pathogen-occupied vacuole in *Legionella*-infected cells and localization of Rab1 to this compartment is mediated by a bacterial Rab1 GEF (Fig. 1). DrrA is a *Legionella* effector that localizes and recruits Rab1 to the *Legionella*-containing vacuole to generate a pool of membrane-associated Rab1 that is distinct from the Rab1 pool residing at the Golgi, which is produced by the host protein complex called TRAPPII (Fig. 1) (Machner and Isberg 2006; Murata et al. 2006; Yamasaki et al. 2009). This example shows how pathogens can utilize the eukaryotic spatial control mechanisms to create an intracellular niche with a distinct molecular code and subvert processes used to recruit host proteins to membrane-bound organelles.

3 Brief Overview of Protein Lipidation in Eukaryotes

In eukaryotic cells, fatty acids and isoprenoids are two of the most common lipids covalently attached to specific residues within distinct protein domains. Bacterial pathogens exploit the lipid transferase enzymes that control these processes to

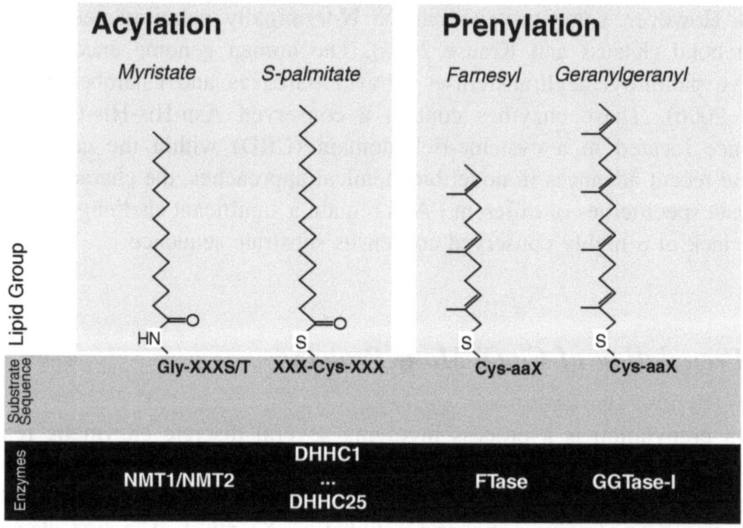

Fig. 2 Enzymology of host lipidation—modifying groups, substrates sequences, and lipidtransferases

spatially control translocated effectors. This section will introduce briefly the enzymes that mediate fatty acylation and prenylation and emphasize important mechanistic aspects of both systems as they pertain to the control of protein localization (Fig. 2).

3.1 Protein Fatty Acylation

Protein acylation refers to the covalent attachment of the saturated fatty acid myristate (C-14) or palmitate (C-16) to proteins by processes known as N-myristoylation and palmitoylation, respectively (Resh 2006). These two processes are distinct. N-myristoylation generally is a co-translational event whereby the initiating methionine is removed by methionine aminopeptidases to expose the neighboring glycine residue that is lipidated by N-myristoyltransferase (NMT) resulting in an amide bond that covalently attaches the two moieties (Farazi et al. 2001). The NMT motif sequence (MGXXXS/T) is present at the N-terminus of the substrate polypeptide. N-myristoylation is irreversible and persists for the life of the modified protein. Up to now, there are only two NMT proteins discovered in vertebrates, which play essential role in early development (Yang et al. 2005). NMT substrates can be predicted with high confidence by algorithm-based analysis due to target sequence conservation. Unlike myristoylation, palmitoylation is a reversible covalent attachment of the palmitic acid on an internal cysteine (S-palmitoylation) through a thioester bond (Smotrys and Linder 2004; Resh

2006). However, some proteins can be N-terminally palmitoylated through an amide bond (Kleuss and Krause 2003). The human genome encodes over 20 putative palmitoyl acyltransferases (PATs) (Greaves and Chamberlain; Fukata et al. 2006). These enzymes contain a conserved Asp-His-His-Cys (DHHC) sequence located in a cysteine-rich domain (CRD) within the catalytic core. Despite recent advances in novel biochemical approaches, the characterization of substrate specificities of different PATs remain a significant challenge, in part due to the lack of a highly conserved consensus substrate sequence.

3.2 Prenylation of CaaX Motif Proteins

Protein prenylation is a process involving several discrete enzymatic reactions performed by multiple enzymes that irreversibly conjugate an isoprenoid moiety through a thioether bond to one or more cysteine residues located within a C-terminal 'CaaX' (Cys—aliphatic—aliphatic—X) motif. The first step in the reaction is the covalent addition of a farnesyl (C-15) or a geranylgeranyl (C-20) isoprenoid moiety by one of two distinct prenyltransferase complexes in the cytosol—farnesyltransferase (FTase) or geranylgeranyltransferase I (GGTase-I) (Sebti 2005). The amino acid sequence of the CaaX motif as well as the neighboring residues determines enzyme binding and thus dictates substrates specificity. The next step is the proteolysis of the three terminal residues from the CaaX domain by the endopeptidase Rce1 (Schmidt et al. 1998; Kim et al. 1999) followed by the methylation of the exposed prenylated Cys by isoprenylcysteine carboxyl methyltransferase (ICMT) (Bergo et al. 2000) at the endoplasmic reticulum. The additional methylation step further increases membrane affinity of the prenylated protein (Kim et al. 1999; Bergo et al. 2000).

3.3 Membrane Affinity of Lipidated Proteins

Although lipidation increases protein hydrophobicity, binding studies clearly demonstrate that membrane affinity for singly modified polypeptides is not sufficient to promote stable membrane association (Peitzsch and McLaughlin 1993; Silvius and l'Heureux 1994). Thus, additional membrane-targeting determinants that increase membrane affinity are required for anchoring. Frequently, proteins encode tandem lipidation sites, such as prenylation/palmitoylation or myristoylation/palmitoylation, which increase membrane affinity sufficiently to provide stable association with a lipid bilayer (Sebti 2005; Resh 2006). A polybasic stretch of residues neighboring a lipidated amino acid also can enhance membrane affinity by providing an electrostatic interaction with negatively charged phospholipids (Sebti 2005; Resh 2006; Yeung et al. 2008). The composition of phospholipids in the bilayer provides a distinct electrostatic potential for each membrane-bound organelle in eukaryotes, which contributes to the spatial distribution of proteins in cells (Yeung et al. 2008; Bohdanowicz and Grinstein 2013). Thus, multiple factors determine membrane affinity and spatial compartmentalization of lipidated proteins.

3.4 Mechanisms of Dynamic Membrane Association of Lipidated Proteins

Several distinct mechanisms have evolved to dynamically control the presence of lipidated proteins on membrane interfaces. Central to all these mechanisms is the low membrane affinity of single lipidated substrates that allow for sampling of membrane leaflets until affinity is sufficiently increased to generate stable association.

3.4.1 Kinetic Trapping of Lipidated Proteins

Single lipidated proteins rapidly sample membrane interfaces with similar 'on/off' rates. However, the addition of a second lipid moiety on an already lipidated substrate significantly increases the membrane affinity essentially "trapping" the dual-lipidated protein on the membrane interface until the second modification is removed (Rocks et al. 2005; Resh 2006). Proteins that are regulated by kinetic trapping encode tandem irreversible and reversible lipidation sites. In this fashion, membrane affinity of an irreversibly prenylated or myristoylated substrate can be controlled by cycles of reversible palmitoylation. Spatial organization is driven by a palmitoylation/depalmitoylation cycle established by the presence of a lipid-modifying enzyme that results in substrates being trapped on organelle membranes containing specific palmitoyl transferases and released from compartments at which thioesterases are present. For example, farnesylated H-Ras is anchored at the Golgi compartment as a result of palmitoylation by a resident PAT, travels via the secretory pathway to the plasma membrane where an esterase removes the palmitate causing membrane dissociation and cycling of H-Ras (Rocks et al. 2005). In this manner, cells generate distinct pools of H-Ras via reversible lipidation mediated by spatially restricted modifying enzymes using the kinetic trapping mechanism.

3.4.2 Dynamic Membrane Association of Lipidated Proteins by Modulation of Electrostatic Potential

The electrostatic potential varies among the membranes of different intracellular compartments due to their phospholipids composition, which cells exploit to localize lipidated proteins. Stretches of polybasic residues encoded by singly lipidated proteins significantly stabilize membrane association by providing electrostatic interactions with negatively charged head groups of membrane phospholipids (Yeung et al. 2008). To control membrane association dynamically, cells can transiently decrease the charge of the polybasic regions by phosphorylation (Bivona et al. 2006) or protein–protein interactions (Fivaz and Meyer 2005). Prenylated K-Ras4B efficiently dissociates from the plasma membrane after protein kinase C phosphorylation of a single serine residue within the polybasic stretch thereby decreasing the electrostatic interactions with membrane-resident phospholipids (Bivona et al. 2006). This type of electrostatic switches can exploit different mechanisms to decrease the protein electrostatic potential including protein–protein interactions and post-translational modifications.

4 Host Prenyltransferases Facilitate Membrane Localization of *Legionella pneumophila* Effectors

Legionella pneumophila is a facultative intracellular bacterium that infects unicellular protozoa in the environment. Pulmonary infections with *Legionella* can produce severe atypical pneumonia in immunocompromised individuals. In infected cells *Legionella* subverts host trafficking pathways to create an ER-like membrane-bound compartment that supports bacterial replication. This process is mediated by a cohort of over 300 bacterial effector proteins translocated in the cytosol of infected cells by a Type IV secretion system (T4SS) system known as the Dot/Icm apparatus. *Legionella* effectors are translocated into the host cytosol at very low levels, which requires robust spatial control mechanisms to ensure that effectors localize to their substrates (Ensminger and Isberg 2009; Hubber and Roy 2010). Recently, a number of *Legionella* effectors have been shown to encode substrate domains for eukaryotic prenyltransferases that modify these effectors to promote their association with host membranes.

4.1 Identification of Legionella Effectors Containing Prenylation Domains

Multiple *Legionella* effectors encode membrane-targeting domains, including putative transmembrane domains, phospholipid-binding domains (Ragaz et al. 2008; Brombacher et al. 2009) and most recently discovered prenylation substrate domains (Ivanov et al. 2010; Price et al. 2010a, b). A bioinformatics approach led to the discovery of ten *Legionella* proteins containing putative CaaX motifs (Fig. 3). All *Legionella* CaaX motif proteins (CMPs) have been validated experimentally as Dot/Icm effectors, with the exception of Lpl2477 and Lpp1863, which have not been tested (Ivanov et al. 2010; Price et al. 2010b). Six of the bacterial

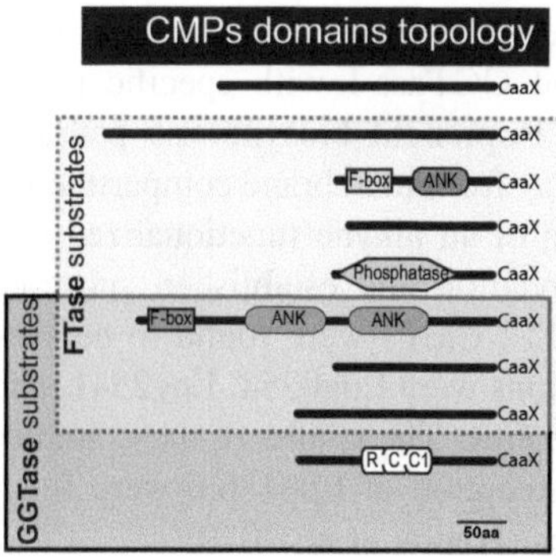

Philadelphia	Lens	Paris	Corby	130b	Lorraine	Alcoy
XXXXX	Lpl1858*	Lpp1863	Lpc_1344*	Lpw19321*	Lpo_1964	Lpa_02733*
XXXXX	Lpl2806	XXXXX	XXXXX	Lpw31621	XXXXX	XXXXX
Lpg2144	Lpl2072	Lpp2082*	Lpc_1593	Lpw23181	Lpo_2207	Lpa_03071
Lpg2375	XXXXX	Lpp2440	Lpc_2117	XXXXX	Lpo_2549	XXXXX
XXXXX	Lpl2477	XXXXX	XXXXX	XXXXX	XXXXX	XXXXX
Lpg2525	XXXXX	XXXXX	XXXXX	XXXXX	XXXXX	XXXXX
Lpg0254	Lpl0307	Lpp0324	Lpc_0331	Lpw03431	XXXXX	XXXXX
Lpg2541	Lpl2462	Lpp2607	Lpc_1927	Lpw27701	Lpo_2749	Lpa_03710
Lpg1976	Lpl1953	Lpp1959	Lpc_1462	Lpw20351	Lpo_2073	Lpa_02889

** Asterisk indicates a homologue that lacks a CaaX motif*

Fig. 3 Phylogeny and enzyme specificities for the Legionella CaaX motif effectors

CMPs are conserved with homologs present in the genomes of multiple *Legionella* strains (Fig. 3). Within this group of proteins, the CaaX substrate motifs were also highly conserved with two notable exceptions—*lpp2082* and *lpp1863*. Lpp2082 is a highly conserved homolog of Lpg2144 from the *Paris* strain that lacks the last ~20 amino acid residues, which include the CaaX motif. The loss of the CaaX domain by Lpp2082 raises the intriguing possibility that Lpp2082 and Lpg2144 function at distinct intracellular localizations in infected cells and might represent an example of functional diversification by loss of a membrane targeting domain. The other exception is Lpp1863, which has gained a putative CaaX motif by point mutation that is absent in other homologs. The *Legionella* CMPs encode several eukaryotic-like domains with diverse functions suggesting that these effectors might play distinct roles in *Legionella* pathogenesis. In this review we refer to the gene alleles from the different *Legionella* strains by their respective strain-specific nomenclature.

4.2 The CaaX Motifs Encoded by the Legionella CMPs are Membrane-Targeting Domains that Control Protein Localization

Convincing biochemical and pharmacological studies using transient transfections in eukaryotic cells demonstrate that the *Legionella* CaaX motifs identified in silico are substrates for host prenyltransferases that control subcellular localization by increasing membrane affinity (Ivanov et al. 2010; Price et al. 2010b). *Legionella* CMPs were found to associate with membranes and to localize at distinct membrane-bound compartments, such as the Golgi and the plasma membrane (Ivanov et al. 2010). Mutagenesis of the substrate cysteine in the CaaX motifs resulted in loss of membrane association and redistribution of the mutant alleles from membranes to the cytosol as shown by biochemical fractionation and immunofluorescence (Ivanov et al. 2010; Price et al. 2010a). Also, inactivation of the CaaX motif lipidating enzymes FTase and GGTase-I with specific peptidomimetic inhibitors or interference with the isoprenoid biosynthesis pathway dissociated and relocalized *Legionella* CMPs away from membrane compartments. Moreover, experiments measuring incorporation of an alkyne-functionalized isoprenol reporter have demonstrated direct prenylation and established substrate specificities for FTase or GGTase-I (Fig. 3). Most CMPs were found to be substrates of either FTase or GGTase-I. The exceptions were Lpg0254, Lpg2541, and Lpg2525, which could be modified by either enzyme. The results of these studies validated that all *Legionella* CMPs, with the exception of Lpp1863, were functional substrates of eukaryotic prenyltransferases (Ivanov et al. 2010).

Importantly, not all CMPs required prenylation for membrane association and intracellular localization. Despite being prenylated by a CaaX motif-dependent mechanism, Lpg2525 and Lpl2806 retained their localization and membrane affinity when their CaaX motifs were mutagenized or prenyltransferases were inhibited (Ivanov et al. 2010). One likely scenario, which remains to be investigated, is that prenylation regulates protein functions independent of localization. Such alternative regulatory mechanisms have been shown for G-protein coupled receptors and postsynaptic density-95 (PSD) protein, which mediate lipidation-induced changes in protein conformation or promote clustering within specific membrane micro-domains (El-Husseini Ael et al. 2002; Qanbar and Bouvier 2003).

Interestingly, several *Legionella* CMPs contain an upstream polybasic amino acid stretch indicating that electrostatic interactions might contribute to localization. Consistent with this idea, CMPs encoding a long polybasic region accumulated at the plasma membrane, which has the most negatively charged membrane leaflet, whereas, proteins with shorter regions localized to the Golgi (Ivanov et al. 2010). An electrostatic regulator has been validated for Lpg2144 using point mutations in the polybasic region to reduce the electrostatic potential, which did not affect membrane association but relocalized the protein from the plasma membrane to the Golgi compartment (Ivanov et al. 2010). It is intriguing to speculate that evolutional pairing of an electrostatic patch with a CaaX motif is advantageous for host-range expansion of a *Legionella* CMP because an electrostatic interaction should be functional in most, if not all, eukaryotic cells.

Despite the strong evidence for spatial control of ectopically produced *Legionella* CMPs by prenylation, demonstrating a similar role for lipidation on a CMP translocated into the host cytosol during infection presented a challenge because *Legionella* effectors are translocated at a very low levels during infection and thus it is difficult to monitor their subcellular distribution by standard fractionation or immunofluorescence methods. An elegant way to circumvent this obstacle was to tag a *Legionella* CMPs enzymatically with the adenylate cyclase domain of the *Bordetella pertussis* CyaA protein (Ladant and Ullmann 1999), which allowed the presence of the fusion protein within biochemical fractions of infected cells to be quantified by the enzymatic production of cAMP (Ivanov et al. 2010). This methodology was used to demonstrate that, at least for Lpg2144, the

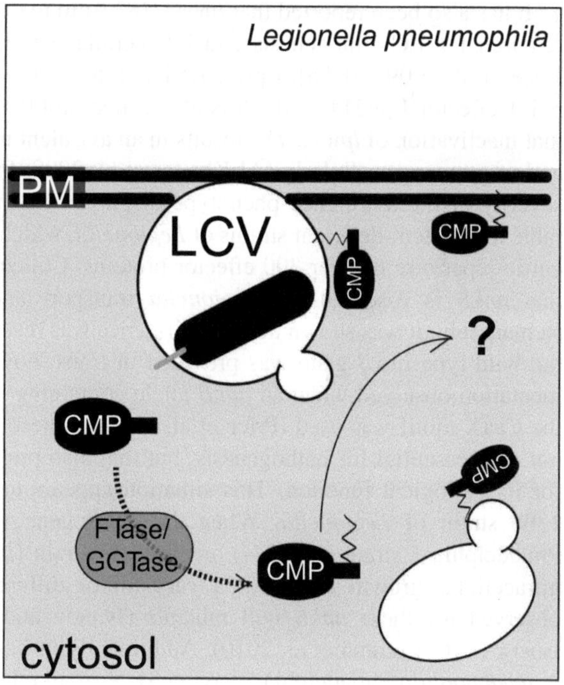

Fig. 4 Spatial regulation of Legionella CMPs

association with host membranes during infection required farnesylation of the cysteine residue of the CaaX motif, similar to the results obtained with the ectopically produced protein (Ivanov et al. 2010). However, on which membrane-bound compartments CMPs reside during infection is still unclear as illustrated in Fig. 4. Frequently, the localizations of ectopically produced and translocated effectors are different (Murata et al. 2006; Ninio et al. 2009). Thus, one should be cautious when interpreting spatial information obtained from ectopically expressed proteins.

4.3 Does Prenylation Regulate the Legionella CMPs Function?

The biochemical functions of individual CMPs remain largely unknown, which makes it difficult to determine whether the host prenylation machinery is required for effector functions. Based on evidence from many eukaryotic proteins, one can imagine that eliminating the spatial restriction imposed by lipidation would result in protein mislocalization, which should negatively affect function. This remains to be demonstrated formally for the *Legionella* CMPs. There is evidence, however, in support of this hypothesis: Inhibitors that disrupt the function of FTase or GGTase-I have been shown to interfere with the ability of *Legionella* to block endosomal maturation (Ivanov et al. 2010).

It has also been reported that *lpw-23181* from the *L. pneumophila* strain 130b is required for LCV biogenesis and intracellular survival (Al-Khodor et al. 2008; Price et al. 2009, 2010a). Lpw-23181 is a homolog of *L. pneumophila* Philadelphia1 effector Lpg2144, which is also called AnkB (Fig. 3). It has been reported that inactivation of *lpw-23181* results in an avirulent mutant that is unable to traffic and to grow intracellularly (Al-Khodor et al. 2008; Price et al. 2009, 2010a). The severity of the avirulence phenotype displayed by the *ankB* mutant was comparable to Dot/Icm-deficient strains of *Legionella*, which are unable to translocate the entire repertoire of over 300 effector proteins. Collectively, these studies indicate that *ankB* is essential for *Legionella* transport and replication in host cells. Remarkably, it was shown that full virulence was restored to an *ankB* mutant when the wild type *ankB* gene was provided in trans, however, there was no complementation observed when an *ankB* allele containing a point mutation inactivating the CaaX motif was used (Price et al. 2010a). These studies suggest that AnkB is not only essential for pathogenesis, but that also prenylation of AnkB is required for its biological function. This situation appears to be unique, however, to the 130b strain of *Legionella*. When the *ankB* gene was deleted from either the Philadelphia-1 strain (*lpg2144*) or the Paris strain (*lpp2082*), there was either no intracellular growth defect or a very minor difference in intracellular growth observed for these *ankB* null mutants (Ivanov and Roy 2009; Ensminger and Isberg 2010; Lomma et al. 2010). Additionally, it has been shown that a strain of *L. pneumophila* Philadelphia 1 having 31 % of all effector genes deleted, including *lpg2144*, retains the ability to replicate intracellularly (O'Connor et al. 2011). Thus, it is astonishing that a single mutation that eliminates the ability of AnkB to be prenylated is sufficient to render the *130b* strain avirulent, but this highlights the potential importance of host prenylation in controlling the function of *Legionella* effector proteins.

5 Mimicry of Eukaryotic Lipidation Substrates by Bacterial Effectors is a Conserved Strategy for Spatial Organization

Similar to *Legionella*, other bacterial pathogens can translocate effectors proteins into the cytosol of infected cells via specialized secretion apparatuses to manipulate host functions (Galan and Wolf-Watz 2006; Grant et al. 2006; Ensminger and Isberg 2009; Voth et al. 2012). Tight spatial control is required to ensure that bacterial effectors execute their functions at the appropriate subcellular compartment. Recent studies have discovered that *Salmonella* and *Pseudomonas* effectors also encode substrates motifs of various eukaryotic lipid transferases. The fact that several pathogens have evolved to subvert host lipidation highlights the importance of host lipid transferases for spatial control of translocated effectors. The principles of regulation uncovered for *Salmonella* and *Pseudomonas* effectors are likely common to *Legionella* effectors as well. For example, substrate promiscuity has been shown for *Legionella* CMPs as well as *Salmonella* effectors. Considering the large repertoire of uncharacterized effectors encoded by *Legionella*, it is possible that additional lipidation motifs similar to the ones found in *Salmonella* and *Pseudomonas* are present. Thus, the spatial control strategies uncovered for *Salmonella* and *Pseudomonas* effectors might also be conserved in *Legionella*.

5.1 Regulation of *Salmonella enterica* Effectors by Lipidation

The intracellular vacuolar pathogen *Salmonella enterica* is an enteric pathogen that expresses two type III secretion systems (T3SS), which deliver more than 60 effector proteins into the cytosol of infected cells to create a lysosomal-like vacuolar niche, known as the *Salmonella*-containing vacuole (SCV) that support bacterial proliferation (Galan and Wolf-Watz 2006; Bakowski et al. 2008). Similar to *Legionella* CMPs several *Salmonella* effectors require modification by host lipidtransferases for membrane association and correct subcellular localization (Reinicke et al. 2005; Hicks et al. 2011).

Three Salmonella T3SS effectors have been found to be substrates of host lipidtransferases—SseI, SspH2, and SifA (Reinicke et al. 2005; Hicks et al. 2011). These effectors perform distinct functions in the host cytosol during infection. SseI inhibits cell migration of immune cells by interacting with the scaffold protein IQ motif containing GTPase activating protein 1 (IQGAP1) (McLaughlin et al. 2009). SifA is a WxxxE motif effector, which functions together with the effector SseJ and the host proteins SKIP and RhoA to promote tubulation of the SCV membrane (Ohlson et al. 2008; McGourty et al. 2012). SspH2 is a novel type of E3 ubiquitin ligase enzyme that modifies yet unknown substrate(s) during infection (Quezada et al. 2009).

The membrane association of SifA is conferred by dual lipidation when the protein is expressed in eukaryotic cells (Reinicke et al. 2005). A prenylation site within a CaaX motif modified by GGTase-I and an adjacent palmitoylation site lipidated by unknown PAT have been identified. Although mutagenesis studies demonstrated that both modifications were required for strong membrane anchoring, the functional implications of the spatial organization brought by SifA lipidation has not been investigated systematically (Reinicke et al. 2005). Because SifA functions at a membrane interface to activate RhoA in a GEF-like fashion, one can expect that lipidation would be crucial for SifA function, however, the ability of a lipidation-deficient SifA mutant to activate RhoA or to promote SCV tubulation has not been investigated.

Unlike SifA, the role of lipidation in regulating effectors SseI and SspH2 is better characterized. SseI and SspH2 are palmitoylated on a common palmitoyltransferase substrate sequence (-HIGSGCLPA) at their N-terminus, which is required for membrane association (Hicks et al. 2011). Mutagenesis of the substrate cysteine completely abolished membrane affinity and redistributed the proteins from the plasma membrane to the cytosol (Hicks et al. 2011). SseI and SspH2 translocated in the host cytosol by *Salmonella* during infection have been shown to be palmitoylated. Importantly, palmitoylation controlled spatial organization of the effectors in vivo demonstrating that by mimicking host lipid transferases substrates *Salmonella* effectors acquired the capacity to associate with host membranes (Hicks et al. 2011). What are the functional implications of palmitoylation for SseI and SspH2? This question remains unanswered for SspH2 because its function remains unknown. However, palmitoylation was required for SseI-dependent inhibition of cell migration, providing convincing evidence that the spatial regulation of bacterial effector by host lipidtransferases is required for their function.

The minimum membrane targeting domain for SseI and SspH2 has been mapped to the N-terminal ~130 amino acids that encompass the palmitoylation motif. When ectopically expressed in polarized cells, these domains localized to the apical side of the plasma membrane. Full-length SspH2 maintained apical localization, however, full-length SseI redistributed to the basolateral side of the plasma membrane, reinforcing the idea that protein localization is ultimately a combination of multiple determinants (Hicks et al. 2011).

How is palmitoylation of the *Salmonella* effectors regulated? Several PATs appear to modify ectopically produced SspH2, of which DHHC-3 and DHHC-7 were demonstrated to lipidate SspH2 directly in vitro (Hicks et al. 2011). DHHC-3 and DHHC-7 have broad-spectrum substrate specificities and reside at the Golgi (Fernandez-Hernando et al. 2006) raising the possibility that lipidated SspH2 travels to the plasma membrane via the secretory pathway after it is modified at the Golgi as is speculated in Fig. 5. Another intriguing unexplored possibility is that a PAT might be recruited to the SCV to ensure efficient lipidation of bacterial effectors as they are translocated by the T3SS. Interestingly, palmitoylated SseI and SspH2 remain acylated for hours (Hicks et al. 2011) during infection raising the possibility that *Salmonella* effectors are somehow protected from host palmitoyl esterases and less likely to be regulated by kinetic trapping mechanisms. The long turnover of SspH2 and SseI lipidation contrasts the minutes scale rate of eukaryotic substrates (Rocks et al. 2005). Therefore, it might be advantageous for *Salmonella* effectors to stably maintain lipidation in order to prevent cycling off the lipid bilayer at their final destination compartment.

5.2 *Membrane Tethering of Pseudomonas syringae Effectors by Dual-Acylation*

Pseudomonas syringae is a phytopathogen that encodes a T3SS and translocates over 30 effector proteins into plant cells to suppress host defenses (Grant et al. 2006). Recently, a family of bacterial cysteine protease effectors has been shown to utilize an unusual spatial control mechanism (Dowen et al. 2009), which is best

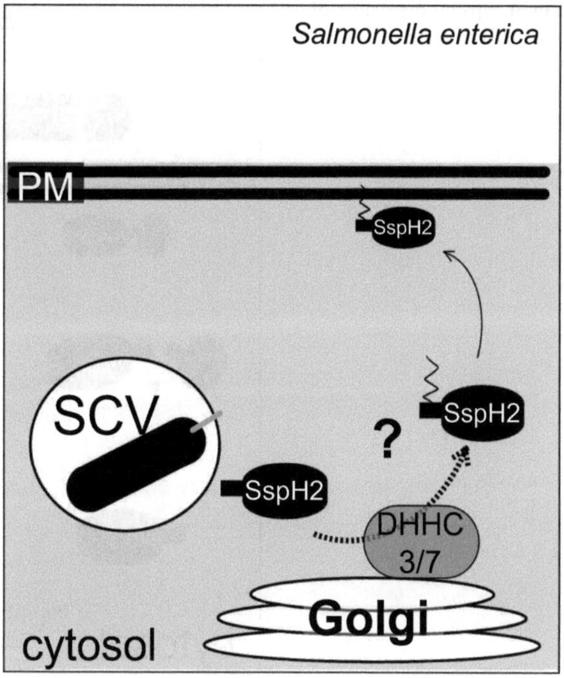

Fig. 5 Spatial regulation of SspH2 in Salmonella-infected cells

characterized for the effector AvrPphB (Fig. 6). When AvrPphB is translocated in the cytosol of plants cells, it cleaves the host serine-threonine protein kinase PBS1 at the plasma membrane (Shao et al. 2002, 2003). In resistant plants, cleavage of PBS1 by AvrPphB is detected by the host, which triggers a form of immune response known as hypersensitivity response (HR) that is manifested by localized programmed cell death at the infection site (Shao et al. 2003). AvrPphB encodes internal myristoylation and palmitoylation sites that are revealed by autoproteolysis upon translocation in the host cytosol (Nimchuk et al. 2000; Dowen et al. 2009). The mature AvrPphB becomes dually acylated and is tethered to the plasma membrane of the host cell where the AvrPphB substrate PBS1 resides. Autoproteolysis is required for host-dependent acylation of AvrPphB, which in turn localizes the translocated effector at the plasma membrane of plant cells (Dowen et al. 2009). Importantly, AvrPphB function is strictly dependent on the effector plasma membrane localization. Acylation-deficient mutant failed to proteolytically process PBS1 and failed to trigger an HR response. As indicated in Fig. 6 it is still unclear which host lipid transferases modify AvrPphB and where they are localized. Mechanism of AvrPphB-mediated proteolysis of PBS1 highlights the important role of host lipid transferases in localizing lipidated bacterial effectors.

AvrPphB autoproteolysis-dependent lipidation mechanism is unique for the AvrPphB-like effector family of bacterial proteases and represents an elegant solution for elimination of the T3SS secretion signal from the N-terminus of the protein once translocation into the host cell is completed and generation of an

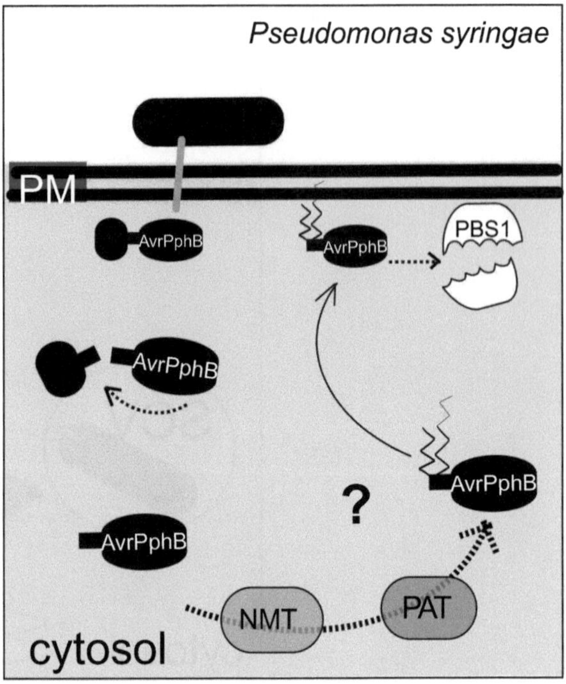

Fig. 6 Spatial regulation of AvrPphB in plant cells

N-terminal glycine required for lipidation by the host N-myristoyltransferase. This mechanism for spatial regulation, however, is not conserved in all members of the AvrPphB-like effector family. ORF4 from *P. syringae* and NopT from *Rhizobium* were both acylated after autoprocessing, whereas RipT from *Ralstonia* was not acylated even though autoproteolysis occurs and putative substrate sites are present (Dowen et al. 2009). Some family members, like HopC1 and HopN1, lacked auto-processing sites and were not acylated (Dowen et al. 2009). Despite the lack of a lipid moiety, all three effectors RipT, HopC1, and HopN1 localized to host membranes in plant cells indicating that additional mechanisms spatially regulate these effectors.

6 Conclusions and Perspectives

In this review, we highlight several examples from a diverse set of pathogenic bacteria producing polypeptides that become substrates for different host lipid transferases after they are translocated into the cytosol of infected cells. Host-dependent lipidation increases membrane affinity and primes bacterial effectors for coincidence detection-based compartmentalization. Such spatial regulatory mechanism is critical to localize the bacterial effector at the appropriate compartment where its target resides. Thus, perturbation of effector lipidation should be detrimental for its function. Although this paradigm has been formally demonstrated for AvrPphB and SseI, the functions of many effectors remain unknown. In the absence of known effector functions it remains difficult to address the biological significance of host-dependent lipidation. It is also likely that other pathogens apart from *Legionella*, *Salmonella*, and *Pseudomonas* exploit host lipidation to localize effectors in infected cells.

The capacity of bacterial effectors to utilize host-dependent lipidation machineries raises several interesting questions about how the acquisition and assembly of different protein domains drives the evolution of effector functions. Gain or loss of a lipidation domain should drastically affect the effector subcellular localization and as a result its function. The highly conserved *Legionella* homologs Lpg2144 and Lpp2082 represent one example where a CaaX motif required for membrane anchoring of Lpg2144 is absent from Lpp2082. Both effectors encode a eukaryotic-like F-box domain predicted to function by targeting substrate(s) for ubiquitinylation. Does the loss of the CaaX motif provide evolutionary diversification of effector functions by altering protein localization allowing the effector to gain access to alternative substrates? Or perhaps the loss of membrane affinity allows Lpp2082 to target a non-membrane substrate pool. Systematic analysis of the Lpg2144 and Lpp2082 substrates and their intracellular localization could yield some of these answers in the future.

Another issue we do not fully understand is the dynamics of the spatial regulation of lipidated effectors. It appears that effectors are either permanently lipidated or remain modified for very long period of time in case of palmitoylated proteins. There is evidence that at least some *Legionella* CMPs are wired for dynamic control based on electrostatic switches, however, the molecular regulators and the mechanisms of action would have to be elucidated. The *Salmonella* effectors SseI and SspH2 have a surprisingly prolonged palmitoylation cycle unsuitable for a rapid membrane cycling by a kinetic trapping mechanism. Thus, it remains to be elucidated whether SseI and SspH2 are generally very poor substrates of palmitoyl esterases or additional mechanisms are in place to maintain the effectors palmitoylated.

Now that it has been established clearly that host-dependent lipidation spatially regulates bacterial effectors, future efforts will have to focus on characterizing the role of these effectors in pathogenesis and elucidating the mechanisms of spatial control.

References

Al-Khodor S, Price CT, Habyarimana F, Kalia A, Abu Kwaik Y (2008) A Dot/Icm-translocated ankyrin protein of Legionella pneumophila is required for intracellular proliferation within human macrophages and protozoa. Mol Microbiol 70:908–923

Alexandrov K, Horiuchi H, Steele-Mortimer O, Seabra MC, Zerial M (1994) Rab escort protein-1 is a multifunctional protein that accompanies newly prenylated rab proteins to their target membranes. EMBO J 13:5262–5273

Bakowski MA, Braun V, Brumell JH (2008) Salmonella-containing vacuoles: directing traffic and nesting to grow. Traffic 9:2022–2031

Bergo MO, Leung GK, Ambroziak P, Otto JC, Casey PJ, Young SG (2000) Targeted inactivation of the isoprenylcysteine carboxyl methyltransferase gene causes mislocalization of K-Ras in mammalian cells. J Biol Chem 275:17605–17610

Bivona TG, Quatela SE, Bodemann BO, Ahearn IM, Soskis MJ, Mor A, Miura J, Wiener HH, Wright L, Saba SG, Yim D, Fein A, Perez de Castro I, Li C, Thompson CB, Cox AD, Philips MR (2006) PKC regulates a farnesyl-electrostatic switch on K-Ras that promotes its association with Bcl-XL on mitochondria and induces apoptosis. Mol Cell 21:481–493

Blumer J, Rey J, Dehmelt L, Mazel T, Wu YW, Bastiaens P, Goody RS, Itzen A (2013) RabGEFs are a major determinant for specific Rab membrane targeting. J Cell Biol 200:287–300

Bohdanowicz M, Grinstein S (2013) Role of phospholipids in endocytosis, phagocytosis, and macropinocytosis. Physiol Rev 93:69–106

Brombacher E, Urwyler S, Ragaz C, Weber SS, Kami K, Overduin M, Hilbi H (2009) Rab1 guanine nucleotide exchange factor SidM is a major phosphatidylinositol 4-phosphate-binding effector protein of Legionella pneumophila. J Biol Chem 284:4846–4856

Carlton JG, Cullen PJ (2005) Coincidence detection in phosphoinositide signaling. Trends Cell Biol 15:540–547

Cho W, Stahelin RV (2005) Membrane-protein interactions in cell signaling and membrane trafficking. Annu Rev Biophys Biomol Struct 34:119–151

Di Paolo G, De Camilli P (2006) Phosphoinositides in cell regulation and membrane dynamics. Nature 443:651–657

Dowen RH, Engel JL, Shao F, Ecker JR, Dixon JE (2009) A family of bacterial cysteine protease type III effectors utilizes acylation-dependent and -independent strategies to localize to plasma membranes. J Biol Chem 284:15867–15879

El-Husseini Ael D, Schnell E, Dakoji S, Sweeney N, Zhou Q, Prange O, Gauthier-Campbell C, Aguilera-Moreno A, Nicoll RA, Bredt DS (2002) Synaptic strength regulated by palmitate cycling on PSD-95. Cell 108:849–863

Ensminger AW, Isberg RR (2009) Legionella pneumophila Dot/Icm translocated substrates: a sum of parts. Curr Opin Microbiol 12:67–73

Ensminger AW, Isberg RR (2010) E3 ubiquitin ligase activity and targeting of BAT3 by multiple Legionella pneumophila translocated substrates. Infect Immun 78:3905–3919

Farazi TA, Waksman G, Gordon JI (2001) The biology and enzymology of protein N-myristoylation. J Biol Chem 276:39501–39504

Fernandez-Hernando C, Fukata M, Bernatchez PN, Fukata Y, Lin MI, Bredt DS, Sessa WC (2006) Identification of Golgi-localized acyl transferases that palmitoylate and regulate endothelial nitric oxide synthase. J Cell Biol 174:369–377

Fivaz M, Meyer T (2005) Reversible intracellular translocation of KRas but not HRas in hippocampal neurons regulated by Ca2 +/calmodulin. J Cell Biol 170:429–441

Fukata Y, Bredt DS, Fukata M (2006) Protein palmitoylation by DHHC protein family. In: Kittler JT, Moss SJ (eds) The dynamic synapse: molecular methods in ionotropic receptor biology. CRC Press, Boca Raton

Galan JE, Wolf-Watz H (2006) Protein delivery into eukaryotic cells by type III secretion machines. Nature 444:567–573

Grant SR, Fisher EJ, Chang JH, Mole BM, Dangl JL (2006) Subterfuge and manipulation: type III effector proteins of phytopathogenic bacteria. Annu Rev Microbiol 60:425–449

Greaves J, Chamberlain LH (2011) DHHC palmitoyl transferases: substrate interactions and (patho)physiology. Trends Biochem Sci 36:245–253

Hicks SW, Charron G, Hang HC, Galan JE (2011) Subcellular targeting of Salmonella virulence proteins by host-mediated S-palmitoylation. Cell Host Microbe 10:9–20

Hubber A, Roy CR (2010) Modulation of host cell function by Legionella pneumophila type IV effectors. Annu Rev Cell Dev Biol 26:261–283

Hutagalung AH, Novick PJ (2011) Role of Rab GTPases in membrane traffic and cell physiology. Physiol Rev 91:119–149

Ivanov SS, Charron G, Hang HC, Roy CR (2010) Lipidation by the host prenyltransferase machinery facilitates membrane localization of Legionella pneumophila effector proteins. J Biol Chem 285:34686–34698

Ivanov SS, Roy CR (2009) Modulation of ubiquitin dynamics and suppression of DALIS formation by the Legionella pneumophila Dot/Icm system. Cell Microbiol 11:261–278

Jean S, Kiger AA (2012) Coordination between RAB GTPase and phosphoinositide regulation and functions. Nat Rev Mol Cell Biol 13:463–470

Kim E, Ambroziak P, Otto JC, Taylor B, Ashby M, Shannon K, Casey PJ, Young SG (1999) Disruption of the mouse Rce1 gene results in defective Ras processing and mislocalization of Ras within cells. J Biol Chem 274:8383–8390

Kleuss C, Krause E (2003) Galpha(s) is palmitoylated at the N-terminal glycine. EMBO J 22:826–832

Konstantinopoulos PA, Karamouzis MV, Papavassiliou AG (2007) Post-translational modifications and regulation of the RAS superfamily of GTPases as anticancer targets. Nat Rev Drug Discov 6:541–555

Krauss M, Haucke V (2007) Phosphoinositides: regulators of membrane traffic and protein function. FEBS Lett 581:2105–2111

Kutateladze TG (2010) Translation of the phosphoinositide code by PI effectors. Nat Chem Biol 6:507–513

Ladant D, Ullmann A (1999) Bordatella pertussis adenylate cyclase: a toxin with multiple talents. Trends Microbiol 7:172–176

Lomma M, Dervins-Ravault D, Rolando M, Nora T, Newton HJ, Sansom FM, Sahr T, Gomez-Valero L, Jules M, Hartland EL, Buchrieser C (2010) The Legionella pneumophila F-box protein Lpp 2082 (AnkB) modulates ubiquitination of the host protein parvin B and promotes intracellular replication. Cell Microbiol 12:1272–1291

Machner MP, Isberg RR (2006) Targeting of host Rab GTPase function by the intravacuolar pathogen Legionella pneumophila. Dev Cell 11:47–56

McGourty K, Thurston TL, Matthews SA, Pinaud L, Mota LJ, Holden DW (2012) Salmonella inhibits retrograde trafficking of mannose-6-phosphate receptors and lysosome function. Science 338:963–967

McLaughlin LM, Govoni GR, Gerke C, Gopinath S, Peng K, Laidlaw G, Chien YH, Jeong HW, Li Z, Brown MD, Sacks DB, Monack D (2009) The Salmonella SPI2 effector SseI mediates long-term systemic infection by modulating host cell migration. PLoS Pathog 5:e1000671

Mizuno-Yamasaki E, Rivera-Molina F, Novick P (2012) GTPase networks in membrane traffic. Annu Rev Biochem 81:637–659

Murata T, Delprato A, Ingmundson A, Toomre DK, Lambright DG, Roy CR (2006) The Legionella pneumophila effector protein DrrA is a Rab1 guanine nucleotide-exchange factor. Nat Cell Biol 8:971–977

Nimchuk Z, Marois E, Kjemtrup S, Leister RT, Katagiri F, Dangl JL (2000) Eukaryotic fatty acylation drives plasma membrane targeting and enhances function of several type III effector proteins from Pseudomonas syringae. Cell 101:353–363

Ninio S, Celli J, Roy CR (2009) A Legionella pneumophila effector protein encoded in a region of genomic plasticity binds to Dot/Icm-modified vacuoles. PLoS Pathog 5:e1000278

O'Connor TJ, Adepoju Y, Boyd D, Isberg RR (2011) Minimization of the Legionella pneumophila genome reveals chromosomal regions involved in host range expansion. Proc Natl Acad Sci U S A 108:14733–14740

Ohlson MB, Huang Z, Alto NM, Blanc MP, Dixon JE, Chai J, Miller SI (2008) Structure and function of Salmonella SifA indicate that its interactions with SKIP, SseJ, and RhoA family GTPases induce endosomal tubulation. Cell Host Microbe 4:434–446

Peitzsch RM, McLaughlin S (1993) Binding of acylated peptides and fatty acids to phospholipid vesicles: pertinence to myristoylated proteins. Biochemistry 32:10436–10443

Price CT, Al-Khodor S, Al-Quadan T, Santic M, Habyarimana F, Kalia A, Kwaik YA (2009) Molecular mimicry by an F-box effector of Legionella pneumophila hijacks a conserved polyubiquitination machinery within macrophages and protozoa. PLoS Pathog 5:e1000704

Price CT, Al-Quadan T, Santic M, Jones SC, Abu Kwaik Y (2010a) Exploitation of conserved eukaryotic host cell farnesylation machinery by an F-box effector of Legionella pneumophila. J Exp Med 207:1713–1726
Price CT, Jones SC, Amundson KE, Kwaik YA (2010b) Host-mediated post-translational prenylation of novel dot/icm-translocated effectors of legionella pneumophila. Front Microbiol 1:131
Qanbar R, Bouvier M (2003) Role of palmitoylation/depalmitoylation reactions in G-protein-coupled receptor function. Pharmacol Ther 97:1–33
Quezada CM, Hicks SW, Galan JE, Stebbins CE (2009) A family of Salmonella virulence factors functions as a distinct class of autoregulated E3 ubiquitin ligases. Proc Natl Acad Sci U S A 106:4864–4869
Ragaz C, Pietsch H, Urwyler S, Tiaden A, Weber SS, Hilbi H (2008) The Legionella pneumophila phosphatidylinositol-4 phosphate-binding type IV substrate SidC recruits endoplasmic reticulum vesicles to a replication-permissive vacuole. Cell Microbiol 10:2416–2433
Reinicke AT, Hutchinson JL, Magee AI, Mastroeni P, Trowsdale J, Kelly AP (2005) A Salmonella typhimurium effector protein SifA is modified by host cell prenylation and S-acylation machinery. J Biol Chem 280:14620–14627
Resh MD (2006) Trafficking and signaling by fatty-acylated and prenylated proteins. Nat Chem Biol 2:584–590
Rocks O, Peyker A, Kahms M, Verveer PJ, Koerner C, Lumbierres M, Kuhlmann J, Waldmann H, Wittinghofer A, Bastiaens PI (2005) An acylation cycle regulates localization and activity of palmitoylated Ras isoforms. Science 307:1746–1752
Schmidt WK, Tam A, Fujimura-Kamada K, Michaelis S (1998) Endoplasmic reticulum membrane localization of Rce1p and Ste24p, yeast proteases involved in carboxyl-terminal CAAX protein processing and amino-terminal a-factor cleavage. Proc Natl Acad Sci U S A 95:11175–11180
Sebti SM (2005) Protein farnesylation: implications for normal physiology, malignant transformation, and cancer therapy. Cancer Cell 7:297–300
Shao F, Golstein C, Ade J, Stoutemyer M, Dixon JE, Innes RW (2003) Cleavage of Arabidopsis PBS1 by a bacterial type III effector. Science 301:1230–1233
Shao F, Merritt PM, Bao Z, Innes RW, Dixon JE (2002) A Yersinia effector and a Pseudomonas avirulence protein define a family of cysteine proteases functioning in bacterial pathogenesis. Cell 109:575–588
Silvius JR, l'Heureux F (1994) Fluorimetric evaluation of the affinities of isoprenylated peptides for lipid bilayers. Biochemistry 33:3014–3022
Smotrys JE, Linder ME (2004) Palmitoylation of intracellular signaling proteins: regulation and function. Annu Rev Biochem 73:559–587
Voth DE, Broederdorf LJ, Graham JG (2012) Bacterial Type IV secretion systems: versatile virulence machines. Future Microbiol 7:241–257
Yamasaki A, Menon S, Yu S, Barrowman J, Meerloo T, Oorschot V, Klumperman J, Satoh A, Ferro-Novick S (2009) mTrs130 is a component of a mammalian TRAPPII complex, a Rab1 GEF that binds to COPI-coated vesicles. Mol Biol Cell 20:4205–4215
Yang SH, Shrivastav A, Kosinski C, Sharma RK, Chen MH, Berthiaume LG, Peters LL, Chuang PT, Young SG, Bergo MO (2005) N-myristoyltransferase 1 is essential in early mouse development. J Biol Chem 280:18990–18995
Yeung T, Gilbert GE, Shi J, Silvius J, Kapus A, Grinstein S (2008) Membrane phosphatidylserine regulates surface charge and protein localization. Science 319:210–213

Phosphoinositide Lipids and the *Legionella* Pathogen Vacuole

Ina Haneburger and Hubert Hilbi

Abstract Subversion of vesicle trafficking is vital for intracellular survival of *Legionella pneumophila* within host cells. *L. pneumophila* produces several type IV-translocated effector proteins that modify components of the phagosomal membrane, in particular the phosphoinositide (PI) lipids. Within eukaryotic cells PIs co-define subcellular compartments and membrane dynamics. The generation, half-life, and localization of PI lipids are not only tightly regulated by the host cell, but also targeted and modulated by a number of *L. pneumophila* effectors. These effectors either anchor to PIs, directly modify the lipids, or recruit PI-metabolizing enzymes to the LCV membrane. Together, PI-subverting *L. pneumophila* effectors act jointly to promote the formation of a replication-permissive niche inside the host.

Abbreviations

AMPylase	Adenylyltransferase
ANTH domain	AP180 N-terminal homology domain
DAG	Diacylglycerol
ENTH domain	Epsin N-terminal homology domain
ER	Endoplasmic reticulum
FYVE domain	Domain occurring in Fab 1 (yeast orthologue of PIKfyve), YOTB, Vac 1 (vesicle transport protein), and EEA1
GDF	GDI displacement factor
GDI	Guanine nucleotide dissociation inhibitor
GEF	Guanine nucleotide exchange factor
LCV	*Legionella*-containing vacuole
LVA domain	*Legionella* vacuole association domain
MVB	Multivesicular body
PH domain	Pleckstrin homology domain

I. Haneburger · H. Hilbi (✉)
Medical Faculty, Max von Pettenkofer Institute, Ludwig-Maximilians University, Munich, Pettenkoferstrasse 9a, 80336 Munich, Germany
e-mail: hilbi@mvp.uni-muenchen.de

Current Topics in Microbiology and Immunology (2013) 376: 155–173
DOI: 10.1007/82_2013_341

Published Online: 6 August 2013

PHOX domain	Phagocyte NADPH oxidase domain
PI	Phosphoinositide
PI3K	Phosphoinositide 3-kinase
PtdIns	Phosphatidylinositol
PX domain	PHOX homology domain
SCV	*Salmonella*-containing vacuole
T2SS	Type II secretion system
T4SS	Type IV secretion system
TGN	*Trans*-Golgi network

Contents

1 Introduction ... 156
2 Regulation of Membrane Dynamics by Phosphoinositide Lipids ... 157
3 Subversion of Phosphoinositide Metabolism by Intracellular Bacterial Pathogens ... 159
4 Phosphoinositides and Intracellular Replication of *L. pneumophila* ... 162
4.1 PI Lipids as Membrane Anchors for L. pneumophila Effectors ... 162
4.2 Modulation of the LCV PI Pattern ... 165
4.3 Role of PI 3-kinases for L. pneumophila Uptake and Replication ... 167
5 Conclusions ... 168
References ... 169

1 Introduction

Intracellular bacteria benefit from a number of advantages that their eukaryotic cellular niche can provide. The host cell represents a potentially rich source of nutrients, it protects the bacteria from free-living environmental predators and warrants that only few bacterial species compete for resources (Hilbi et al. 2007; Cosson and Soldati 2008). However, these advantages come with a price, and in order to thrive within their host cells the bacteria had to evolve intricate and distinct survival strategies. Intracellular bacteria adapt by either (i) acclimate to the acidic bactericidal lysosome, (ii) escape the phagosome and grow inside the host cell cytoplasm, or (iii) evade the degradative pathway and create a replication-permissive organelle.

Legionella pneumophila is a facultative intracellular bacterium that establishes inside phagocytic host cells a membrane-bound, replication-permissive compartment termed the *Legionella*-containing vacuole (LCV). In order to create this replicative niche, *L. pneumophila* manipulates host cell vesicle trafficking pathways to circumvent fusion of the phagosome with lysosomes and to establish a replicative organelle with characteristics of early and late endosomes as well as the endoplasmic reticulum (ER) (Kagan et al. 2004; Urwyler et al. 2009; Hilbi and Haas 2012). In the course of this process, LCVs also undergo a morphological transition from tight to spacious vacuoles, where the phagosome membrane detaches from the bacteria and only stays connected at the bacterial poles (Li et al. 2005; Lu and Clarke 2005). To promote LCV formation, *L. pneumophila* employs a type IV secretion system (T4SS), the so-called Icm/Dot (intracellular multiplication/defective in organelle transport) machinery (Marra et al. 1992; Berger and Isberg 1993), which translocates approximately 300 different effector proteins into host cells (Hubber and Roy 2010; Zhu et al. 2011; Lifshitz et al. 2013). While the Icm/Dot T4SS is essential for intracellular replication, most effectors are dispensable due to a high redundancy. Indeed, simultaneous deletion of up to 31 % of the *L. pneumophila* Icm/Dot substrates had only minor effects on growth in mouse macrophages (O'Connor et al. 2011). The molecular function of the majority of these effectors is not understood so far, but several of them have been shown to interact with different host phosphoinositide (PI) lipids and to modulate the PI pattern of the LCV. This topic will be discussed in detail in this chapter.

2 Regulation of Membrane Dynamics by Phosphoinositide Lipids

In eukaryotic cells, the different subcellular compartments are defined by their particular lipid and protein composition. Trafficking between distinct organelles is tightly regulated to ensure organelle identity and the delivery of cargo from a donor to the correct recipient compartment. This organelle "branding" is partly accomplished by different PI lipids. PIs occur with a low abundance in cellular membranes and constitute less than 10 % of total cellular phospholipids. PI lipids are derivatives of phosphatidylinositol (PtdIns) comprising a diacylglycerol (DAG) moiety and a D-*myo*-inositol 1-phosphate head group, which can be phosphorylated and dephosphorylated at positions 3′, 4′, and/or 5′ (Payrastre et al. 2001; Di Paolo and De Camilli 2006; Saarikangas et al. 2010) (Fig. 1). PIs are highly interconvertable, and every organelle is equipped with a distinct array of PI-metabolizing enzymes (PI kinases and PI phosphatases) to establish compartmentalization within the cell (De Matteis and Godi 2004; Sasaki et al. 2009). The spatial and temporal production and consumption of PIs is precisely balanced in order to establish and maintain the differential cellular and subcellular patterns.

PIs localize to the cytoplasmic leaflet of membranes and, together with small GTPases, recruit specific downstream effector proteins. The majority of PI-related enzymes is cytosolic and the mechanisms how they are targeted to the membrane are only partially understood (De Matteis and Godi 2004). PI-metabolizing and PI-interacting enzymes can bind to their targets via different PI lipid-binding modules (Lemmon 2008), such as the FYVE (Simonsen et al. 1998), PHOX

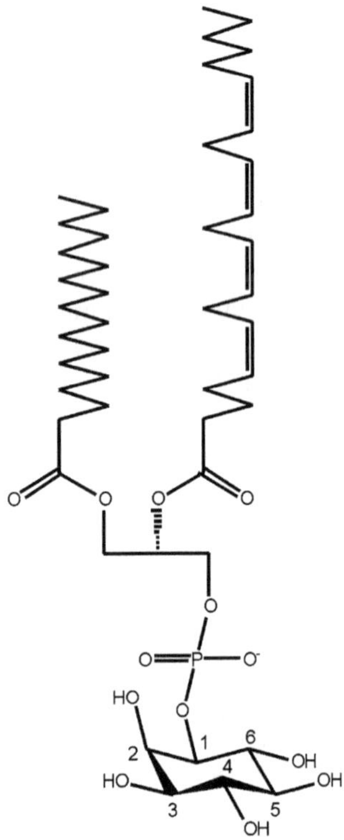

Fig. 1 Chemical structure of phosphoinositides. Phosphoinositide (PI) lipids are derivatives of phosphatidylinositol (PtdIns) comprising a diacylglycerol (DAG) moiety and a D-*myo*-inositol 1-phosphate head group, which can be phosphorylated by kinases at the positions 3, 4, and/or 5 of the D-*myo*-inositol ring, dephosphorylated by phosphatases, or hydrolyzed by phospholipases. The lipid moieties may vary, but mostly stearic acid and arachidonic acid are found. PI lipids can function as second messengers or as membrane anchors for endogenous eukaryotic proteins or bacterial effectors

homology (PX) (Wishart et al. 2001), pleckstrin homology (PH) (Wang and Shaw 1995), ENTH (De Camilli et al. 2002), and ANTH (Ford et al. 2001) domains. Peripheral membrane proteins harboring these domains, together with the PI composition of a cellular membrane, determine to a large part the identity of organelles, as well as the vesicle trafficking routes within eukaryotic cells (Behnia and Munro 2005; Di Paolo and De Camilli 2006).

Often, PI-metabolizing enzymes themselves are recruited to subcellular compartments via the action of peripheral membrane proteins, such as small GTPases. For example Rab5, a small GTPase localizing to endosomes, activates the endosomal type III phosphoinositide 3-kinase (PI3K) generating PtdIns(3)*P* (Christoforidis et al. 1999; Murray et al. 2002), whereas Arf6 recruits phosphoinositide 5-kinase (PI5K) to the plasma membrane (Krauss et al. 2003). Small GTPases localize to their target membrane through the concerted action of guanine nucleotide dissociation inhibitor (GDI), GDI displacement factor (GDF), and guanine nucleotide exchange factor (GEF). Distinct GEFs activate specific small GTPases by exchanging GDP with GTP. The hydrolysis of GTP, catalysed by the interaction with GAP proteins (GTPase activating protein), inactivates the small GTPase. The interaction of most GTPases with the target membrane is realized through the addition of hydrophobic prenyl moieties to the enzyme.

Different PI lipids are functionally coupled, as the lipid product produced by a given PI-metabolizing enzyme at the membrane of a certain compartment can serve as a substrate for "downstream" lipid-converting enzymes. In such a way, a framework is created that is essential for efficient and correct membrane trafficking. At the plasma membrane, PtdIns(4,5)P_2 and PtdIns(4)P are enriched, and during signal transduction or phagocytosis PtdIns(3,4,5)P_3 transiently accumulates. PtdIns(4)P predominantly occurs at the Golgi apparatus and on secretory vesicles as well as at the ER (Behnia and Munro 2005; Blumental-Perry et al. 2006; Di Paolo and De Camilli 2006). PtdIns(3)P is characteristic for phagosomes, early endosomes, and multivesicular bodies (MVBs). In addition to PtdIns(3)P, the membranes of MVBs and late endosomes also contain PtdIns(3,5)P_2.

3 Subversion of Phosphoinositide Metabolism by Intracellular Bacterial Pathogens

As outlined above, PIs are crucial for the temporal and spatial regulation of many host cell processes and as such are often targeted by intracellular bacteria. In order to generate a distinct replication-permissive organelle, intracellular bacterial pathogens such as *Mycobacterium tuberculosis*, *Salmonella enterica* serovar Typhimurium, or *Shigella flexneri* manipulate the PI pattern of cellular membranes and subcellular compartments. By mimicking the lipid identity of host cell organelles, the bacteria avoid being degraded by the host lysosomal or autophagosomal defense machinery, employing a strategy that has been termed "identity theft" (Behnia and Munro 2005). For an overview, see also Table 1 and earlier reviews (Pizarro-Cerda and Cossart 2004; Hilbi 2006; Weber et al. 2009b; Ham et al. 2011).

During uptake, *M. tuberculosis* impedes the maturation of the phagosome into a phagolysosome by the secretion of the PtdIns(3)P phosphatase SapM (Vergne et al. 2004). SapM was originally described to be a secreted acid phosphatase that dephosphorylates a wide variety of organic phosphoesters (Saleh and Belisle 2000). However, SapM also dephosphorylates PtdIns(3)P to PtdIns and thus reduces the amount of PtdIns(3)P on the *Mycobacterium*-containing phagosome (MCV). Due to lower amounts of PtdIns(3)P, the fate of the phagosome is altered and fusion with lysosomes is prevented (Vergne et al. 2005). A second secreted phosphatase from *M. tuberculosis*, MptpB, is characterized by a broad substrate range and promotes intracellular growth and MCV formation (Beresford et al. 2007; Beresford et al. 2009).

S. enterica Typhimurium employs a PI phosphatase to invade, survive, and replicate inside host cells. The type III-secreted phosphatase SopB (also known as SigD) is required for fission of *Salmonella*-triggered invaginations by reducing the concentration of PtdIns(4,5)P_2, and consequently, weakening membrane-cytoskeleton interactions (Terebiznik et al. 2002). After phagosome closure, SopB

Table 1 Bacterial effectors subverting the PI pattern of pathogen-containing vacuoles

Effector	Substrate(s)/ ligand(s)	Product(s)	Activity and cellular functions	References
L. pneumophila				
LidA	PtdIns(3)*P*, PtdIns(4)*P*	Not modified	Rab interactor; activation of Rab GTPases	(Conover et al. 2003; Brombacher et al. 2009; Schoebel et al. 2011; Neunuebel et al. 2012)
LpdA, LecE	PI binding not assessed	PtdIns(4)*P* (indirect)	Phospholipase D; production of DAG Activates PA phosphatase	(Viner et al. 2012)
LpnE	PtdIns(3)*P*	Not modified	Binds OCRL1; promotes uptake and intracellular growth	(Newton et al. 2006, 2007; Weber et al. 2009a)
SetA	PtdIns(3)*P*	Not modified	Glycosyltransferase; interference with vesicle trafficking	(Jank et al. 2012)
SidC/ SdcA	PtdIns(4)*P* (PtdIns(3)*P*)	Not modified	Adaptor (?); recruitment of ER vesicles	(Luo and Isberg 2004; Weber et al. 2006; Ragaz et al. 2008)
SidF	PtdIns(3,4)P_2, PtdIns(3,4,5)P_3	PtdIns(4)*P*, PtdIns(4,5)P_2	PI phosphatase; modulation of LCV PI pattern	(Banga et al. 2007; Hsu et al. 2012)
SidM (DrrA)	PtdIns(4)*P*	Not modified	Rab1 GDF/GEF, Rab1 AMPylase; activation of Rab GTPases	(Machner and Isberg 2006; Ingmundson et al. 2007; Brombacher et al. 2009; Schoebel et al. 2010)
RalF	No PI binding	PtdIns(4)*P* (indirect)	Arf 1 GEF; recruitment of Arf1 to LCV	(Nagai et al. 2002; Brombacher et al. 2009)
M. tuberculosis				
MptpB	PtdIns(3)*P*, PtdIns(4)*P*, PtdIns(5)*P*, Phospho-Tyr	PtdIns, Tyr	PI/protein phosphatase; phagosome maturation arrest	(Beresford et al. 2007; Beresford et al. 2009)
SapM	PtdIns(3)*P*	PtdIns	PI phosphatase; phagosome maturation arrest	(Saleh and Belisle 2000; Vergne et al. 2005)

(continued)

Table 1 (continued)

Effector	Substrate(s)/ligand(s)	Product(s)	Activity and cellular functions	References
S. Typhimurium				
SopB (SigD)	PtdIns(4,5)P_2 (PtdIns(3,4)P_2, PtdIns(3,4,5)P_3)	PtdIns(5)P	PI phosphatase; membrane fission during uptake, phagosome maturation arrest	(Terebiznik et al. 2002; Hernandez et al. 2004; Mason et al. 2007; Mallo et al. 2008; Bakowski et al. 2010)
S. flexneri				
IpgD	PtdIns(4,5)P_2	PtdIns(5)P	PI phosphatase; membrane fission during uptake	(Niebuhr et al. 2002)

modulates the trafficking of the *Salmonella*-containing vacuole (SCV) by maintaining high levels of PtdIns(3)*P* (Hernandez et al. 2004), which is achieved by recruiting Rab5 and the PI3K Vps34 (Mallo et al. 2008). In addition, SopB interacts with the essential host cell Rho GTPase Cdc42 by mimicking a host GDI (Burkinshaw et al. 2012). Cdc42 regulates vital steps in eukaryotic cytoskeletal organization and membrane trafficking. By diminishing the amount of PtdIns(4,5)P_2 and phosphatidylserine, SopB also controls the overall surface charge of SCVs, which leads to dissociation of numerous host cell proteins from the membrane and prevents fusion of the SCV with lysosomes (Bakowski et al. 2010).

The enteroinvasive bacterium causing bacillary dysentery, *Shigella flexneri*, also possesses a type III-secreted PI phosphatase. The enzyme, termed IpgD, is injected into the host cell during the early stages of infection. IpgD is a PtdIns(4,5)P_2 phosphatase that generates PtdIns(5)*P* upon hydrolysis of the substrate. The decrease in PtdIns(4,5)P_2 at the plasma membrane reduces cortical tethering forces and weakens the connection between the cytoskeleton and the membrane, eventually causing membrane blebbing (Niebuhr et al. 2002). The decline in PtdIns(4,5)P_2 also influences the distribution of membrane-cytoskeleton protein crosslinkers, which interferes with cell polarization and thus inhibits the migration of T cells (Konradt et al. 2011).

4 Phosphoinositides and Intracellular Replication of *L. pneumophila*

4.1 PI Lipids as Membrane Anchors for L. pneumophila Effectors

L. pneumophila translocates approximately 300 different effector proteins through its Icm/Dot T4SS into host cells, some of which localize to the cytoplasmic face of LCVs. Several of these effectors anchor to the LCV membrane by binding to specific PI lipids, namely PtdIns(4)*P* or PtdIns(3)*P* (Weber et al. 2009b; Hilbi et al. 2011). The 105 kDa Icm/Dot substrate SidC and its paralogue SdcA localize to the LCV membrane (Luo and Isberg 2004) and represent the first *Legionella* effectors shown to bind to PtdIns(4)*P* (and much weaker also to PtdIns(3)*P*) (Weber et al. 2006). The association of SidC with PtdIns(4)*P* is realized through a unique 20 kDa C-terminal binding domain (P4C) that does not show similarity to eukaryotic PI-binding motifs. The 70 kDa N-terminal domain, which is predicted to form coiled-coils, interacts with ER-derived vesicles and recruits ER membranes to the LCV (Table 1, Fig. 2) (Ragaz et al. 2008). LCVs harboring an *L. pneumophila* mutant strain lacking *sidC* and *sdcA* interact slower and to a smaller extent with the ER and do not undergo the transition from tight to spacious vacuoles (Weber et al. 2006).

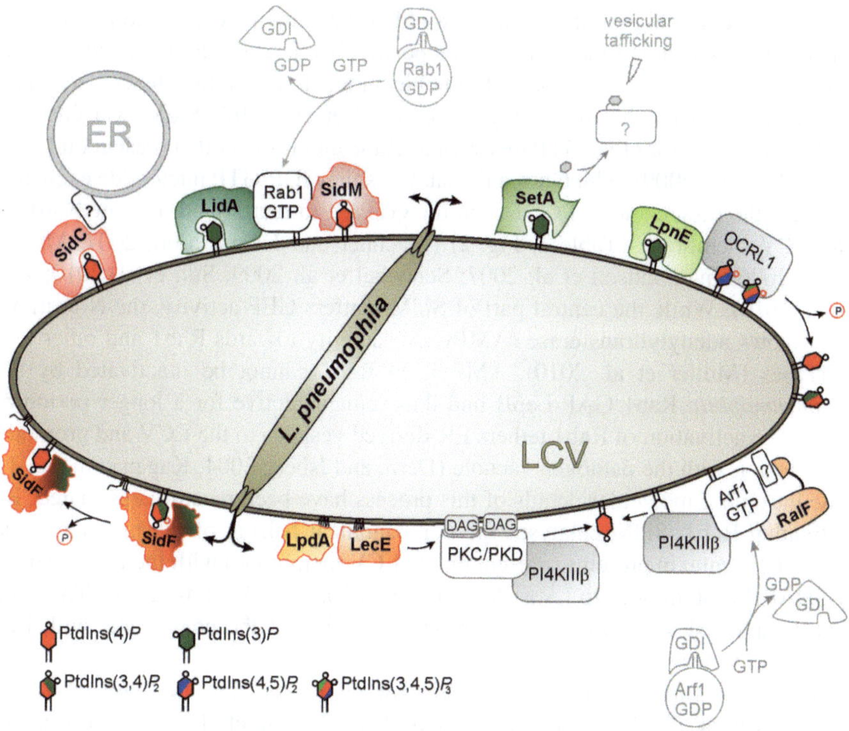

Fig. 2 *L. pneumophila* effector proteins anchoring to and manipulating host PI lipids. The *L. pneumophila* Icm/Dot substrates SidC and SidM bind to PtdIns(4)*P* and recruit ER-derived vesicles or the small GTPase Rab1, respectively. SidM activates Rab1 through its GEF and AMPylase activities. The Icm/Dot substrate LidA anchors to LCVs by binding to PtdIns(3)*P* as well as to PtdIns(4)*P* and stabilizes activated Rab1. The glycosyltransferase SetA is another PtdIns(3)*P*-binding Icm/Dot substrate. LpnE is secreted by an unknown mechanism and also binds PtdIns(3)*P*. The LCV PI pattern, i.e., an accumulation of PtdIns(4)*P*, is catalysed by *L. pneumophila* effectors that directly modulate PIs or recruit host PI-metabolizing enzymes. SidF is an Icm/Dot-translocated PI 3-phosphatase that associates with the LCV via transmembrane regions and dephosphorylates PtdIns(3,4)P_2 and PtdIns(3,4,5)P_3 to PtdIns(4)*P* or PtdIns(4,5)P_2. The Icm/Dot substrates LpdA and LecE bind to LCVs and promote the production of phosphatidic acid and diacylglycerol(DAG), respectively, which through the recruitment of protein kinase C and D (PKC, PKD), and finally binding of PI 4-kinase IIIβ (PI4KIIIβ) produces PtdIns(4)*P*. The type IV-translocated Arf1 GEF RalF anchors through an unknown factor to the LCV and recruits Arf1, which in turn also interacts with PI4KIIIβ. Finally, LpnE binds OCRL1, a PI 5-phosphatase that dephosphorylates PtdIns(4,5)P_2 to yield PtdIns(4)*P*. (?) denotes unknown factors. See text for details

Another PI-binding *L. pneumophila* effector is SidM (also termed DrrA), a 73 kDa Icm/Dot substrate that localizes to the LCV membrane already early during infection (Ingmundson et al. 2007). SidM binds to PtdIns(4)*P* through a C-terminal domain (P4M), which is unrelated to P4C; yet the effector competes with SidC for binding to LCVs (Brombacher et al. 2009). PtdIns(4)*P* is bound by

SidM with a very high affinity (K_d = 18 nM) via electrostatic interactions between the negatively charged membrane and a positively charged arginine residue in the P4M domain (R544) and via hydrophobic interactions of two leucine residues (L610, L614) with the lipid bilayer (Schoebel et al. 2010). SidM is a GEF for Rab1, as well as for other GTPases such as Rab8 and Rab14 (also present on LCVs (Urwyler et al. 2009), which not only catalyses the GDP-GTP nucleotide exchange but also the dissociation of GDI from the GTPase to recruit and activate Rab1 on the LCV membrane (Table 1, Fig. 2) (Machner and Isberg 2006, 2007; Murata et al. 2006; Ingmundson et al. 2007; Schoebel et al. 2009; Suh et al. 2010; Zhu et al. 2010). While the central part of SidM confers GEF activity, the N-terminal part shows adenylyltransferase (AMPylase) activity towards Rab1 and other Rab GTPases (Müller et al. 2010). AMPylated Rab1 cannot be inactivated by the *L. pneumophila* Rab1 GAP LepB und thus remains active for a longer period of time. The activation of Rab1 tethers ER-derived vesicles to the LCV and promotes their fusion with the pathogen vacuole (Derre and Isberg 2004; Kagan et al. 2004). The intriguing molecular details of this process have been recently elucidated. By activating Rab1, SidM catalyses the non-canonical pairing of plasma membrane t-SNARE syntaxin proteins (present on the LCV membrane) with the ER-localized v-SNARE protein Sec22b (Arasaki and Roy 2010; Arasaki et al. 2012). Thus, the SidM-catalysed activation of Rab1 promotes fusion of the phagosome with ER-derived vesicles.

LidA is a further Icm/Dot substrate that modulates Rab1 activity on LCVs (Conover et al. 2003; Machner and Isberg 2006; Murata et al. 2006) and directly binds to PIs. Through its central domain, LidA preferentially binds to PtdIns(3)*P* but to a lesser extent also to PtdIns(4)*P* (Brombacher et al. 2009; Neunuebel et al. 2012). The 83 kDa effector LidA not only targets Rab1 but also interacts with several other host Rab GTPases, including Rab1, Rab8, and Rab6A (Machner and Isberg 2006; Schoebel et al. 2011; Cheng et al. 2012; Chen and Machner 2013). The effector appears early after infection on the LCV membrane and supports SidM-dependent recruitment of Rab1 (Machner and Isberg 2006). LidA binds with equal affinity to the GDP- and GTP- bound state of Rab1, in either case stabilizing the active conformation of the GTPase (Machner and Isberg 2006; Cheng et al. 2012) and also interacts with the AMPylated form of Rab1, thus preventing inactivation by GAPs (Müller et al. 2010; Schoebel et al. 2011).

The 73 kDa Icm/Dot substrate SetA has been identified in a screen for *L. pneumophila* proteins that interfere with vesicular trafficking of eukaryotic cells (Heidtman et al. 2009). Ectopically expressed SetA localizes to endosomes and LCVs through a C-terminal domain comprising amino acids 401-644 (Heidtman et al. 2009; Jank et al. 2012). This C-terminal domain specifically binds to PtdIns(3)*P* that is enriched on early endosomal membranes (Table 1, Fig. 2) (Jank et al. 2012). The N-terminal portion of SetA shows similarity to glycosyl transferases, and indeed, UDP-glucose serves as a substrate for glycosylation of histones in vitro. Since only the full-length protein is toxic for host cells, both PI-binding and glycosylation are required for its cellular function.

Finally, LpnE is an *L. pneumophila* virulence factor that also rather selectively binds to PtdIns(3)*P* (Weber et al. 2009a). The 41 kDa protein is secreted into the culture supernatant by an unknown mechanism, as translocation is apparently independent of the Icm/Dot T4SS or the Lsp T2SS (Newton et al. 2006, 2007). *L. pneumophila* lacking LpnE is impaired for infection of *Acanthamoeba castellanii* and entry into human macrophage-like cells. Moreover, intracellular trafficking as well as virulence in the A/J mouse strain is compromised in the absence of *lpnE*. LpnE contains a so-called Sel1 repeat motif that is thought to mediate protein–protein interactions (Newton et al. 2007). Indeed, LpnE was shown to interact with the N-terminus of the human PI 5-phosphatase OCRL1 ($OCRL1_{1-236}$) heterologously produced in *D. discoideum* and in turn, LpnE was precipitated from *L. pneumophila* lysates by purified GST-$OCRL1_{1-236}$ (Weber et al. 2009a).

4.2 Modulation of the LCV PI Pattern

L. pneumophila effector proteins might modulate the LCV PI pattern in different ways. Icm/Dot-translocated factors possibly (i) represent bacterial PI phosphatases or kinases, (ii) directly bind (and recruit, activate, or inhibit) host PI-metabolizing enzymes, (iii) activate small host GTPases or other host factors, thus indirectly modulating the activity of PI phosphatases or kinases, or (iv) titrate (mask) LCV PIs (Hilbi et al. 2011). Moreover, *L. pneumophila* T4SS-translocated effectors might act either *in cis* (on the LCV membrane) or *in trans* (in a distance from the LCV). Given the intricate spatial and temporal control of PI metabolism, *L. pneumophila* likely subverts the PI pattern primarily on LCVs. Thus far, evidence has been obtained that *L. pneumophila* modulates the LCV PI pattern through Icm/Dot substrates, which (i) are a PI phosphatase (SidF), (ii) bind and recruit a host PI phosphatase (LpnE), or (iii) activate small GTPases or other factors, which in turn recruit PI-metabolizing host enzymes (RalF, SidM, LpdA, and LecE).

Among the *L. pneumophila* effectors described so far, SidF is the only one that directly metabolizes PI lipids (Hsu et al. 2012). Originally, the Icm/Dot substrate SidF has been described as a suppressor of host cell death (Banga et al. 2007). The phosphatase activity of SidF was identified by screening *L. pneumophila* effectors for the presence of a "CX_5R" motif (Hsu et al. 2012) that is characteristic for PI phosphatases (Norris et al. 1998). The 102 kDa SidF possesses a large N-terminal domain (amino acids 1-760) and two predicted C-terminal transmembrane domains that localize the protein to the LCV membrane upon translocation. Crystal structure analysis of the N-terminal catalytic domain in complex with its substrate PtdIns(3,4)P_2 revealed that like in other PI phosphatases, the catalytic center comprises a positively charged groove (Hsu et al. 2012). Two hydrophobic loops protrude from the protein, putatively inserting into the membrane bilayer and thus positioning the catalytic groove close to the membrane. Upon binding of PtdIns(3,4)P_2, a conformational shift is induced that leads to the formation of a highly cationic pocket able to accommodate the D4 phosphate group of the PI.

SidF specifically hydrolyses PtdIns(3,4)P_2 and PtdIns(3,4,5)P_3 typically occurring on early phagosomes. Thus, through the action of SidF and perhaps in combination with host cell enzymes the LCV lipid identity is shifted towards a PtdIns(4)P-rich compartment that resembles the *cis*-Golgi prone to interact with ER vesicles (Table 1, Fig. 2). Indeed, deletion of SidF decreased the recruitment of PtdIns(4)P-anchoring effectors, in agreement with a SidF-dependent accumulation of PtdIns(4)P on the LCV (Hsu et al. 2012).

The *L. pneumophila* PtdIns(3)P-binding virulence factor LpnE binds the human PI 5-phosphatase OCRL1 (Oculocerebrorenal syndrome of Lowe 1). OCRL1 and its *Dictyostelium* homologue Dd5P4 (*D. discoideum* 5-phosphatase 4) are PI 5-phosphatases and hydrolyse PtdIns(4,5)P_2 as well as PtdIns(3,4,5)P_3 to yield PtdIns(4)P and PtdIns(3,4)P_2, respectively (Zhang et al. 1995; Loovers et al. 2007). ORCL1 can functionally restore the lack of Dd5P4 in a *D. discoideum* mutant strain (Loovers et al. 2007). Dd5P4 restricts intracellular growth of *L. pneumophila* and localizes to the LCV via a 132 amino acid N-terminal domain termed *Legionella* vacuole association (LVA) domain (Weber et al. 2009a). Dd5P4 is catalytically active on LCVs and thus increases the PtdIns(4)P available for binding of effectors such as SidC or SidM (Weber et al. 2009a). Therefore, by binding to and recruiting OCRL1/Dd5P4, LpnE might modulate the PI pattern on LCVs such that the concentration of PtdIns(4)P is increased and anchoring of effectors is enhanced.

The *L. pneumophila* effectors RalF and SidM possibly contribute to the modulation of the LCV PI pattern indirectly through the recruitment and activation of small host GTPases. RalF, the first identified Icm/Dot substrate, activates and recruits the small GTPase Arf1 to the LCV (Nagai et al. 2002). Arf1 acts as key regulator in vesicle trafficking from the ER to the Golgi [reviewed in (Donaldson and Jackson 2000)]. RalF was identified based on its similarity to eukaryotic Arf1 GEFs harboring a Sec7 domain. Indeed, RalF shows Arf1 GEF activity and is required for Arf1 localization on the LCV as well as fusion of the LCV with the ER (Nagai et al. 2002). After translocation, RalF is found at the cytoplasmic leaflet of the LCV although structural data did not reveal any membrane interaction determinants (Amor et al. 2005) and no indication of an interaction with PIs was obtained (Weber et al. 2006). RalF contains a C-terminal globular "capping" domain that auto-inhibits the Arf1 docking sites of the effector (Alix et al. 2012), thus possibly also masking its membrane binding determinants. PI 4-kinase IIIβ (PI4KIIIβ) is recruited by activated Arf1 to the *trans*-Golgi network (TGN) (Godi et al. 1999), and thus, RalF might indirectly increase the PtdIns(4)P concentration on LCVs (Table 1, Fig. 2). In agreement with a role for PI4KIIIβ during LCV formation, the depletion by RNA interference of PI4KIIIβ, but not PI4KIIIα or PI4KIIα decreased SidC on LCVs in replication-permissive *Drosophila* Kc167 phagocytes, suggesting that in the absence of PI4KIIIβ PtdIns(4)P was reduced (Brombacher et al. 2009).

Analogously, SidM recruits and activates Rab1 on LCVs. Activated Rab1 (Hyvola et al. 2006) as well as Arf1 (Lichter-Konecki et al. 2006) has been shown to recruit OCRL1 to endosomal membranes. Therefore, SidM as well as RalF might indirectly contribute to an increase of PtdIns(4)*P* on LCV membranes. If the translocation of RalF or SidM indeed indirectly modulates the PtdIns(4)*P* level on the LCV membrane, the concentration of this PI lipid on LCVs should decrease in the absence of the GEFs, and lower amounts of SidC should bind to LCVs. However, in the absence of RalF or SidM, SidC on LCVs remained constant or even decreased (Brombacher et al. 2009). Consequently, *L. pneumophila* probably does not (or at least not exclusively) control the decoration of LCVs with PIs by the indirect recruitment of host PI kinases or phosphatases through GEFs.

Lastly, two *L. pneumophila* effectors have recently been identified that activate the DAG biosynthetic pathway and thus might also indirectly modulate the LCV PI pattern (Viner et al. 2012). LpdA is a 50 kDa Icm/Dot substrate, which possesses a phospholipase D (PLD) domain and catalyses the reaction from phosphatidylcholine (PC) to phosphatidic acid (PA). LecE, identified in a search for genes specifically present in *Legionella* and *Coxiella* species and analysed through yeast genetics, is a 61 kDa Icm/Dot substrate that enhances the activity of the eukaryotic PA phosphatase Pah1, which leads to the conversion of PA into DAG. Both LpdA and LecE localize to LCVs, and therefore, their combined activity likely results in the production of DAG on the LCV membrane. DAG is a second messenger that recruits protein kinase D (PKD) (Fu and Rubin 2011) and its activator protein kinase C (PKC) (Almena and Merida 2011) to membranes. Activated PKD then interacts with PI4KIIIβ, thereby possibly contributing to an increase in PtdIns(4)*P* on the LCV membrane (Viner et al. 2012) (Table 1, Fig. 2).

4.3 Role of PI 3-kinases for L. pneumophila Uptake and Replication

While a number of *L. pneumophila* effectors targeting PtdIns(4)*P* or PtdIns(3)*P* have been characterized, thus far no *L. pneumophila* factors subverting PtdIns(4,5)P_2 or PtdIns(3,4,5)P_3 have been identified. Yet, both PI lipids play a crucial role in the regulation of phagocytosis, since PtdIns(4,5)P_2 links the cortical actin to the plasma membrane and PtdIns(3,4,5)P_3 is transiently formed at the inner leaflet of the phagocytic cup (Botelho et al. 2000; Marshall et al. 2001; Vieira et al. 2001). PtdIns(3,4,5)P_3 is produced by type I PI3Ks which can be inhibited by the structurally unrelated inhibitors, wortmannin or LY294002 (Araki et al. 1996; Cox et al. 1999).

An analysis of the role of PI3Ks for the uptake of *L. pneumophila* yielded conflicting results. Phagocytosis of wild type *L. pneumophila* by replication-permissive human U937 macrophage-like cells apparently occurred through a wortmannin-insensitive pathway, while the uptake of an Icm/Dot mutant was strongly reduced by the addition of PI3K inhibitors (Khelef et al. 2001; Harada et al. 2012). In contrast, the uptake of *L. pneumophila* by murine J774A.1 macrophage-like cells was inhibited by wortmannin or LY294002 (Tachado et al. 2008; Charpentier et al. 2009; Peracino et al. 2010). However, J774A.1 macrophages do not support intracellular growth of *L. pneumophila*, and hence, the evasion of PI3K signaling during uptake might be essential for the formation of a replication-permissive vacuole (Weber et al. 2009b).

PI3Ks are also largely dispensable for the uptake of wild-type *L. pneumophila* by *D. discoideum*, but required for the uptake of an Icm/Dot mutant strain (Weber et al. 2006; Peracino et al. 2010). A role for PI3Ks during uptake of *L. pneumophila* by *D. discoideum* would be in agreement with a PI3K-dependent macropinocytotic process, rather than a phagocytic mechanism that proceeds independently of PI3Ks (Peracino et al. 2010). In any case, the pharmacological inhibition of PI phospholipase C (PLC), an enzyme that hydrolyses PtdIns(4,5)P_2 to yield DAG and inositol-1,4,5-trisphosphate abolished the uptake of *L. pneumophila* (Peracino et al. 2010). Taken together, the function of PI3Ks during uptake of *L. pneumophila* and the role of the Icm/Dot T4SS in this process seem to be to a certain degree dependent on the bacterial strain and the host phagocyte.

Upon deletion of two or five class I PI3Ks in *D. discoideum*, respectively, wild-type *L. pneumophila* replicated more efficiently in the amoeba (Weber et al. 2006; Peracino et al. 2010). It is currently unclear how a decrease in the cellular level of PtdIns(3,4,5)P_3 promotes intracellular bacterial replication. Since mono-phosphorylated PtdIns(3)*P* largely determines the endocytic maturation of a phagosome, a decrease in the cellular content of this PI lipid would readily explain a more efficient LCV formation and intracellular replication. Perhaps, the *L. pneumophila* PI 3-phosphatase SidF also directly contributes to a reduction of PtdIns(3)*P* on the LCV membrane, thereby diverting the pathogen vacuole from the endocytic pathway.

5 Conclusions

In order to avoid degradation by the host cell, *L. pneumophila* manipulates cell signaling in an intricate manner. One major aspect determining the identity, fate, and interactions of a cellular compartment is its PI lipid pattern. *L. pneumophila* subverts the LCV PI composition by Icm/Dot-translocated effector proteins that (i) anchor to distinct PIs, (ii) catalyse PI dephosphorylation, (iii) directly bind and recruit host PI-metabolizing enzymes, and (iv) activate small host GTPases or other factors, which in turn recruit PI kinases or phosphatases. Most of the effectors known to modulate the LCV PI pattern shift the LCV into a Golgi/ER-similar and PtdIns(4)*P*-rich compartment. Thus, interference with the LCV PI composition appears to be crucial to form a replication-permissive compartment, wherein the bacteria are protected from degradation and fit to replicate. Future studies should aim at a time-resolved analysis of the complex interplay of *L. pneumophila* effectors targeting host PI lipid metabolism.

References

Alix E, Chesnel L, Bowzard BJ, et al. (2012) The capping domain in RalF regulates effector functions. PLoS Path 8:e1003012

Almena M, Merida I (2011) Shaping up the membrane: diacylglycerol coordinates spatial orientation of signaling. Trends Biochem Sci 36:593–603

Amor JC, Swails J, Zhu X, et al. (2005) The structure of RalF, an ADP-ribosylation factor guanine nucleotide exchange factor from *L. pneumophila*, reveals the presence of a cap over the active site. J Biol Chem 280:1392–1400

Araki N, Johnson MT, Swanson JA (1996) A role for phosphoinositide 3-kinase in the completion of macropinocytosis and phagocytosis by macrophages. J Cell Biol 135:1249–1260

Arasaki K, Roy CR (2010) *L. pneumophila* promotes functional interactions between plasma membrane syntaxins and Sec22b. Traffic 11:587–600

Arasaki K, Toomre DK, Roy CR (2012) The *L. pneumophila* effector DrrA is sufficient to stimulate SNARE-dependent membrane fusion. Cell Host Microbe 11:46–57

Bakowski MA, Braun V, Lam GY, et al. (2010) The phosphoinositide phosphatase SopB manipulates membrane surface charge and trafficking of the *Salmonella*-containing vacuole. Cell Host Microbe 7:453–462

Banga S, Gao P, Shen X, Fiscus V, Zong WX, Chen L, Luo ZQ (2007) *Legionella pneumophila* inhibits macrophage apoptosis by targeting pro-death members of the Bcl2 protein family. Proc Nat Acad Sci U S A 104:5121–5126

Behnia R, Munro S (2005) Organelle identity and the signposts for membrane traffic. Nature 438:597–604

Beresford N, Patel S, Armstrong J, Szoor B, Fordham-Skelton AP, Tabernero L (2007) MptpB, a virulence factor from *M. tuberculosis*, exhibits triple-specificity phosphatase activity. Biochem J 406:13–18

Beresford NJ, Mulhearn D, Szczepankiewicz B et al (2009) Inhibition of MptpB phosphatase from *M. tuberculosis* impairs mycobacterial survival in macrophages. J Antimicrob Chemother 63:928–936

Berger KH, Isberg RR (1993) Two distinct defects in intracellular growth complemented by a single genetic locus in *L. pneumophila*. Mol Microbiol 7:7–19

Blumental-Perry A, Haney CJ, Weixel KM, Watkins SC, Weisz OA, Aridor M (2006) Phosphatidylinositol 4-phosphate formation at ER exit sites regulates ER export. Dev Cell 11:671–682

Botelho RJ, Teruel M, Dierckman R et al (2000) Localized biphasic changes in phosphatidylinositol-4,5-bisphosphate at sites of phagocytosis. J Cell Biol 151:1353–1368

Brombacher E, Urwyler S, Ragaz C, Weber SS, Kami K, Overduin M, Hilbi H (2009) Rab1 guanine nucleotide exchange factor SidM is a major phosphatidyl-inositol 4-phosphate-binding effector protein of *L. pneumophila*. J Biol Chem 284:4846–4856

Burkinshaw BJ, Prehna G, Worrall LJ, Strynadka NC (2012) Structure of *Salmonella* effector protein SopB N-terminal domain in complex with host Rho GTPase Cdc42. J Biol Chem 287:13348–13355

Charpentier X, Gabay JE, Reyes M et al (2009) Chemical genetics reveals bacterial and host cell functions critical for type IV effector translocation by *L. pneumophila*. PLoS Path 7:e1000501

Chen Y, Machner MP (2013) Targeting of the small GTPase Rab6A' by the *L. pneumophila* effector LidA. Infect Immun 81:2226–35

Cheng W, Yin K, Lu D et al (2012) Structural insights into a unique *L. pneumophila* effector LidA recognizing both GDP and GTP bound Rab1 in their active state. PLoS Path 8:e1002528

Christoforidis S, Miaczynska M, Ashman K et al (1999) Phosphatidylinositol-3-OH kinases are Rab5 effectors. Nat Cell Biol 1:249–252

Conover GM, Derre I, Vogel JP, Isberg RR (2003) The *L. pneumophila* LidA protein: a translocated substrate of the Dot/Icm system associated with maintenance of bacterial integrity. Mol Microbiol 48:305–321

Cosson P, Soldati T (2008) Eat, kill or die: when amoeba meets bacteria. Curr Op Microbiol 11:271–276

Cox D, Tseng CC, Bjekic G, Greenberg S (1999) A requirement for phosphatidyl-inositol 3-kinase in pseudopod extension. J Biol Chem 274:1240–1247

De Camilli P, Chen H, Hyman J, Panepucci E, Bateman A, Brunger AT (2002) The ENTH domain. FEBS Lett 513:11–18

De Matteis MA, Godi A (2004) PI-loting membrane traffic. Nat Cell Biol 6:487–492

Derre I, Isberg RR (2004) *Legionella pneumophila* replication vacuole formation involves rapid recruitment of proteins of the early secretory system. Inf Immun 72:3048–3053

Di Paolo G, De Camilli P (2006) Phosphoinositides in cell regulation and membrane dynamics. Nature 443:651–657

Donaldson JG, Jackson CL (2000) Regulators and effectors of the ARF GTPases. Curr Op Cell Biol 12:475–482

Ford MG, Pearse BM, Higgins MK et al (2001) Simultaneous binding of PtdIns(4,5)P_2 and clathrin by AP180 in the nucleation of clathrin lattices on membranes. Science 291:1051–1055

Fu Y, Rubin CS (2011) Protein kinase D: coupling extracellular stimuli to the regulation of cell physiology. EMBO Rep 12:785–796

Godi A, Pertile P, Meyers R et al (1999) ARF mediates recruitment of PtdIns-4–OH kinase-beta and stimulates synthesis of PtdIns(4,5)P_2 on the Golgi complex. Nat Cell Biol 1:280–287

Ham H, Sreelatha A, Orth K (2011) Manipulation of host membranes by bacterial effectors. Nat Rev Microbiol 9:635–646

Harada T, Tanikawa T, Iwasaki Y, Yamada M, Imai Y, Miyake M (2012) Phagocytic entry of *L. pneumophila* into macrophages through phosphatidyl-inositol 3,4,5-trisphosphate-independent pathway. Biol Pharma Bull 35:1460–1468

Heidtman M, Chen EJ, Moy MY, Isberg RR (2009) Large-scale identification of *L. pneumophila* Dot/Icm substrates that modulate host cell vesicle trafficking pathways. Cell Microbiol 11:230–248

Hernandez LD, Hueffer K, Wenk MR, Galan JE (2004) *Salmonella* modulates vesicular traffic by altering phosphoinositide metabolism. Science 304:1805–1807

Hilbi H (2006) Modulation of phosphoinositide metabolism by pathogenic bacteria. Cell Microbiol 8:1697–1706

Hilbi H, Haas A (2012) Secretive bacterial pathogens and the secretory pathway. Traffic 13:1187–1197

Hilbi H, Weber S, Finsel I (2011) Anchors for effectors: subversion of phosphoinositide lipids by *Legionella*. Front Microbiol 2:91

Hilbi H, Weber SS, Ragaz C, Nyfeler Y, Urwyler S (2007) Environmental predators as models for bacterial pathogenesis. Env Microbiol 9:563–575

Hsu F, Zhu W, Brennan L, Tao L, Luo ZQ, Mao Y (2012) Structural basis for substrate recognition by a unique *Legionella* phosphoinositide phosphatase. Proc Nat Acad Sci U S A 109:13567–13572

Hubber A, Roy CR (2010) Modulation of host cell function by *L. pneumophila* type IV effectors. Ann Rev Cell Dev Biol 26:261–283

Hyvola N, Diao A, McKenzie E, Skippen A, Cockcroft S, Lowe M (2006) Membrane targeting and activation of the Lowe syndrome protein OCRL1 by Rab GTPases. EMBO J 25:3750–3761

Ingmundson A, Delprato A, Lambright DG, Roy CR (2007) *Legionella pneumophila* proteins that regulate Rab1 membrane cycling. Nature 450:365–369

Jank T, Bohmer KE, Tzivelekidis T, Schwan C, Belyi Y, Aktories K (2012) Domain organization of *Legionella* effector SetA. Cell Microbiol 14:852–868

Kagan JC, Stein MP, Pypaert M, Roy CR (2004) *Legionella* subvert the functions of Rab1 and Sec22b to create a replicative organelle. J Exp Med 199:1201–1211

Khelef N, Shuman HA, Maxfield FR (2001) Phagocytosis of wild-type *L. pneumophila* occurs through a wortmannin-insensitive pathway. Inf Immun 69:5157–5161

Konradt C, Frigimelica E, Nothelfer K et al (2011) The *S. flexneri* type three secretion system effector IpgD inhibits T cell migration by manipulating host phosphoinositide metabolism. Cell Host Microbe 9:263–272

Krauss M, Kinuta M, Wenk MR, De Camilli P, Takei K, Haucke V (2003) Arf6 stimulates clathrin/AP-2 recruitment to synaptic membranes by activating phosphatidylinositol phosphate kinase type I gamma. J Cell Biol 162:113–124

Lemmon MA (2008) Membrane recognition by phospholipid-binding domains. Nature Rev Mol Cell Biol 9:99–111

Li Z, Solomon JM, Isberg RR (2005) *D. discoideum* strains lacking the RtoA protein are defective for maturation of the *L. pneumophila* replication vacuole. Cell Microbiol 7:431–442

Lichter-Konecki U, Farber LW, Cronin JS, Suchy SF, Nussbaum RL (2006) The effect of missense mutations in the RhoGAP-homology domain on Ocrl1 function. Mol Genetics Metabol 89:121–128

Lifshitz Z, Burstein D, Peeri M et al (2013) Computational modeling and experimental validation of the *Legionella* and *Coxiella* virulence-related type-IVB secretion signal. Proc Nat Acad Sci U S A 110:E707–E715

Loovers HM, Kortholt A, de Groote H, Whitty L, Nussbaum RL, van Haastert PJ (2007) Regulation of phagocytosis in *Dictyostelium* by the inositol 5-phosphatase OCRL homolog Dd5P4. Traffic 8:618–628

Lu H, Clarke M (2005) Dynamic properties of *Legionella*-containing phagosomes in *Dictyostelium* amoebae. Cell Microbiol 7:995–1007

Luo ZQ, Isberg RR (2004) Multiple substrates of the *L. pneumophila* Dot/Icm system identified by interbacterial protein transfer. Proc Nat Acad Sci U S A 101:841–846

Machner MP, Isberg RR (2006) Targeting of host Rab GTPase function by the intravacuolar pathogen *L. pneumophila*. Dev Cell 11:47–56

Machner MP, Isberg RR (2007) A bifunctional bacterial protein links GDI displacement to Rab1 activation. Science 318:974–977

Mallo GV, Espina M, Smith AC et al (2008) SopB promotes phosphatidylinositol 3-phosphate formation on *Salmonella* vacuoles by recruiting Rab5 and Vps34. J Cell Biol 182:741–752

Marra A, Blander SJ, Horwitz MA, Shuman HA (1992) Identification of a *Legionella pneumophila* locus required for intracellular multiplication in human macrophages. Proc Nat Acad Scie U S A 89:9607–9611

Marshall JG, Booth JW, Stambolic V et al (2001) Restricted accumulation of phosphatidylinositol 3-kinase products in a plasmalemmal subdomain during Fc gamma receptor-mediated phagocytosis. J Cell Biol 153:1369–1380

Mason D, Mallo GV, Terebiznik MR et al (2007) Alteration of epithelial structure and function associated with PtdIns(4,5)P_2 degradation by a bacterial phosphatase. J Gen Physiol 129:267–283

Müller MP, Peters H, Blumer J, Blankenfeldt W, Goody RS, Itzen A (2010) The *Legionella* effector protein DrrA AMPylates the membrane traffic regulator Rab1b. Science 329:946–949

Murata T, Delprato A, Ingmundson A, Toomre DK, Lambright DG, Roy CR (2006) The *Legionella pneumophila* effector protein DrrA is a Rab1 guanine nucleotide-exchange factor. Nat Cell Biol 8:971–977

Murray JT, Panaretou C, Stenmark H, Miaczynska M, Backer JM (2002) Role of Rab5 in the recruitment of hVps34/p150 to the early endosome. Traffic 3:416–427

Nagai H, Kagan JC, Zhu X, Kahn RA, Roy CR (2002) A bacterial guanine nucleotide exchange factor activates ARF on *Legionella* phagosomes. Science 295:679–682

Neunuebel MR, Mohammadi S, Jarnik M, Machner MP (2012) *Legionella pneumophila* LidA affects nucleotide binding and activity of the host GTPase Rab1. J Bact 194:1389–1400

Newton HJ, Sansom FM, Bennett-Wood V, Hartland EL (2006) Identification of *L. pneumophila*-specific genes by genomic subtractive hybridization with *L. micdadei* and identification of *lpnE*, a gene required for efficient host cell entry. Inf Immun 74:1683–1691

Newton HJ, Sansom FM, Dao J, McAlister AD, Sloan J, Cianciotto NP, Hartland EL (2007) Sel1 repeat protein LpnE is a *L. pneumophila* virulence determinant that influences vacuolar trafficking. Infect Immun 75:5575–5585

Niebuhr K, Giuriato S, Pedron T et al (2002) Conversion of PtdIns(4,5)P_2 into PtdIns(5)P by the *S. flexneri* effector IpgD reorganizes host cell morphology. EMBO J 21:5069–5078

Norris FA, Wilson MP, Wallis TS, Galyov EE, Majerus PW (1998) SopB, a protein required for virulence of *S. dublin*, is an inositol phosphate phosphatase. Proc Nat Acad Sci U S A 95:14057–14059

O'Connor TJ, Adepoju Y, Boyd D, Isberg RR (2011) Minimization of the *L. pneumophila* genome reveals chromosomal regions involved in host range expansion. Proc Nat Acad Sci U S A 108:14733–14740

Payrastre B, Missy K, Giuriato S, Bodin S, Plantavid M, Gratacap M (2001) Phosphoinositides: key players in cell signalling, in time and space. Cell Signal 13:377–387

Peracino B, Balest A, Bozzaro S (2010) Phosphoinositides differentially regulate bacterial uptake and Nramp1-induced resistance to *Legionella* infection in *Dictyostelium*. J Cell Sci 123:4039–4051

Pizarro-Cerda J, Cossart P (2004) Subversion of phosphoinositide metabolism by intracellular bacterial pathogens. Nat Cell Biol 6:1026–1033

Ragaz C, Pietsch H, Urwyler S, Tiaden A, Weber SS, Hilbi H (2008) The *L. pneumophila* phosphatidylinositol-4 phosphate-binding type IV substrate SidC recruits endoplasmic reticulum vesicles to a replication-permissive vacuole. Cell Microbiol 10:2416–2433

Saarikangas J, Zhao H, Lappalainen P (2010) Regulation of the actin cytoskeleton-plasma membrane interplay by phosphoinositides. Physiol Rev 90:259–289

Saleh MT, Belisle JT (2000) Secretion of an acid phosphatase (SapM) by *M. tuberculosis* that is similar to eukaryotic acid phosphatases. J Bacteriol 182:6850–6853

Sasaki T, Takasuga S, Sasaki J, Kofuji S, Eguchi S, Yamazaki M, Suzuki A (2009) Mammalian phosphoinositide kinases and phosphatases. Prog Lipid Res 48:307–343

Schoebel S, Blankenfeldt W, Goody RS, Itzen A (2010) High-affinity binding of phosphatidylinositol 4-phosphate by *L. pneumophila* DrrA. EMBO Rep 11:598–604

Schoebel S, Cichy AL, Goody RS, Itzen A (2011) Protein LidA from *Legionella* is a Rab GTPase supereffector. Proc Nat Acad Sci U S A 108:17945–17950

Schoebel S, Oesterlin LK, Blankenfeldt W, Goody RS, Itzen A (2009) RabGDI displacement by DrrA from *Legionella* is a consequence of its guanine nucleotide exchange activity. Mol Cell 36:1060–1072

Simonsen A, Lippe R, Christoforidis S et al (1998) EEA1 links PI3K function to Rab5 regulation of endosome fusion. Nature 394:494–498

Suh HY, Lee DW, Lee KH et al (2010) Structural insights into the dual nucleotide exchange and GDI displacement activity of SidM/DrrA. EMBO J 29:496–504

Tachado SD, Samrakandi MM, Cirillo JD (2008) Non-opsonic phagocytosis of *L. pneumophila* by macrophages is mediated by phosphatidylinositol 3-kinase. Plos One 3:e3324

Terebiznik MR, Vieira OV, Marcus SL et al (2002) Elimination of host cell PtdIns(4,5)P_2 by bacterial SigD promotes membrane fission during invasion by *Salmonella*. Nat Cell Biol 4:766–773

Urwyler S, Brombacher E, Hilbi H (2009) Endosomal and secretory markers of the *Legionella*-containing vacuole. Comm Integrat Biol 2:107–109

Vergne I, Chua J, Lee HH, Lucas M, Belisle J, Deretic V (2005) Mechanism of phagolysosome biogenesis block by viable *M. tuberculosis*. Proc Nat Acad Sci U S A 102:4033–4038

Vergne I, Chua J, Singh SB, Deretic V (2004) Cell biology of *M. tuberculosis* phagosome. Ann Rev Cell Dev Biol 20:367–394

Vieira OV, Botelho RJ, Rameh L et al (2001) Distinct roles of class I and class III phosphatidylinositol 3-kinases in phagosome formation and maturation. J Cell Biol 155:19–25

Viner R, Chetrit D, Ehrlich M, Segal G (2012) Identification of two *L. pneumophila* effectors that manipulate host phospholipids biosynthesis. PLoS Path 8:e1002988

Wang DS, Shaw G (1995) The association of the C-terminal region of beta I sigma II spectrin to brain membranes is mediated by a PH domain, does not require membrane proteins, and coincides with a inositol-1,4,5 triphosphate binding site. Biochem Biophys Res Commun 217:608–615

Weber SS, Ragaz C, Hilbi H (2009a) The inositol polyphosphate 5-phosphatase OCRL1 restricts intracellular growth of *Legionella*, localizes to the replicative vacuole and binds to the bacterial effector LpnE. Cell Microbiol 11:442–460

Weber SS, Ragaz C, Hilbi H (2009b) Pathogen trafficking pathways and host phosphoinositide metabolism. Mol Microbiol 71:1341–1352

Weber SS, Ragaz C, Reus K, Nyfeler Y, Hilbi H (2006) *L. pneumophila* exploits PI(4)*P* to anchor secreted effector proteins to the replicative vacuole. PLoS Path 2:e46

Wishart MJ, Taylor GS, Dixon JE (2001) Phoxy lipids: revealing PX domains as phosphoinositide binding modules. Cell 105:817–820

Zhang X, Jefferson AB, Auethavekiat V, Majerus PW (1995) The protein deficient in Lowe syndrome is a phosphatidylinositol-4,5-bisphosphate 5-phosphatase. Proc Nat Acad Sci U S A 92:4853–4856

Zhu W, Banga S, Tan Y, Zheng C, Stephenson R, Gately J, Luo ZQ (2011) Comprehensive identification of protein substrates of the Dot/Icm type IV transporter of *L. pneumophila*. Plos One 6:e17638

Zhu Y, Hu L, Zhou Y, Yao Q, Liu L, Shao F (2010) Structural mechanism of host Rab1 activation by the bifunctional *Legionella* type IV effector SidM/DrrA. Proc Nat Acad Sci U S A 107:4699–4704

Legionella Phospholipases Implicated in Virulence

Katja Kuhle and Antje Flieger

Abstract Phospholipases are diverse enzymes produced in eukaryotic hosts and their bacterial pathogens. Several pathogen phospholipases have been identified as major virulence factors acting mainly in two different modes: on the one hand, they have the capability to destroy host membranes and on the other hand they are able to manipulate host signaling pathways. Reaction products of bacterial phospholipases may act as secondary messengers within the host and therefore influence inflammatory cascades and cellular processes, such as proliferation, migration, cytoskeletal changes as well as membrane traffic. The lung pathogen and intracellularly replicating bacterium *Legionella pneumophila* expresses a variety of phospholipases potentially involved in disease-promoting processes. So far, genes encoding 15 phospholipases A, three phospholipases C, and one phospholipase D have been identified. These cell-associated or secreted phospholipases may contribute to intracellular establishment, to egress of the pathogen from the host cell, and to the observed lung pathology. Due to the importance of phospholipase activities for host cell processes, it is conceivable that the pathogen enzymes may mimic or substitute host cell phospholipases to drive processes for the pathogen's benefit. The following chapter summarizes the current knowledge on the *L. pneumophila* phospholipases, especially their substrate specificity, localization, mode of secretion, and impact on host cells.

Abbreviations

1,2-DG	1,2-diacylglycerol
3-HB	3-hydroxybutyrate
FFA	Free fatty acid/s
GCAT	Glycerophospholipid:cholesterol acyltransferase
IL-8	Interleukin-8

K. Kuhle · A. Flieger (✉)
FG 11 - Division of Enteropathogenic Bacteria and Legionella, Robert Koch-Institut, Burgstr. 37 38855 Wernigerode, Germany
e-mail: FliegerA@rki.de

Current Topics in Microbiology and Immunology (2013) 376: 175–209
DOI: 10.1007/82_2013_348

Published Online: 8 August 2013

LCV	*Legionella*-containing vacuole
LPA	Lysophosphatidic acid
LPC	Lysophosphatidylcholine
LPG	Lysophosphatidylglycerol
LPLA	Lysophospholipase A
PA	Phosphatidic acid
PC	Phosphatidylcholine
PC-PLC	PC-preferring PLC
PE	Phosphatidylethanolamine
PHB	Poly-3-hydroxybutyrate
PI	Phosphatidylinositol
PI-PLC	Phosphatidylinositol-specific PLC
PKC	Protein kinase C
PLA	Phospholipase A
PLB	Phospholipase B
PLC	Phospholipase C
PLD	Phospholipase D
PLP	Patatin-like protein
pNP	para-nitrophenol
pNPPC	para-nitrophenylphosphorylcholine
PS	Phosphatidylserine
SCV	*Salmonella*-containing vacuole
Sifs	*Salmonella*-induced filaments
SM	Sphingomyelin
SMase	Sphingomyelinase
VPS	Vacuole protein sorting

Contents

1 Introduction 177
2 Bacterial Phospholipases as Virulence Factors 178
  2.1 Classification and General Properties of Lipases/Phospholipases 178
  2.2 Role of Phospholipases in Virulence 183
3 Manifold Phospholipases of *L. pneumophila* 186
  3.1 Phospholipases A: GDSL Lipase Family, PlaB, Patatin-Like Protein Family 187
  3.2 *Legionella* Phospholipases C: PlcA, PlcB, and PlcC 192
  3.3 Phospholipase D: LpdA 193
4 Importance of *Legionella* Phospholipases for Host Cell Infection 194
5 Conclusion 198
References 199

1 Introduction

Legionella pneumophila is a Gram-negative, facultative intracellular pathogen which replicates primarily in free-living protozoans, such as amoebae. These environmental hosts provide nutrients, protect the bacteria from adverse conditions, and also render the bacteria more virulent. Via contaminated aerosols, *L. pneumophila* gets access to the human lung and subsequently colonizes lung macrophages and epithelial cells in a similar way like environmental hosts. Bacterial replication may lead to lung destruction and consequently to a severe form of pneumonia, Legionnaires' disease (Albert-Weissenberger et al. 2007; Newton et al. 2010). Histopathologic features of infected lungs show distention of the alveolar spaces by macrophages and neutrophil exudate, fibrin-rich proteinaceous exudates, and diffuse alveolar damage. The lymphatic and hematogenous spread of the bacteria in the host has been described and occurs mainly in liver and spleen (Hernandez et al. 1980; Hicklin et al. 1980; Winn and Myerowitz 1981).

Within a host cell, *L. pneumophila* exhibits a biphasic life cycle consisting of a replicative and a transmissive phase characterized by distinct physiologies and morphologies. Replicative phase bacteria are not cytotoxic and the gene expression pattern mirrors nutrient usage essential for efficient multiplication. When nutrients cease, the bacteria switch to a cytotoxic stage where gene expression targets allocation of new nutrient sources and egress from the host cell. In the different infection stages, effector proteins are transferred from the bacterium to the phagosomal space or to specific places in the host cell allowing both intracellular establishment and subsequent escape (Albert-Weissenberger et al. 2007; Brüggemann et al. 2006; Byrne and Swanson 1998; Molofsky and Swanson 2004; Weissenmayer et al. 2011).

In this regard, two protein secretion systems, the Lsp type II and the Dot/Icm type IVB systems, are mainly associated with *L. pneumophila* virulence (Albert-Weissenberger et al. 2007; Cianciotto 2009; Hubber and Roy 2010; Zhu et al. 2011). The Lsp secretion machinery exports many degradative enzymes facilitating infection of protozoan as well as macrophage host cells and growth within the mammalian lung (Cianciotto 2009; DebRoy et al. 2006; Rossier and Cianciotto 2001; Rossier et al. 2004). Recently, McCoy-Simandle et al. showed that the Lsp system also promotes infection of lung epithelial cells and dampens cytokine/chemokine release from macrophages and epithelial cells, for example by restricting cytokine transcript levels (McCoy-Simandle et al. 2011). More than 25 type II-dependently secreted proteins are yet identified including some phospholipases (Cianciotto 2009; DebRoy et al. 2006; Tyson et al. 2013). The type IVB secretion system Dot/Icm injects a plethora of about 300 effector proteins into the host cell cytosol or to the phagosomal membrane and strongly promotes *L. pneumophila* virulence (Ensminger and Isberg 2009; Hubber and Roy 2010; Isberg et al. 2008; Segal et al. 1998; Vogel et al. 1998; Zhu et al. 2011). In addition, *L. pneumophila* possesses other putative protein secretion systems, such as the type I Lss system, several type VIA machineries (Lvh, Trb-1, Trb-2), and a type V autotransporter (Albert-Weissenberger et al. 2007; Brassinga et al. 2003; Cazalet et al. 2004; Glöckner et al. 2008; Jacobi and Heuner 2003; Schroeder et al. 2010; Segal et al. 1999).

Important virulence factors of pathogenic bacteria are phospholipases supporting a variety of processes including invasion and colonization of the host. *L. pneumophila* for example possesses a multitude of phospholipases (Table 1 and Figure 2). Some might assist nutrient acquisition in the *Legionella*-containing vacuole (LCV) as well as suppression of host defense allowing intracellular establishment in the replicative phase, while others might promote bacterial release from the LCV and host cell as well as impairment of lung function in the transmissive phase. The following chapter summarizes the current knowledge on bacterial phospholipases, their general properties, and impact on host cells.

2 Bacterial Phospholipases as Virulence Factors

2.1 Classification and General Properties of Lipases/Phospholipases

Phospholipases, a multifacetted subclass of lipases/esterases, are diverse and ubiquitous enzymes which are widespread in eukaryotic and prokaryotic organisms fulfilling a variety of cellular functions, for example in membrane synthesis and composition, in secondary messenger generation, in inflammatory response, and in cellular turnover (Dennis et al. 1991; Ibarguren et al. 2010; Linkous and Yazlovitskaya 2010; Sreenivas et al. 1998). In many cases, the catalytic active residues and surrounding sequence motifs are conserved in lipases/phospholipases of evolutionary distinct species (Akoh et al. 2004; Arpigny and Jaeger 1999; Banerji and Flieger 2004; Messaoudi et al. 2010; Upton and Buckley 1995). They may therefore resemble and replace each other upon host-pathogen contact and pathogen phospholipases may then hijack cellular processes to serve the pathogen's needs. Phospholipases act on phospholipid substrates, which are amphipathic molecules usually assembling into subcellular structures, such as bilayers or monolayers. Phosphatidylcholine (PC), phosphatidylethanolamine (PE), phosphatidylinositol (PI), and phosphatidylserine (PS) are the most common in mammalian cells (Kikkawa et al. 1975; Mason and Williams 1980; Rooney et al. 1977; Schmiel and Miller 1999; Virtanen et al. 1998).

Phospholipases hydrolyze phospholipids at different positions and four major specificities termed A, B, C, and D have been described. The capital letters indicate the specific targeted ester bond. Common phospholipids consist of a glycerol backbone with two esterified fatty acids and one phosphate group (Fig. 1). The nature of the fatty acids and the alcohol attached to the phosphate group plays an important role in phospholipase substrate specificity. Acylhydrolases, such as phospholipases A (PLA), release free fatty acids (FFA) at sn-1 (PLA_1) and/or sn-2

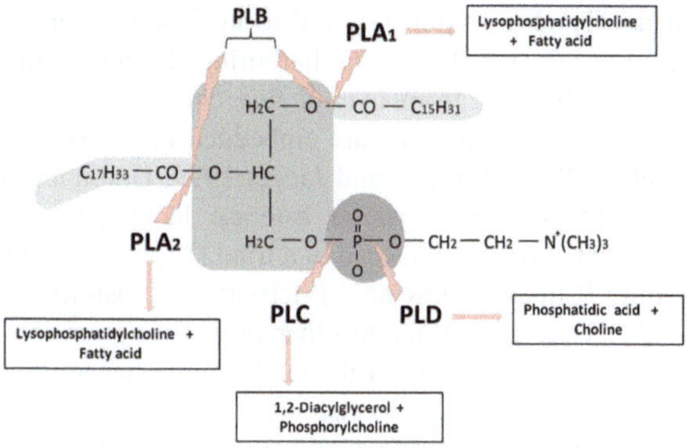

Fig. 1 Cleavage of phosphatidylcholine by different phospholipases and resulting reaction products. Important parts of a phospholipid, such as phosphatidylcholine, are the central glycerol backbone (*gray*) with two esterified fatty acids (*medium gray*) and one phosphate group (*dark gray circle*). The phosphate is further bonded to a polar head group such as choline (*light gray*). See text for further details

(PLA_2) position and thereby, in the case of PC hydrolysis, generate toxic lysophosphatidylcholine (LPC) (Huang et al. 2005; Masamune et al. 2001). Enzymes with a lysophospholipase A activity (LPLA) cleave the remaining FFA from the glycerol backbone and thereby detoxify LPC and generate glycerophosphorylcholine. Phospholipase B (PLB) cleaves both sn-1 and sn-2 acyl chains of the phospholipid. Phosphodiesterases, such as phospholipase C (PLC) and D (PLD), hydrolyze either the glycerol-oriented or the alcohol-oriented phosphoester of their substrate, releasing 1,2-diacylglycerol (1,2-DG) and a phosphoryl alcohol or phosphatidic acid (PA) and an alcohol, respectively (Fig. 1).

Phospholipases A Arpigny and Jäger proposed a bacterial acylhydrolase/lipase classification scheme based on comparison of amino acid sequence and fundamental biological properties, and identified eight different families (Arpigny and Jaeger 1999). These families usually utilize a characteristic catalytic triad of Ser-Asp-His and show the typical fold of the α/β-hydrolase superfamily. We here first focus on two of the eight defined families with at least one member conferring PLA activity, specifically family I and II.

Lipolytic enzymes of **family I** are designated true lipases hydrolyzing neutral lipids and are subdivided into eleven subfamilies. However, substrate preferences of some lipases are not very strict and they may also cleave amphipathic phospholipids, such as observed for *Staphylococcus hyicus* lipase which is rare among such true family I lipases (Van Kampen et al. 1998; van Kampen et al. 2001; Van Oort et al. 1989). Members of this group share the conserved Gly-X-Ser-X-Gly motif, enclosing the nucleophilic Ser, and the two other catalytic triad residues, Asp and His. This family is the most abundant and enzymes are widely distributed among Gram-positive and Gram-negative bacteria (Arpigny and Jaeger 1999; Messaoudi et al. 2010).

Members of **family II** are often referred to as **GDSL esterases** because they comprise a Gly-Asp-Ser-(Leu/Ile) motif harboring the active-site Ser, which is located in homology block I from overall five conserved blocks. The putative catalytic triad members Asp and His are embedded in an Asp-X-X-His motif in block V (Akoh et al. 2004; Arpigny and Jaeger 1999; Upton and Buckley 1995). As an exception, *Streptomyces scabies* esterase lacks the catalytic Asp and therefore harbors a catalytic dyad instead of a triad (Arpigny and Jaeger 1999; Wei et al. 1995). Family II lipases show an α/β-tertiary fold distinct from the classical α/β-hydrolase fold. They also differ in other properties from the well-known α/β-hydrolases, such as the proximity of the nucleophilic Ser to the N-terminus and the close localization of the catalytic Asp and His (Akoh et al. 2004; Mølgaard et al. 2000; Upton and Buckley 1995). GDSL enzymes are found in many bacteria as well as in higher plants and a further related eukaryotic protein is the α1 subunit of the brain platelet-activating factor acetylhydrolase (α1 PAF-AH) (Akoh et al. 2004; Arai et al. 2002; Arpigny and Jaeger 1999; Upton and Buckley 1995). GDSL enzymes show in addition to their lipase/acetylhydrolase activities PLA, LPLA as well as hemolytic and acyltransferase activities (Arpigny and Jaeger 1999; Banerji et al. 2008; Lang et al. 2012; Shinoda et al. 1991; Upton and Buckley 1995). A paradigm bacterial GDSL protein is the glycerophospholipid: cholesterol acyltransferase (GCAT) SatA of *Aeromonas salmonicida* (Buckley et al. 1982), transferring fatty acids to cholesterol, a sterol usually not present in bacteria (van der Meer-Janssen et al. 2010). *L. pneumophila* GDSL enzymes will be discussed in more detail in Sect. 3.1.

A new family named **patatin-like proteins** (PLP) according to protein sequence similarity to patatin, the most abundant protein in the potato tuber, showing lipid acyl hydrolase activity, has been defined (Anderson et al. 2002; Andrews et al. 1988; Banerji and Flieger 2004; Kienesberger et al. 2009; Shewry 2003). Members of the patatin-like protein family exhibit four conserved sequence blocks. Block II comprises the active site Ser embedded in the well-known Gly-X-Ser-X-Gly motif. Block IV contains the second amino acid of a catalytic dyad, an Asp embedded in an Asp-Gly motif, unique among known bacterial lipases. Also a stretch of glycines close to the N-terminus within block I characterizes PLPs (Banerji and Flieger 2004; Hirschberg et al. 2001; Rydel et al. 2003; Schrag and Cygler 1997). The eukaryotic and the bacterial PLPs show different motifs embedding a conserved Pro in block III. Bacterial PLPs harbor a conserved Ala-Ser-X-X-X-Pro motif, while eukaryotic patatin homologs possess an Ala-Ala-Pro sequence. Following block IV, a conserved Ser is only present in eukaryotic patatins, including human $cPLA_2$ (Banerji and Flieger 2004; Hirschberg et al. 2001). At present, about 10,000 proteins possessing a patatin/phospholipase A_2-related domain are encoded in sequenced bacterial genomes (http://www.ebi.ac.uk/interpro). *Pseudomonas aeruginosa* ExoU represents the first characterized bacterial member of this family and its crystal structure was solved recently (Banerji and Flieger 2004; Gendrin et al. 2012; Phillips et al. 2003; Sato and Frank 2004; Sato et al. 2003). Until now only a few additional bacterial PLPs have been experimentally addressed, for example PlpD of *P. aeruginosa*, YvdO of

Bacillus subtilis as well as VipD/PatA, VpdA/PatC, VpdB/PatG, VpdC/PatF, and PatD of *L. pneumophila* (Aurass et al. 2009; Kato et al. 2010; Ku et al. 2012; Salacha et al. 2010; Shohdy et al. 2005; VanRheenen et al. 2006). The *L. pneumophila* PLPs will be discussed in more detail in Sect. 3.1.

Recently, another novel group of lipolytic enzymes was discovered by screening hemolytic *Escherichia coli* clones expressing an *L. pneumophila* gene library. The identified enzyme, designated **PlaB**, showed hemolytic and also PLA and LPLA activities and shares no significant homology to previously described phospholipases and lipolytic enzymes, except to uncharacterized and hypothetical proteins of some water-associated bacteria (Flieger et al. 2004). PlaB utilizes a typical triad of Ser-Asp-His embedded in a new amino acid consensus motif unique for this protein family. The active site Ser is located within an unusual pentapeptide Thr-X-Ser-X-Gly as opposed to the other lipase families described above. Furthermore, the catalytic Asp and His are found in an uncommon Gly-Ser-Asp-Gly-Val-Val motif and in a Ser-His-Ser motif, respectively (Bender et al. 2009; Flieger et al. 2004). *L. pneumophila* PlaB will be further discussed in Sect. 3.1.

Phospholipases C Although intensely investigated already from the 1980s due to their importance as major bacterial toxins, there are not many bacterial PLC families described to date. The **Zn^{2+}-dependent PC-preferring PLCs (PC-PLC) of Gram-positive bacteria**, such as *Bacillus cereus* PC-PLC, *Listeria monocytogenes* PLC-B, and *Clostridium perfringens* α-toxin, represent a well-characterized major class of PLCs (Geoffroy et al. 1991; González-Bulnes et al. 2010; Hough et al. 1989; Krug and Kent 1984; Sakurai et al. 1993; Slepkov et al. 2010; Titball 1993). All of the characterized enzymes hydrolyze PC and other phospholipids, for example PE, PS, or sphingomyelin (SM), with various efficiencies and require a cation, such as zinc, for activity (Geoffroy et al. 1991; Moreau et al. 1988; Otnaess et al. 1977; Songer 1997; Titball 1993). They also show differences in their toxicity, which may correlate with their hemolytic activity (Titball 1993). Crystallographic and chemical modification studies of *B. cereus* PC-PLC supplied an insight into possible tertiary structures of homologous regions in other family members. The Zn^{2+}-coordinating His, Glu, Trp, and Asn residues are located at similar positions in the other members (Hough et al. 1989; Titball 1993). The importance of the His residues for Zn^{2+}-coordination was experimentally confirmed for *C. perfringens* α-toxin (Titball 1993; Titball et al. 1999; Titball and Rubidge 1990). Furthermore, the putative active site of the α-toxin including the three catalytically essential Zn^{2+} ions, coordinated by Trp, Glu, two Asp, and five His residues, is located within the N-terminal domain (Naylor et al. 1998). Mutagenic studies suggest that the N-terminal domain is essential for all observed activities, such as phospholipase C toward PC, lethal, and hemolytic activity (Titball et al. 1999).

Members of another PLC family, the **PLC/acid phosphatase family**, show in addition to their PC-PLC activity also sphingomyelinase activity (SMase), except PlcN, which hydrolyses PC and PS (Ostroff et al. 1990; Stonehouse et al. 2002), or phosphatase activity and are found in several bacterial pathogens, such as

Mycobacterium tuberculosis, Bordetella pertussis, Francisella tularensis, Burkholderia pseudomallei, and *Xanthomonas campestris* (Stonehouse et al. 2002). Paradigm members of this PLC branch are PlcH and PlcN, the hemolytic and non-hemolytic enzymes of *P. aeruginosa*, respectively (Costas et al. 2010; Ostroff et al. 1990). Enzymes of this family showed no sequence homology and are further structurally unrelated to the above-mentioned PC-PLCs of Gram-positive bacteria as well as the PI-specific PLCs (PI-PLCs) of *B. cereus* and *L. monocytogenes* (Costas et al. 2010; Stonehouse et al. 2002). Furthermore, the catalytic mechanism for hydrolysis of phosphodiester bonds (PC-PLC, SMase) or phosphomonoester bonds (phosphatase) here is remarkably distinct from the above described PC-PLCs. For example, extracellular PC-PLC/SMase, PlcH, of *P. aeruginosa,* forming a heterodimeric complex with a chaperon protein, and *F. tularensis* acidic phosphatase, AcpA, are strongly inhibited by Zn^{2+} ions and are not affected in their enzymatic activity when the cation chelator EDTA is added (Reilly et al. 1996; Stonehouse et al. 2002). The family contains proteins ranging from plain phosphatases to more complex heterodimeric and multifunctional enzymes. A recently discovered member of this family is *Pseudomonas fluorescens* CGDEase, a pyrophosphatase and glycerophosphodiesterase, showing low activity on choline derivates and inactivity for phospholipids with long-chain fatty acids (Costas et al. 2010).

P. fluorescens PC-PLC however did not show significant amino acid sequence homology to the above described PLCs of Gram-positive and Gram-negative bacteria, suggesting presence of **another subclass of PLCs** (Preuss et al. 2001; Rossignol et al. 2008). And indeed, homologous proteins are also found in different *Legionella* species but not in other bacteria. Most interestingly genes coding for related proteins are present in fungi, including plant, insect, and human pathogenic species, such as *Gibberella zeae, Cordiceps militaris,* and *Trichophyton rubrum*, respectively (Aurass et al. 2013). The enzymatic activity of the *P. fluorescens* and three *L. pneumophila* proteins have been characterized. They require Zn^{2+} or Ca^{2+} ions for activity which distinguishes those enzymes from the above described PC-PLC/SMase/phosphatase family and they prefer PC or phosphatidylglycerol (PG) rather than PI as a substrate (Aurass et al. 2013; Preuss et al. 2001). Overall the novel PLC family shares six blocks of amino acid homology containing a variety of amino acids essential for enzymatic activity, including the amino acids Asp, His, Phe, Glu, and Arg (Aurass et al. 2013). Interestingly, central block III shows homology to a portion of the Zn^{2+}-binding motif of the PC-PLCs from *B. cereus, C. perfringens,* and *L. monocytogenes* (Aurass et al. 2013; Geoffroy et al. 1991; Hansen et al. 1993; Hough et al. 1989; Naylor et al. 1998; Titball 1993). With exception of block V which is also present in *C. perfringens* PC-PLC, conservation of further protein stretches was not observed. Although block III might be involved in Zn^{2+} binding in the novel PLC family as well, a close relationship between these and the above described PLCs from Gram-positive bacteria is not evident. The *L. pneumophila* PLCs will be discussed in more detail in Sect. 3.2.

Phospholipases D Although PLA and PLC are the most common bacterial, some bacterial PLDs were identified in pathogens. Members of the PLD superfamily share a conserved catalytic His-X-Lys-X4-Asp-X6-Gly-Ser-X-Asn motif (HKD) and are proposed to hydrolyze phosphodiester bonds via a similar reaction mechanism (Koonin 1996; Ponting and Kerr 1996; Zhao et al. 1997). Nearly all of the members have a duplication of the HKD motif (Zhao et al. 1997). Known PLDs of this superfamily are found in Gram-negative bacteria, such as *Neisseria gonorrhoeae* NgPLD (Edwards and Apicella 2006; Edwards et al. 2003) and *Yersinia pestis* toxin Ymt (Rudolph et al. 1999). Less well characterized PLDs are *Chlamydia ssp.* pzPLD (Nelson et al. 2006) and *P. aeruginosa* PLD (Wilderman et al. 2001). The recently identified *L. pneumophila* PLD, LpdA, contains the catalytic core of the PLD domain (Viner et al. 2012) and will be discussed in Sect. 3.3.

PLD enzymes lacking the HKD motif are referred to as non-HKD PLD enzymes (Selvy et al. 2011) and are mainly found in Gram-positive bacteria. The well-characterized non-HKD PLD of *Streptomyces chromofuscus* shows phosphodiesterase and phosphatase activities. This enzyme utilizes Fe^{3+} and Mn^{2+} ions for its catalysis and is activated by PA (Geng et al. 1999; Zambonelli et al. 2003). Other non-HKD PLDs are found in *Corynebacterium pseudotuberculosis* and *Arcanobacterium haemolyticum* showing significant sequence identity. Both reveal limited homology to a substrate-binding domain of a dehydrogenase (Cuevas and Songer 1993; McNamara et al. 1995). Furthermore, they exhibit divalent cation-dependent activities and hydrolyze SM rather than PC (Hodgson et al. 1990; Lucas et al. 2010; Selvy et al. 2011).

2.2 *Role of Phospholipases in Virulence*

The number of phospholipases discovered in prokaryotic pathogens has been steadily increasing. These enzymes, with demonstrated and putative roles in virulence, have been identified in intracellular and extracellular bacteria and characterized predominantly in the following species, such as different *Clostridium* and *Bacillus* species, *L. monocytogenes, P. aeruginosa, Staphylococcus aureus, M. tuberculosis, C. pseudotuberculosis, Salmonella enterica, Helicobacter pylori, Yersinia enterocolitica,* and *L. pneumophila.* The different classes of bacterial phospholipases play a multitude of roles in bacterial pathogenesis and therefore are involved in different disease-promoting processes (Higgins et al. 1989; Istivan and Coloe 2006; Popoff and Bouvet 2009; Schmiel and Miller 1999; Singh et al. 2010; Sitkiewicz et al. 2007; Songer 1997; Titball 1998; Titball 1993). Those processes basically fall into two categories (a) phospholipase action via massive cytolytic and destructive properties and (b) via host cell manipulation upon interference with signal transduction processes.

Phospholipases cause membrane and cell destruction Phospholipase toxicity is linked to its cytolytic activity. Hydrolysis of phospholipids is either caused directly by the bacterial phospholipases or together with other host degradative enzymes induced during infection. This leads to a depletion of structural lipids, causes changes in membrane constituents, or generates lytic/membrane-remodeling reaction products, such as LPC. Both processes result in a loss of membrane integrity and therefore cytotoxicity (Schmiel and Miller 1999; Weltzien 1979). For example, the two distinct PLCs of *L. monocytogenes*, a broad range PLC (PC-PLC) and a PI-PLC, and an additional pore-forming cytolysin, listeriolysin O, are required for efficient lysis of the pathogen-containing vacuole and subsequent escape into the cytosol. Further, the bacterial phospholipases are sufficient to trigger escape from double-membrane compartments and therefore mediate cell-to-cell spread (Alberti-Segui et al. 2007; Camilli et al. 1991; Camilli et al. 1993; Goldfine et al. 1995; Smith et al. 1995; Vazquez-Boland et al. 1992). The extracellular lung pathogen *P. aeruginosa* injects ExoU, a PLP displaying PLA/LPLA activities with broad substrate specificity, directly into the host cell via the type III secretion system and therefore induces cytotoxic effects leading to necrotic cell death as a consequence of membrane destruction (Finck-Barbançon et al. 1997; Saliba et al. 2006; Sato and Frank 2004; Sato et al. 2003; Tamura et al. 2004). Disruption of epithelial and endothelial barriers allows *P. aeruginosa* to disseminate into the bloodstream (Hauser 2009). Accordingly, *C. pseudotuberculosis* PLD hydrolyzes SM in host cell membranes causing endothelial membrane leakage and cytolysis, leading to enhanced vascular permeability (McNamara et al. 1994).

Notably, phospholipases may also regulate bacteria-induced membrane extensions. The effector protein SseJ of *S.* Typhimurium, a GDSL enzyme exhibiting PLA and GCAT activities, regulates the level of tubular extensions of the *Salmonella*-containing vacuole (SCV), known as *Salmonella*-induced filaments (Sifs). After injection into the host cell via the *Salmonella* pathogenicity island II (SPI II)-associated type III secretion system, SseJ localizes to the SCV membrane as well as to Sifs and antagonizes the stimulatory effect of SifA, an essential protein for Sif formation. Both thereby regulate the dynamics of the SCV membrane (Lossi et al. 2008; Ruiz-Albert et al. 2002).

Phospholipases manipulate/modulate host signaling events *P. aeruginosa* ExoU on the one hand is a potent cytotoxin, and on the other hand shares homology with eukaryotic $cPLA_2$ and indeed triggers an arachidonic acid-dependent inflammatory cascade in vivo. Thereby, ExoU causes increased production of proinflammatory eicosanoids and induces specific transcription factors regulating cytokine production and subsequent neutrophil recruitment (Cuzick et al. 2006; McMorran et al. 2003; Saliba et al. 2005; Sutterwala et al. 2007). Evidently, phospholipase reaction products may act as secondary messengers and therefore support pathogen-driven manipulation of host signaling events and modulate inflammatory responses.

PC hydrolysis by PLA_2 in addition to fatty acid release generates LPC, exhibiting a variety of functions in eukaryotic cells. For example, LPC enhances superoxide anion production by activation of the NADH/NADPH oxidase system, increases expression of chemokines such as interleukin-8 (IL-8), and activates the small GTPase RhoA in a protein kinase C (PKC) α-dependent manner and therefore influences endothelial permeability, mediates apoptosis, and the inflammatory response (Huang et al. 2005; Masamune et al. 2001; Murugesan et al. 2003; Takeshita et al. 2000). LPC accordingly is a proinflammatory mediator and may be involved in the disruption of membrane barrier functions not only by its lytic capacity at higher concentrations (Weltzien 1979).

1,2-DG, a reaction product of PLCs, was shown to activate PKC which regulates a variety of cellular processes and growth (Newton 2010; Oliva et al. 2005; Titball 1993). Bacterial PLCs indeed trigger signal transduction pathways of eukaryotic cytoplasmic PLCs (Exton 1990; Sakurai et al. 2004). For example, *C. perfringens* α-toxin, one of the most toxic PLCs and one of the best characterized toxins, leads to the production of 1,2-DG and additionally PA by incubation together with neutrophils. PA formation in that case is due to the activation of endogenous PI-PLC and PLD (Ochi et al. 2002; Sakurai et al. 1993, 2004; Titball et al. 1999). The α-toxin activates PKC and various signal transduction pathways such as the arachidonic acid pathway, which causes uncontrolled production of several intracellular mediators such as leukotrienes and prostaglandins as well as intracellular adhesion molecules, IL-8, TFN-α, finally leading to hemolysis, O^{2-} production, muscle contraction, inflammation, vascular permeability, membrane damage, and platelet aggregation among others (Bryant and Stevens 1996; Popoff and Bouvet 2009; Sakurai et al. 2004; Titball et al. 1999).

Lysophosphatidic acid (LPA) is another product, which can be generated from PA by PLA_1 and PLA_2 activities or by PLD-mediated cleavage of lysophospholipids, such as LPC. This signaling molecule acts through G-protein-coupled receptors and thereby activates signal pathways, including the pathways initiated by the GTPases Ras, Rac as well as Rho and therefore alters different cellular responses, such as cell proliferation, survival, migration, cytoskeletal changes, cytokine and chemokine secretion (Lin et al. 2010; Moolenaar et al. 2004; van Leeuwen et al. 2003; Zhao and Natarajan 2012). The PLD/SMaseD of *C. pseudotuberculosis* for example generates LPA. The degradation of circulated LPC to LPA in blood plasma, which activates LPA receptors, may play an important role in pathogenesis (van Meeteren et al. 2004). Furthermore, the exogenous PLD of *S. chromofuscus* was found to stimulate the accumulation of Ras, to induce Ca^{2+} mobilization, membrane depolarization, and Rho-mediated neurite retraction (cytoskeletal contraction). PLD-generated bioactive LPA presented in the membrane mediates these activities and therefore may be causative for the observed effects (Van Dijk et al. 1998). The data suggest that PLD activity and especially the PLD reaction products are involved in the regulation of cytoskeleton rearrangement and cellular movement.

PI is a substrate of several bacterial PLCs and hydrolysis facilitates vacuolar membrane damage and therefore, as in the case of *L. monocytogenes* PI-PLC, escape of the intracellular pathogen from the vacuole is essential for successful survival (Alberti-Segui et al. 2007; Camilli et al. 1991, 1993; Smith et al. 1995; Titball 1998). Furthermore, the phosphorylated variants of PI, the so-called phosphoinositides, represent an important class of signaling molecules. They also play a role in regulation of receptor-mediated endocytosis and phagocytosis and are involved in the recruitment of cytoskeletal elements. Pathogens therefore have developed mechanisms to interfere with phosphoinositide metabolism to affect phagosomal maturation or uptake processes (Di Paolo and De Camilli 2006; van der Meer-Janssen et al. 2010). Bacterial PLCs may alter normal cell behavior by mimicking the function of endogenous PLCs or may stimulate endogenous PLAs or PI-PLCs to generate secondary messengers such as 1,2-DG or inositol triphosphate (IP_3), known to activate PKC and Ca^{2+} channels. Activation of eukaryotic membrane-bound phospholipases was shown for the *C. perfringens* α-toxin conferring PLC activity (Titball 1993, 1998).

As it has been discussed from other bacterial pathogens, phospholipases are important promoters of a variety of destruction and modulation processes. Therefore, it is very interesting that the lung pathogen and intracellular bacterium *L. pneumophila* possesses such a variety of phospholipases which will be discussed in the following (Table 1).

3 Manifold Phospholipases of *L. pneumophila*

The genus *Legionella* comprises 57 different previously characterized species. However, only *L. pneumophila* is responsible for more than 90 % of the disease cases (Campocasso et al. 2012; Diederen 2008; Fields et al. 2002; Hilbi et al. 2010; Muder and Victor 2002; Pearce et al. 2012; http://www.bacterio.cict.fr/l/legionella.html). Therefore, *L. pneumophila*-specific proteins, which do not occur in the other species or which do not show comparable functional properties in non-*pneumophila* species, are of particular interest. *L. pneumophila* expresses a variety of phospholipases potentially involved in disease-promoting processes and development of pneumonia. To address the impact of *L. pneumophila* phospholipases on bacterial pathogenicity, the identification of these enzymes at the protein and gene level started more than 10 years ago. Up to now, it is known that the genomes of several *L. pneumophila* strains contain at least 15 genes coding for potential PLAs. They can be classified into three different families, the GDSL lipase family, the PlaB family, and the PLP family (Fig. 2) (Banerji et al. 2008; Lang and Flieger 2011). Further, there are at least three known PLC enzymes (Aragon et al. 2002; Aurass et al. 2013; McCoy-Simandle et al. 2011) and one PLD (Fig. 2) (Viner et al. 2012). Paralogs of these phospholipases are mostly found in other yet genome-sequenced *Legionella* species, such as *L. drancourtii* or *L. longbeachae* (Cazalet et al. 2010; Kozak et al. 2010; Moliner et al. 2009). However, some are so far only detected in *L. pneumophila*, such as VipD/PatA, PlcA, and LpdA (Aurass et al. 2013; Lang and Flieger 2011; Viner et al. 2012). The following sections summarize the current knowledge on *L. pneumophila* phospholipases A, C, and D.

3.1 Phospholipases A: GDSL Lipase Family, PlaB, Patatin-Like Protein Family

GDSL Lipase Family Three different GDSL enzymes, designated PlaA, PlaC, and PlaD, are coded in the *L. pneumophila* genomes (Fig. 2). The proteins are homologous to SatA of *A. salmonicida* and contain the five characteristic blocks conserved in GDSL enzymes (Akoh et al. 2004; Banerji et al. 2008; Cazalet et al. 2004; Chien et al. 2004; D'Auria et al. 2010; Glöckner et al. 2008; Lang et al. 2012; Upton and Buckley 1995). Orthologs of the three GDSL enzymes are also encoded in non-*pneumophila* species such as *L. drancourtii* (strain LLAP12) and *L. longbeachae* (strains NSW150 and D-4968) (Cazalet et al. 2010; Kozak et al. 2010; Lang and Flieger 2011; Moliner et al. 2009). Surprisingly, PlaA and PlaD are each represented by two homologs within *L. drancourtii*. Furthermore, *L. longbeachae* comprises two PlaD homologs (Lang and Flieger 2011).

PlaA, the first characterized GDSL enzyme of *L. pneumophila,* was identified by N-terminal sequencing of a purified LPLA from culture supernatant (Flieger et al. 2001). Furthermore, the supernatant of a generated *plaA* mutant lost more than 80 % of its lysophosphatidylglycerol- (LPG) and LPC-hydrolyzing LPLA activities and possesses reduced PLA and lipase activity. Those data implied that PlaA is the most prominent secreted LPLA of *L. pneumophila* with additional less prominent PLA and lipase activities. Since, the *plaA* mutant still showed high PLA and some LPLA activities, presence of other secreted PLA/LPLA enzymes was expected (Flieger et al. 2002).

PlaC, the closest relative of PlaA in *L. pneumophila*, possesses predominantly PLA and some LPLA activities. *L. pneumophila* also exhibits secreted GCAT activity (Banerji et al. 2005; Flieger et al. 2002) and a *plaC* mutant as opposed to *plaA* and *plaD* mutants lost its ability to transfer long-chain fatty acids from dipalmitoylphospholipids to cholesterol. Therefore, PlaC is the major secreted GCAT of *L. pneumophila* with additional PLA/LPLA activities (Banerji et al. 2005; Lang et al. 2012). Recently, it was demonstrated that PlaC is not only able to transfer fatty acids to cholesterol but also to ergosterol, a typical membrane lipid of protozoa, fungi, and microalgae (Lang et al. 2012; Raederstorff and Rohmer 1985; Volkman 2003). Surprisingly, PlaA and PlaD also contribute to sterol acylation. However, these two enzymes only transfer short chain fatty acids, whereas PlaC is able to transfer both short and long-chain fatty acids (Lang et al. 2012). PlaC requires an activating factor for development of GCAT activity which is present in the culture supernatant. ProA, a secreted zinc metalloproteinase of *L. pneumophila,* is essential for development of GCAT activity (Banerji et al. 2005; Lang et al. 2012). Elucidation of the activation mechanism determined that ProA directly processes PlaC by cleaving a disulfide loop region and thereby activates the enzyme. Furthermore, site-directed mutagenesis revealed that throughout the GDSL enzymes conserved residues, Ser37, Asp398, and His401, form the typical catalytic triad in PlaC and are responsible for both PLA and GCAT activities (Lang et al. 2012).

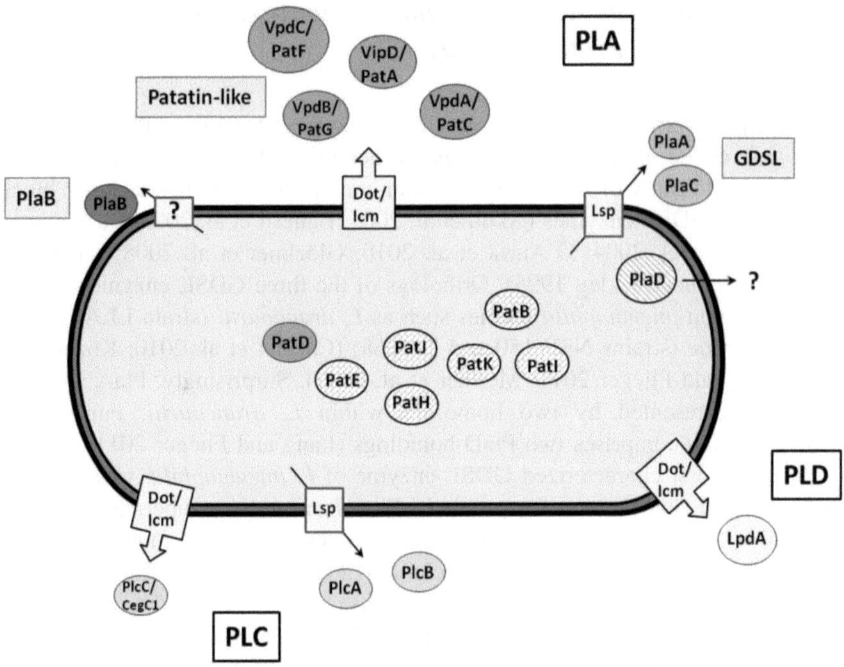

Fig. 2 Overview of *L. pneumophila* phospholipases and their secretion type. The Dot/Icm secretion system injects proteins into the host cells. The question mark represents unknown secretion types. Enzymes with light background (PatB, PatE, PatH-PatK) are not yet experimentally characterized. For references please refer to the manuscript

It is not much known about the third and largest member of the GDSL family in *L. pneumophila*. A special feature of PlaD is a C-terminal appendage of about 170 amino acids after the last homology block missing in PlaA and PlaC (Banerji et al. 2008). In addition to LPLA activity, PlaD also displays some PLA and acyltransferase activities which are reduced upon addition of ProA suggesting independence from activation by ProA. Similarly to PlaC, PlaA acyltransferase activity was only found after activation by ProA (Lang et al. 2012).

In terms of secretion and localization, it is known, that PlaA and PlaC are secreted proteins containing a predicted signal peptide. Whereas the fact that PlaD seems to have no predicted N-terminal signal peptide implies a cell-associated localization (Banerji et al. 2005, 2008). PlaA and PlaC are found in the culture supernatant of *L. pneumophila* and the latter also associates with outer membrane vesicles (Galka et al. 2008). To clarify the mode of PlaA and PlaC secretion, type II secretion mutant culture supernatants were tested for LPLA and GCAT activities and further for protein presence by secretome analysis. The data indicate that both enzymes are type II-dependently secreted (Aragon et al. 2000; Banerji et al. 2005; DebRoy et al. 2006; Flieger et al. 2001, 2002; Rossier and Cianciotto 2001, 2005). It should be noted that the supernatants of type II secretion mutants still show residual PLA/LPLA activities (Flieger et al. 2001; Rossier and Cianciotto 2001). Therefore, it is likely that further lipolytic enzymes are secreted by other secretion mechanisms.

Furthermore, secreted PLA activity is reduced but LPLA activity is increased in *L. pneumophila* mutants of the two-component regulatory system LetA/S and the alternative sigma factor RpoS, important for the switch from the replicative to the transmissive phase (Broich et al. 2006; Hales and Shuman 1999; Hammer et al. 2002; Lynch et al. 2003). The data indicate the induction of a secreted PLA by LetA/S and RpoS (Broich et al. 2006) and PlaC is a possible candidate. Indeed, GCAT activity was reduced and *plaC* mRNA was decreased in *letA* and *rpoS* mutants, implying a direct or indirect regulation of *plaC* expression (Broich et al. 2006). The data further point to a repression of a major LPLA, such as PlaA, by means of LetA/S, and RpoS (Broich et al. 2006). Microarray analysis comparing the expression profile during *Acanthamoeba* infection or during growth in broth revealed that *plaC* and *plaD* are upregulated whereas *plaA* expression was upregulated during amoeba infection but downregulated when the bacteria were grown in broth (Brüggemann et al. 2006; Weissenmayer et al. 2011). These data suggest that the three genes might be important at the later stages of an amoeba infection. Accordingly, Creasey and Isberg showed that PlaA similar to another GDSL enzyme, S. Typhimurium SseJ, promotes vacuole disruption in the absence of the type IVB-secreted effector SdhA (Creasey and Isberg, 2012).

PlaB *L. pneumophila* PlaB, the major cell-associated PLA/LPLA with hemolytic activity, shares no significant homology to previously described phospholipases. This enzyme only shows sequence homology to hypothetical and uncharacterized proteins of water-associated bacteria, such as *P. aeruginosa, Shewanella paleana, Marinobacter algicola,* and *Persephonella marina* (Bender et al. 2009; Flieger et al. 2004). Very minor protein homology was found for the secreted lipase LipB. Whether this protein also possesses PLA activity in addition to lipase activity is not known (Aragon et al. 2002; Flieger et al. 2004). Furthermore, paralogs of *plaB* are also conserved in non-*pneumophila Legionella* species, and some of these, such as *L. gormanii* and *L. spiritensis* showed cell-associated PLA/LPLA activities comparable with *L. pneumophila.* However, PC-hydrolyzing capacity associated with *L. pneumophila* PlaB was most prominent compared to other *Legionella* species (Bender et al. 2009).

Characterization of PlaB enzymatic profile revealed that a *L. pneumophila plaB* mutant showed no quantitative changes in the secreted PLA/LPLA/GCAT activities but instead an almost complete loss of the cell-associated PLA/LPLA activities. Those observations indicate that PlaB is the most prominent cell-associated PLA/LPLA preferentially hydrolyzing PG and PC as well as the respective lysophospholipids with long-chain fatty acids. Notably, these cell-associated PLA/LPLA activities of *L. pneumophila* exceeded about 100-fold the secreted phospholipolytic activities (Bender et al. 2009; Flieger et al. 2004). Furthermore, because PlaB is the first characterized member of a novel lipase family, it was necessary to investigate the residues essential for enzymatic activity. Site-directed mutagenesis revealed a typical catalytic triad of Ser85, Asp203, and His251 embedded into uncommon sequence motifs (see Sect. 2.1) and located within the N-terminal region of PlaB. A Ser129 mutation interestingly causes a ~90 % reduction in PC- but not PG-hydrolyzing PLA activity. The mutated protein was not hemolytic to human red blood cells corroborating the direct linkage between PC hydrolysis and the hemolytic potential of PlaB (Bender et al. 2009). Sole detection of PLA activity therefore may not be sufficient to define a virulence factor. Accordingly, the fact that the enzyme targets important eukaryotic lipids, such as PC or PS, and comprises the major cell-associated PLA/LPLA activity, might be crucial for host cell infection. Indeed, PlaB of *L. spiritensis*, a *Legionella* species so far not associated with known disease cases (Fang et al. 1989; Muder and Victor 2002), showed a less prominent PC-targeting PLA activity and accordingly a 50 % reduced hemolytic activity towards human red blood cells compared with *L. pneumophila* PlaB (Bender et al. 2009).

plaB expression in *L. pneumophila* is most prominent during early exponential growth phase in broth media and decreases afterwards, indicating that expression occurs at an early bacterial life stage. However, the corresponding hemolytic and PC-targeting PLA activities peak during the late-exponential growth phase (Schunder et al. 2010; Weissenmayer et al. 2011). Microarray analysis did not show significant/clear changes of *plaB* expression during amoeba infection, suggesting that mRNA production might be induced early after infection and is kept at this level during the infection cycle (Brüggemann et al. 2006; Weissenmayer et al. 2011). Bacteria-associated PLA/LPLA activities were detected during amoeba and human cell infections with *L. pneumophila* and a *plaB* mutant and demonstrate that PlaB is presumably the most prominent PLA/LPLA of the pathogen during intracellular infection (Bender et al. 2009). To obtain further information about the localization of PlaB, cell fractionation experiments together with detection of hemolytic and PLA activity determined the presence of PlaB in the bacterial outer membrane. Proteinase K accessibility of PlaB on intact *L. pneumophila* cells additionally confirmed this localization. These data indicate a surface-exposition of PlaB and therefore a direct interaction with host cell lipids. However, the process of translocation of PlaB through the inner and outer membranes as well as of anchoring to the bacterial surface is unknown. None of the tested *L. pneumophila* secretion systems (Lss, Tat, Lsp, Dot/Icm) is essential for PlaB export (Schunder et al. 2010). Furthermore, a Sec or Tat signal sequence is not predicted for PlaB which would be necessary to drive translocation for example via type II or type V secretion systems (De Buck et al. 2007; Schunder et al. 2010).

Patatin-Like Protein Family Most yet genome-sequenced *L. pneumophila* strains encode eleven PLPs, designated VipD/PatA, PatB, VpdA/PatC, PatD, PatE, VpdC/PatF, VpdB/PatG, PatH to PatK (Fig. 2) (Aurass et al. 2009; Banerji et al. 2008; Lang and Flieger 2011; Shohdy et al. 2005; VanRheenen et al. 2006). Consequently, *L. pneumophila* is the bacterium with the highest density of PLP genes per number of protein-coding sequences; followed by *M. tuberculosis* with 8 PLPs. The *L. pneumophila* genomes of the strains Lens and Corby interestingly do not comprise VipD/PatA and therefore only encode ten PLPs (Banerji et al. 2008). Five *L. pneumophila* PLPs (VipD/PatA, VpdA/PatC, VpdC/PatF, VpdB/PatG, PatD) were characterized previously (Aurass et al. 2009; Ku et al. 2012; Shohdy et al. 2005; VanRheenen et al. 2006).

One of the more intensely investigated PLP, VipD/PatA, was found by screening a *L. pneumophila* genomic library for vacuole protein sorting (VPS) defects in yeast. Furthermore, VipD lacking the N-terminal half containing the patatin catalytic domain disturbed the late secretory pathway between the endoplasmic reticulum, Golgi membrane, and the vacuole more intensely than the full-length protein (Shohdy et al. 2005). Ku et al. recently presented the crystal structure of VipD and show that the C-terminus of VipD interferes with endosomal trafficking via selective interaction with Rab5 and Rab22, two key regulators of endosomal vesicle trafficking (Ku et al. 2012). Interestingly, the expression of VipD was not toxic to yeast in contrast to the high cytotoxicity of its *P. aeruginosa* homolog, ExoU, depending on the PLP domain (Finck-Barbançon et al. 1997; Sato et al. 2003; Shohdy et al. 2005; VanRheenen et al. 2006). Only overexpression of VipD in yeast resulted in poor growth (Heidtman et al. 2009; VanRheenen et al. 2006; Viner et al. 2012). Further, PLA_2 activity was detected for VipD using a fluorogenic phospholipid substrate (red/green BODIPY PC-A2) but substrate specificity of VipD was not addressed. Mutations of Ser73 and Asp288, conserved among bacterial PLP, abolish activity indicating that these residues form the catalytic dyad (Ku et al. 2012).

Only limited information is available for the three closest homologs of VipD, including that VpdA and VpdB are similar in size to VipD, whereas VpdC contains an additional N-terminal extension of > 200 amino acids (VanRheenen et al. 2006). VpdA and VpdC induce a strong lethal effect when expressed in yeast (Heidtman et al. 2009; Viner et al. 2012) and VipD as well as VpdA cause secretory defects in yeast (Heidtman et al. 2009).

Enzymatic activity toward phospholipid substrates was shown for PatD which in addition to PlaB, is a cell-associated PLA/LPLA contributing to about 40 % of the total cell-associated lipolytic activities (Aurass et al. 2009). *patD* is arranged in an operon together with *bdhA*, encoding for a protein with homology to 3-hydroxybutyrate (3-HB) dehydrogenase of *Sinorhizobium* spp. (Aneja and Charles 2005; Aurass et al. 2009). 3-HB dehydrogenases oxidize depolymerized poly-3-hydroxybutyrate (PHB), a common bacterial carbon and energy storage compound, to acetoacetate, thus supporting metabolization of energy reserves (Anderson and Dawes 1990; Aneja and Charles 2005; Jendrossek and Handrick 2002). A *L. pneumophila bdhA-patD* mutant showed an increased number of PHB granula in comparison to the wild type suggesting an involvement of the operon in the PHB metabolism. However, it is not clarified whether PatD is directly involved in PHB mobilization, for example as a PHB depolymerase, because this enzyme is structurally related to PHB depolymerases (Aurass et al. 2009; Papageorgiou et al. 2008; Rydel et al. 2003). Interestingly, a protein with homology to a classical PHB depolymerase is not evidently encoded in the *L. pneumophila* genome (Aurass et al. 2009).

Only little is known in terms of PLP localization and secretion. PatD is a predicted cytoplasmic membrane protein and does not show a predicted signal peptide. As outlined above, enzymatic assays of *L. pneumophila* wild type and *patD* mutant cell lysates suggested an association with the bacterial cell (Aurass et al. 2009). Since PatD may get in touch with storage lipids, it is conceivable, that this protein is localized in the cytoplasm, at the inner membrane or even at lipid inclusion membranes (Aurass et al. 2009; Lang and Flieger 2011). VipD, VpdA, VdpB, and VpdC are translocated by the Dot/Icm type IVB secretion system into the host cell (Shohdy et al. 2005; VanRheenen et al. 2006; Zhu et al. 2011). Recently, Ku et al. demonstrated that VipD localizes to early endosomes by means of the C-terminal domain and blocks endosomal vesicle trafficking (Ku et al. 2012). However, the targets of VpdA, VdpB, and VpdC have not yet been discovered. Microarray analysis during an infection of *Acanthamoeba castellanii* revealed different expression patterns of the PLP family members in *L. pneumophila*. Brüggemann et al. showed that seven PLP genes (*vipD/patA, patB, vpdA/patC, patD, patE, patI, patK)* are upregulated from the replicative to the transmissive phase. Particularly *vipD/patA, patD, patE,* and *patI* were 8- to 11-fold increasingly expressed (Brüggemann et al. 2006). Weissenmayer et al. detected in addition to *vipD/patA, patD, patI,* and *patK*, also *patH* among the upregulated genes (Weissenmayer et al. 2011). Presumably, some of these PLPs might be very important in the late stage of intracellular growth or for induction of a new infection cycle. This coincides with increased expression of many other effectors in early stationary growth phase, the phase at which *L. pneumophila* is most virulent (Byrne and Swanson 1998; Conover et al. 2003; Luo and Isberg 2004).

3.2 Legionella *Phospholipases C: PlcA, PlcB, and PlcC*

The sequenced *L. pneumophila* genomes encode three phospholipases C, designated PlcA, PlcB, and PlcC/CegC1 (Fig. 2) (Aurass et al. 2013; Cazalet et al. 2004; Chien et al. 2004; D'Auria et al. 2010; Glöckner et al. 2008). *plcA* is so far only found in *L. pneumophila* genomes whereas *plcB* was also conserved in the two *non-pneumophila* strains of *L. longbeachae* (NSW150 and D-4968), and *plcC/cegC1* is present in all so far genome-sequenced strains (Aurass et al. 2013; Cazalet et al. 2010; Kozak et al. 2010; Moliner et al. 2009; Qin et al. 2012). PlcC/CegC1 was described as a toxic type IVB-secreted *L. pneumophila* effector protein when expressed in yeast (Altman and Segal 2008; Heidtman et al. 2009; Huang et al. 2011). Interestingly, the three proteins are transcriptionally induced during host cell infection or growth in broth (Brüggemann et al. 2006; Faucher et al. 2011; Weissenmayer et al. 2011).

Initially, a secreted hydrolytic activity releasing para-nitrophenol (pNP) from the artificial water-soluble substrate para-nitrophenylphosphorylcholine (pNPPC) was found in *L. pneumophila* which may indicate the existence of PLC activity (Aragon et al. 2002; Baine 1985, 1988; Flieger et al. 2000a). The release of the typical PLC reaction products 1,2-DG or phosphorylalcohol from phospholipids was recently shown (Aurass et al. 2013). PlcC expressed in *E. coli* hydrolyzes a broad phospholipid spectrum, including PC, PG, and PI. Addition of Zn^{2+} ions activates, while EDTA inhibits PlcC-derived PLC activity. Protein homology search reveals that the three *L. pneumophila* enzymes and *P. fluorescens* PC-PLC share conserved domains also present in uncharacterized fungal proteins. Fifteen conserved amino acids are essential for enzyme activity as identified via PlcC mutagenesis. Analysis of defined *L. pneumophila* knockout mutants indicates Lsp-dependent export of PG-hydrolyzing PLC activity. PlcA and PlcB exhibit PG-specific activity after activation by *L. pneumophila* culture supernatant and contain a predicted Sec signal sequence. In line with the reported requirement of host cell contact for Dot/Icm-dependent effector translocation, PlcC shows cell-associated PC-specific PLC activity after bacterial growth in broth (Aurass et al. 2013; Preuss et al. 2001; Rossignol et al. 2008).

Twin-arginine signal peptide-containing PlcA is responsible for about 70 % secreted *L. pneumophila* pNPPC-hydrolase activity. Additionally, a *L. pneumophila* Lsp type II secretion mutant showed an 85 % reduction in secreted pNPPC hydrolase activity, which seems to partially depend (about 30 %) on the Tat pathway (Aragon et al. 2002; Rossier and Cianciotto 2005). The detected differences of about 30 % and about 70 % suggest that PlcA might be translocated either by both the Sec and Tat pathways or by the Sec pathway in absence of the Tat system (Rossier and Cianciotto 2005). PlcB also contributes to pNPPC hydrolase activity (Aurass et al. 2013; McCoy-Simandle et al. 2011). In summary, *L. pneumophila* comprises a novel Zn^{2+}-dependent PLC family.

3.3 Phospholipase D: LpdA

The sequenced genomes of *L. pneumophila* and none of the so far sequenced non-*pneumophila* species encode the PLD LpdA, a type IVB-secreted effector (Fig. 2) (Gomez-Valero et al. 2011; Viner et al. 2012; Zhu et al. 2011). LpdA possesses a functional HKD-PLD domain and localizes to the LCV together with another effector, LecE, where they probably manipulate the phagosome phospholipid composition. Furthermore, it was shown, that LpdA generates PA which may be further converted to 1,2-DG and two Lys residues are important for PLD activity (Viner et al. 2012).

Interestingly, microarray analysis of *A. castellanii* infections revealed that the *lpdA* gene is upregulated from the replicative to the transmissive phase (Brüggemann et al. 2006; Weissenmayer et al. 2011).

4 Importance of *Legionella* Phospholipases for Host Cell Infection

As mentioned before, phospholipases are classical virulence factors of pathogenic bacteria, which contribute to a variety of processes such as invasion and modulation of the host to stimulate intracellular survival and escape from the LCV.

The secreted GDSL enzyme PlaA promotes vacuole disruption and host cell death, suggesting that it may directly target vacuole membrane lipids (Creasey and Isberg 2012). However, the tested *plaA* and *plaC* single knockout mutants did not show any defect for host cell infection in either *A. castellanii* or human macrophages (Banerji et al. 2005; Flieger et al. 2002). Similar results were observed for a *plaA* mutant tested in a mouse infection model (DebRoy et al. 2006). Recent studies interestingly determined a replication defect of a *plaC* mutant in *Hartmannella vermiformis* and *Naegleria lovaniensis* (Tyson et al. 2013). It is likely that the three GDSL enzymes may compensate in part for the absence of a specific PLA/LPLA activity in some hosts and therefore may have overlapping or synergistic roles in *L. pneumophila* virulence. In case of *B. anthracis,* disruption of all three PLC genes was necessary to obtain attenuation in a murine model, reduction of bacterial growth as well as survival in macrophages (Heffernan et al. 2006). Furthermore, a more efficient attenuation in cell culture and in a mouse model was obtained by deleting an additional gene encoding anthrolysin O (Heffernan et al. 2007). A similar example is known for the three PLCs, PlcA, PlcB, and PlcC, of *M. tuberculosis*, the causative agent of tuberculosis. Disruption of the three genes impaired the ability of the pathogen to multiply in the lungs and in the spleen of infected mice. Virulence was partially restored by complementing the triple mutant with the single genes (Raynaud et al. 2002).

Further, it has been shown that the Dot/Icm-secreted effector SdhA counteracts PlaA disruptive forces and comparable antagonists may also exist for PlaC and PlaD. Indeed protein homologs of SdhA, such as SdhB and SidH, are encoded by *L. pneumophila* (Creasey and Isberg 2012; Laguna et al. 2006; Luo and Isberg 2004; Ninio et al. 2005). Therefore, the phenotype of the GDSL mutants may only be detected after elimination of the antagonist or at specific time points when the antagonists are not active/present.

The host lipids PC and PG are components of lung surfactant as well as host membranes and are efficiently hydrolyzed by the described *L. pneumophila* PLA activities to lysophospholipids and FFA. LPLA activities thereupon degrade and detoxify the pore-forming agent and signal transducer LPC (Prokazova et al. 1998; Weltzien 1979). The precise adjustment of the GDSL enzyme activities might support bacterial survival by detoxification of LPC and might also guarantee host cell survival and integrity until the end of the infection cycle. Notably, the detoxifying function of LPLA activities promoting bacterial survival was particularly shown for PlaA (Flieger et al. 2002). Furthermore, the release of FFA or the transfer of the fatty acid to an acceptor molecule, such as cholesterol, and the resulting release of LPC might influence host signal transduction. PlaC is known as a secreted *L. pneumophila* enzyme with such a GCAT activity (Lang et al. 2012). The other two GDSL enzymes PlaA and PlaD might also contribute to sterol acylation under in vivo conditions as has been shown in the case of ergosterol as an acceptor. Conceivable, GCAT activity can modify eukaryotic membranes with important sterol-rich regions and therefore may influence membrane organization or receptor presentation (Banerji et al. 2005; Ruiz-Albert et al. 2002).

Moreover, the remaining *L. pneumophila* PLA enzymes (PlaB, PLPs) may also modulate the host cell. An important role as a virulence factor is attributed to PlaB. It was shown that a *plaB* mutant was impaired for replication in the lungs and dissemination to the spleen in an in vivo guinea pig infection model. This was shown by a 400-fold increase of CFU in the lung after 2 days infection for wild type as compared to the only 20-fold increase of the *plaB* mutant. The *plaB* mutant bacteria spread 100-fold less to the spleen than the wild type. Furthermore, distinct destruction of lung tissue and indication of inflammation was evident only for the wild type infection and barely noticeable with the *plaB* mutant (Schunder et al. 2010). These observations can be explained by the cytolytic potential of PlaB, which may facilitate the dissemination of other organs. It is also possible that PlaB contributes to the modulation of the host immune response, because recruitment of macrophages was observed especially in *L. pneumophila* wild type infections but was reduced in those of a *plaB* mutant; an observation which may be influenced by the reduced ability of the mutant to generate lipid secondary messengers, such as the lysophospholipid LPC (Schunder et al. 2010). Another possible mechanism of PlaB action could be the destruction of lung surfactant, because of its substrate specificity for PC and PG (Flieger et al. 2004). This might explain the efficient colonization of the lungs by wild type bacteria in comparison to the *plaB* mutant. The wild type bacteria may therefore have a better accessibility to alveolar macrophages and subsequently may replicate more efficiently (Schunder et al. 2010).

L. pneumophila single knockout mutants for PlaB, VipD, VpdA, VpdB, and VpdC were not attenuated in in vitro host cell infection models of macrophages and amoebae as already mentioned for the GDSL enzyme single mutants (Bender et al. 2009; Flieger et al. 2004; VanRheenen et al. 2006). However, VipD produces trafficking defects when expressed in yeast but does not result in a detectable toxicity (Shohdy et al. 2005). Furthermore, the *L. pneumophila* strain bearing a quadruple mutation for the four PLPs (VipD, VpdA, VpdB, VpdC) was not significantly impaired for growth in macrophages or in *Dictyostelium discoideum* (VanRheenen et al. 2006). The fact that *L. pneumophila* exhibits protein families with a multitude of members such as PLPs increases the possibility of their functional redundancy. Therefore, an essential role in host cell infection might be masked, which does not mean that the single PLPs or the PLPs as a whole might not play a significant role in host cell modification. Furthermore, in vitro models, such as macrophages, are incomplete models and do not reflect the natural complex in vivo conditions which may be more restrictive. Enzymes with PLA activities (including host and bacterial proteins) were also shown to play a significant role in membrane trafficking through the eukaryotic secretory pathway

(Brown et al. 2003; Choukroun et al. 2000; de Figueiredo et al. 2000; Drecktrah and Brown 1999; Ge and Shao 2011; Gendrin et al. 2012; Schmidt et al. 2010). Heidtman et al. identified *L. pneumophila* proteins, including the two PLPs VipD and VpdA that delay trafficking of host cell secretory proteins to the yeast vacuole (Heidtman et al. 2009). Interestingly, VipD deactivates the key endosomal regulators Rab5 and Rab22, in a way that subsequent lysosomal degradation of the bacteria might be prevented (Ku et al. 2012).

Nevertheless, a *bdhA/patD* knockout mutant shows a prominent infection defect in macrophage and amoeba infection models (Aurass et al. 2009). This defect is comparable to a Dot/Icm secretion system mutant and indicates that PHB metabolism seems to be important in the life of *L. pneumophila*. One plausible scenario is that provision of energy by the cleavage of PHB may energize the Dot/Icm secretion system or another important component (Aurass et al. 2009; Lang and Flieger 2011).

The three *L. pneumophila* PLCs may release the typical reaction products such as 1,2-DG or phosphorylalcohol from host cell membranes and thereby contribute to pathogenesis. Therefore it is conceivable that secreted PLCs contribute to host membrane lysis or degradation of lung surfactant by means of hydrolyzing phospholipids, such as PC and PG. PLCs might also influence the interaction of the bacterium with host phagocytes by mimicking the activation of cellular PLCs, which catalyze the hydrolysis of PIP_2 to the secondary messengers IP_3 and 1,2-DG (Dowling et al. 1992). And indeed low level hydrolysis of PI has been detected for PlcC/CegC1 (Aurass et al. 2013). 1,2-DG as a secondary messenger may then activate host cell PKC and various signal transduction pathways such as the arachidonic acid pathway (see Sect. 2.2). *L. pneumophila plcA, plcB,* and *plcC/cegC1* individually, in two or three gene combinations are not required for infection of lung epithelial cells, macrophages, and amoebae. Knockout mutants intracellularly replicate to levels similar to the wild type indicating that these genes are not required for in vitro infections (Aragon et al. 2002; Aurass et al. 2013; McCoy-Simandle et al. 2011). Furthermore, a *plcA* mutant does not show a defect in an in vivo mouse infection (DebRoy et al. 2006). Determination of cytokine output of infected macrophages indicates that PlcA and PlcB are also not required for limitation of the cytokine response as observed for *L. pneumophila* wild type and specifically the secreted zinc metalloproteinase ProA (McCoy-Simandle et al. 2011). Most interestingly, a PLC triple mutant, but not single or double mutants, exhibited reduced host killing in a *Galleria mellonella* infection model, highlighting the importance of the three PLCs in pathogenesis (Aurass et al. 2013).

L. pneumophila effectors likely manipulate phospholipid compositions in many ways to result in a successful infection. PLD activity of *L. pneumophila* LpdA together with the *Legionella* type IV-secreted effector protein LecE affects the in vivo levels and distribution of 1,2-DG and PA in mammalian cells (see also Sect. 3.3); a mechanism by which the intracellular pathogen might change the lipid composition of the phagosome. This process might lead to the recruitment of specific bacterial and host cell factors to the vacuole, such as protein kinase D or PKC (Viner et al. 2012). Interestingly, an intracellular growth defect of an *lpdA*

Table 1 Overview of possible impacts of *L. pneumophila* phospholipases on host cells, for references please refer to manuscript text

Type of phospholipase	*L. pneumophila* phospholipases	Reaction product after phosphatidylcholine cleavage	Possible impacts on host cells
PLA	**GDSL family**:	Fatty acid	Membrane lysis and cell destruction
	PlaA, PlaC, PlaD	(e.g., **arachidonic acid**)	Generation of secondary messenger LPC: activates RhoA in a PKCα-dependent manner, enhances superoxide anion production, increases expression of chemokines and permeability of cell membranes
	PlaB family: PlaB	and	
		Lysophosphatidylcholine	Triggers arachidonic acid-dependent inflammatory cascade
	Patatin-like family: VipD/PatA, VpdA/PatB, PatC, PatD, PatE,	(Glycerophosphorylcholine)	Production of secondary messenger eicosanoids influences innate and acquired immune response
			Production of leukotrienes for example increases vascular permeability
	VpdC/PatF, VpdB/PatG, PatH, PatI, PatJ, PatK		Production of prostaglandins for example causes vasoactive platelet aggregation, regulates Ca^{2+} influx, inflammation, and muscle contraction
			Induction of cytokines recruits neutrophils and macrophages
			Regulation of membrane extensions
PLC	PlcA, PlcB, PlcC/CegC1	**1,2-Diacylglycerol**	Membrane lysis and cell destruction
			Generation of secondary messenger 1,2-DG: activates PKC, arachidonic acid cascade, and endogenous PLA and PI-PLC
		and Phosphorylcholine	Hydrolysis of PI and its phosphorylated variants yields secondary messengers, which activate Ca^{2+} gates and PKC, modulate cell processes and growth
			Production of intracellular mediators, such as adhesion molecules, IL-8, TNF-α, leukotrienes, and prostaglandins
PLD	LpdA	**Phosphatidic acid**	Membrane lysis and cell destruction
		and Choline	Production of a signaling molecule LPA: activates Ras, Rac, and Rho GTPases through G-coupled receptors, stimulates migration, cell proliferation, and cytoskeletal changes

mutant in *A. castellanii* was not detected. Although *lpdA* expressed in wild type yeast does not show any lethal effect, the expression in a *dgk1* (diacylglycerol kinase gene) deletion mutant clearly shows a lethal effect and LpdA furthermore enhances the lethality of LecE (Viner et al. 2012).

All in all, *L. pneumophila* seems to apply a huge variety of proteins modulating cellular functions for its benefit, including the above described phospholipases (Table 1). An interesting and significant role of bacterial phospholipases may involve the mimicking or substitution of host cell phospholipases for advantage of the pathogen, which was also shown for some *L. pneumophila* eukaryotic-like proteins (Gomez-Valero et al. 2011). Therefore, the PLA/LPLA, PLC, and PLD enzymes represent a repertoire of tools for *L. pneumophila* to attack host-cell membranes, release secondary messengers as well as cytotoxic LPC, an important mediator of apoptosis and inflammatory signaling, inducing the release of chemokines (Linkous and Yazlovitskaya 2010; Murugesan et al. 2003; Takahashi et al. 2002). Furthermore, GCAT activity is an important instrument for host cell manipulation, specifically cholesterol modification. The destruction of lipid monolayers such as pulmonary surfactant represents a further essential virulence mechanism of *L. pneumophila* phospholipases, which may lead to complications in patients and additionally explain the development of pneumonia during a *L. pneumophila* infection (Flieger et al. 2000b).

5 Conclusion

L. pneumophila possesses 15 distinct potential as well as confirmed PLA/LPLA enzymes, three additional PLC enzymes, and one PLD which are cell-associated or secreted by at least two different secretion systems, the type IVB Dot/Icm and type II Lsp systems (Fig. 2). The abundance of phospholipases, highlighting functional redundancy in lipid usage and modulation, argues their importance for *L. pneumophila*. The fact, that these phospholipases additionally show on the one hand cytotoxicity and on the other hand signaling properties increases their action potential as virulence factors. So far, PlaB, PlaC, PatD, and the combination of the three PLCs are essential for host cell infections (Aurass et al. 2009, 2013; Schunder et al. 2010; Tyson et al. 2013). Thus, further studies concerning the individual modes of phospholipase action, such as secretion mechanism, localization, cellular targets, functional and biochemical properties, are required to gain a better understanding of their function during host infection. Future studies may answer the question as to whether the mentioned enzymes take part in phagosomal membrane modulation/destruction which may also influence signaling events in the host. It is further interesting to analyze the specific action site of PlaC, the major secreted GCAT of *L. pneumophila*, as well as the other phospholipases in the host cell and consequences in terms of infection. In addition, whether PatD is directly involved in PHB mobilization and delivery of the resulting energy for virulence determinants, for instance the Dot/Icm secretion machinery, has still to be clarified. Further investigation is necessary concerning the seemingly unusual secretion mechanism of PlaB, its attachment to the outer membrane, and more importantly its precise action during host infection. It needs to be resolved, whether VipD supports survival of *L. pneumophila* in macrophages via its signal blocking property within endosomal trafficking. The recently determined crystal structure of VipD helps to figure out the thitherto unknown function of its C-terminal domain (Ku et al. 2012). Such analyses facilitate understanding the function of additional protein domains attached to the catalytic domain of phospholipases, such as also found in the case of PlaD, VpdA, VpdB, VpdC, PlaB, as well as PlcC/CegC1. In addition, crystal structure analysis may confirm genetically acquired data on catalytic residues and moreover may give information about interaction between the proteins and their substrates which will be invaluable for drug discovery. The identification of interaction partners for the secreted proteins may contribute to a better understanding of molecular processes during the infection cycle of *L. pneumophila* and may also help decipher biological functions of the single phospholipase.

References

Akoh CC, Lee G-C, Liaw Y-C, Huang T-H, Shaw J-F (2004) GDSL family of serine esterases/ lipases. Prog Lipid Res 43(6):534

Alberti-Segui C, Goeden KR, Higgins DE (2007) Differential function of Listeria monocytogenes listeriolysin O and phospholipases C in vacuolar dissolution following cell-to-cell spread. Cell Microbiol 9(1):179–195

Albert-Weissenberger C, Cazalet C, Buchrieser C (2007) Legionella pneumophila—a human pathogen that co-evolved with fresh water protozoa. Cell Mol Life Sci 64(4):432–448

Altman E, Segal G (2008) The response regulator CpxR directly regulates expression of several Legionella pneumophila icm/dot components as well as new translocated substrates. J Bacteriol 190(6):1985–1996

Anderson AJ, Dawes EA (1990) Occurrence, metabolism, metabolic role, and industrial uses of bacterial polyhydroxyalkanoates. Microbiol Rev 54(4):450–472

Anderson C, Pinsirodom P, Parkin KL (2002) Hydrolytic selectivity of patatin (lipid acyl hydrolase) from potato (Solanum tuberosum L.) tubers toward various lipids. J Food Biochem 26(1):63–74

Andrews DL, Beames B, Summers M, Park W (1988) Characterization of the lipid acyl hydrolase activity of the major potato (Solanum tuberosum) tuber protein, patatin, by cloning and abundant expression in a baculovirus vector. Biochem J 252(1):199

Aneja P, Charles TC (2005) Characterization of bdhA, encoding the enzyme d-3-hydroxybutyrate dehydrogenase, from Sinorhizobium sp. strain NGR234. FEMS Microbiol Lett 242(1):87–94

Aragon V, Kurtz S, Flieger A, Neumeister B, Cianciotto NP (2000) Secreted Enzymatic Activities of Wild-Type andpilD-Deficient Legionella pneumophila. Infect Immun 68(4):1855–1863

Aragon V, Rossier O, Cianciotto NP (2002) Legionella pneumophila genes that encode lipase and phospholipase C activities. Microbiology 148(7):2223–2231

Arai H, Koizumi H, Aoki J, Inoue K (2002) Platelet-activating factor acetylhydrolase (PAF-AH). J Biochem 131(5):635–640

Arpigny JL, Jaeger K-E (1999) Bacterial lipolytic enzymes: classification and properties. Biochem J 343(Pt 1):177

Aurass P, Pless B, Rydzewski K, Holland G, Bannert N, Flieger A (2009) bdhA-patD operon as a virulence determinant, revealed by a novel large-scale approach for identification of Legionella pneumophila mutants defective for amoeba infection. Appl Environ Microbiol 75(13):4506–4515

Aurass P, Schlegel M, Metwally O, Harding CR, Schroeder GN, Frankel G, Flieger A (2013) The Legionella pneumophila Dot/Icm-secreted effector PlcC/CegC1 together with PlcA and PlcB promotes virulence and belongs to a novel zinc metallophospholipase C family present in bacteria and fungi. J Biolo Chem 288(16):11080–11092

Baine WB (1985) Cytolytic and phospholipase C activity in Legionella species. J Gen Microbiol 131(6):1383–1391

Baine WB (1988) A phospholipase C from the Dallas 1E strain of Legionella pneumophila serogroup 5: purification and characterization of conditions for optimal activity with an artificial substrate. J Gen Microbiol 134(2):489–498

Banerji S, Flieger A (2004) Patatin-like proteins: a new family of lipolytic enzymes present in bacteria? Microbiology 150(3):522–525

Banerji S, Bewersdorff M, Hermes B, Cianciotto NP, Flieger A (2005) Characterization of the major secreted zinc metalloprotease-dependent glycerophospholipid: cholesterol acyltransferase, PlaC, of Legionella pneumophila. Infect Immun 73(5):2899–2909

Banerji S, Aurass P, Flieger A (2008) The manifold phospholipases A of Legionella pneumophila—Identification, export, regulation, and their link to bacterial virulence. Int J Med Microbiol 298(3):169–181

Bender J, Rydzewski K, Broich M, Schunder E, Heuner K, Flieger A (2009) Phospholipase PlaB of Legionella pneumophila represents a novel lipase family protein residues essential for lipolytic activity, substrate specificity and hemolysis. J Biol Chem 284(40):27185–27194

Brassinga AKC, Hiltz MF, Sisson GR, Morash MG, Hill N, Garduno E, Edelstein PH, Garduno RA, Hoffman PS (2003) A 65-kilobase pathogenicity island is unique to Philadelphia-1 strains of Legionella pneumophila. J Bacteriol 185(15):4630–4637

Broich M, Rydzewski K, McNealy TL, Marre R, Flieger A (2006) The global regulatory proteins LetA and RpoS control phospholipase A, lysophospholipase A, acyltransferase, and other hydrolytic activities of Legionella pneumophila JR32. J Bacteriol 188(4):1218–1226

Brown WJ, Chambers K, Doody A (2003) Phospholipase A2 (PLA2) enzymes in membrane trafficking: mediators of membrane shape and function. Traffic 4(4):214–221

Brüggemann H, Hagman A, Jules M, Sismeiro O, Dillies MA, Gouyette C, Kunst F, Steinert M, Heuner K, Coppée JY (2006) Virulence strategies for infecting phagocytes deduced from the in vivo transcriptional program of Legionella pneumophila. Cell Microbiol 8(8):1228–1240

Bryant A, Stevens D (1996) Phospholipase C and perfringolysin O from Clostridium perfringens upregulate endothelial cell-leukocyte adherence molecule 1 and intercellular leukocyte adherence molecule 1 expression and induce interleukin-8 synthesis in cultured human umbilical vein endothelial cells. Infect Immun 64(1):358–362

Buckley JT, Halasa LN, MacIntyre S (1982) Purification and partial characterization of a bacterial phospholipid: cholesterol acyltransferase. J Biol Chem 257(6):3320–3325

Byrne B, Swanson MS (1998) Expression of Legionella pneumophilavirulence traits in response to growth conditions. Infect Immun 66(7):3029–3034

Camilli A, Goldfine H, Portnoy DA (1991) Listeria monocytogenes mutants lacking phosphatidylinositol-specific phospholipase C are avirulent. J Exp Med 173(3):751–754

Camilli A, Tilney LG, Portnoy DA (1993) Dual roles of plcA in Listeria monocytogenes pathogenesis. Mol Microbiol 8(1):143–157

Campocasso A, Boughalmi M, Fournous G, Raoult D, La Scola B (2012) Legionella tunisiensis sp. nov. and Legionella massiliensis sp. nov., isolated from environmental water samples. Int J Syst Evolution Microbiol, 62(Pt 12):3003–3006

Cazalet C, Rusniok C, Brüggemann H, Zidane N, Magnier A, Ma L, Tichit M, Jarraud S, Bouchier C, Vandenesch F (2004) Evidence in the Legionella pneumophila genome for exploitation of host cell functions and high genome plasticity. Nat Genet 36(11):1165–1173

Cazalet C, Gomez-Valero L, Rusniok C, Lomma M, Dervins-Ravault D, Newton HJ, Sansom FM, Jarraud S, Zidane N, Ma L (2010) Analysis of the Legionella longbeachae genome and transcriptome uncovers unique strategies to cause Legionnaires' disease. PLoS Genet 6(2):e1000851

Chien M, Morozova I, Shi S, Sheng H, Chen J, Gomez SM, Asamani G, Hill K, Nuara J, Feder M (2004) The genomic sequence of the accidental pathogen Legionella pneumophila. Science 305(5692):1966–1968

Choukroun GJ, Marshansky V, Gustafson CE, McKee M, Hajjar RJ, Rosenzweig A, Brown D, Bonventre JV (2000) Cytosolic phospholipase A2 regulates Golgi structure and modulates intracellular trafficking of membrane proteins. J Clin Investig 106(8):983–993

Cianciotto NP (2009) Many substrates and functions of type II secretion: lessons learned from Legionella pneumophila. Future Microbiol 4(7):797–805

Conover GM, Derré I, Vogel JP, Isberg RR (2003) The Legionella pneumophila LidA protein: a translocated substrate of the Dot/Icm system associated with maintenance of bacterial integrity. Mol Microbiol 48(2):305–321

Costas MJ, Pinto RM, Cordero PM, Cabezas A, Alves-Pereira I, Cameselle JC, Ribeiro JM (2010) CGDEase, a Pseudomonas fluorescens protein of the PLC/APase superfamily with CDP-ethanolamine and (dihexanoyl) glycerophosphoethanolamine hydrolase activity induced by osmoprotectants under phosphate-deficient conditions. Mol Microbiol 78(6):1556–1576

Creasey EA, Isberg RR (2012) The protein SdhA maintains the integrity of the Legionella-containing vacuole. PNAS 109(9):3481–3486

Cuevas WA, Songer J (1993) Arcanobacterium haemolyticum phospholipase D is genetically and functionally similar to Corynebacterium pseudotuberculosis phospholipase D. Infect Immun 61(10):4310–4316

Cuzick A, Stirling FR, Lindsay SL, Evans TJ (2006) The type III pseudomonal exotoxin U activates the c-Jun NH2-terminal kinase pathway and increases human epithelial interleukin-8 production. Infect Immun 74(7):4104–4113

D'Auria G, Jiménez-Hernández N, Peris-Bondia F, Moya A, Latorre A (2010) Legionella pneumophila pangenome reveals strain-specific virulence factors. BMC Genomics 11(1):181

De Buck E, Anné J, Lammertyn E (2007) The role of protein secretion systems in the virulence of the intracellular pathogen Legionella pneumophila. Microbiology 153(12):3948–3953

de Figueiredo P, Drecktrah D, Polizotto RS, Cole NB, Lippincott-Schwatrz J, Brown WJ (2000) Phospholipase A2 antagonists inhibit constitutive retrograde membrane traffic to the endoplasmic reticulum. Traffic 1(6):504–511

DebRoy S, Dao J, Söderberg M, Rossier O, Cianciotto NP (2006) Legionella pneumophila type II secretome reveals unique exoproteins and a chitinase that promotes bacterial persistence in the lung. PNAS 103(50):19146–19151

Dennis EA, Rhee SG, Billah MM, Hannun Y (1991) Role of phospholipase in generating lipid second messengers in signal transduction. FASEB J 5(7):2068–2077

Di Paolo G, De Camilli P (2006) Phosphoinositides in cell regulation and membrane dynamics. Nature 443(7112):651–657

Diederen B (2008) Legionella spp. and Legionnaires' disease. J Infect 56(1):1–12

Dowling JN, Saha A, Glew R (1992) Virulence factors of the family Legionellaceae. Microbiol Rev 56(1):32

Drecktrah D, Brown WJ (1999) Phospholipase A2 antagonists inhibit nocodazole-induced Golgi ministack formation: evidence of an ER intermediate and constitutive cycling. Mol Biol Cell 10(12):4021–4032

Edwards JL, Apicella MA (2006) Neisseria gonorrhoeae PLD directly interacts with Akt kinase upon infection of primary, human, cervical epithelial cells. Cell Microbiol 8(8):1253–1271

Edwards JL, Entz DD, Apicella MA (2003) Gonococcal phospholipase D modulates the expression and function of complement receptor 3 in primary cervical epithelial cells. Infect Immun 71(11):6381–6391

Ensminger AW, Isberg RR (2009) Legionella pneumophila Dot/Icm translocated substrates: a sum of parts. Curr Opin Microbiol 12(1):67–73

Exton J (1990) Signaling through phosphatidylcholine breakdown. J Biol Chem 265(1):1–4

Fang G-D, YU VL, Vickers RM (1989) Disease due to the Legionellaceae (other than Legionella pneumophila): historical, microbiological, clinical, and epidemiological review. Medicine 68(2):116

Faucher SP, Mueller CA, Shuman HA (2011) Legionella pneumophila transcriptome during intracellular multiplication in human macrophages. Front Microbiol 2:60

Fields BS, Benson RF, Besser RE (2002) Legionella and Legionnaires' disease: 25 years of investigation. Clin Microbiol Rev 15(3):506–526

Finck-Barbançon V, Goranson J, Zhu L, Sawa T, Wiener-Kronish JP, Fleiszig SM, Wu C, Mende-Mueller L, Frank DW (1997) ExoU expression by Pseudomonas aeruginosa correlates with acute cytotoxicity and epithelial injury. Mol Microbiol 25(3):547–557

Flieger A, Gong S, Faigle M, Deeg M, Bartmann P, Neumeister B (2000a) Novel Phospholipase A activity secreted byLegionella species. J Bacteriol 182(5):1321–1327

Flieger A, Gong S, Faigle M, Mayer HA, Kehrer U, Mußotter J, Bartmann P, Neumeister B (2000b) Phospholipase A secreted by Legionella pneumophila destroys alveolar surfactant phospholipids. FEMS Microbiol Lett 188(2):129–133

Flieger A, Gong S, Faigle M, Stevanovic S, Cianciotto NP, Neumeister B (2001) Novel lysophospholipase A secreted byLegionella pneumophila. J Bacteriol 183(6):2121–2124

Flieger A, Neumeister B, Cianciotto NP (2002) Characterization of the gene encoding the major secreted lysophospholipase A of Legionella pneumophila and its role in detoxification of lysophosphatidylcholine. Infect Immun 70(11):6094–6106

Flieger A, Rydzewski K, Banerji S, Broich M, Heuner K (2004) Cloning and characterization of the gene encoding the major cell-associated phospholipase A of Legionella pneumophila, plaB, exhibiting hemolytic activity. Infect Immun 72(5):2648–2658

Galka F, Wai SN, Kusch H, Engelmann S, Hecker M, Schmeck B, Hippenstiel S, Uhlin BE, Steinert M (2008) Proteomic characterization of the whole secretome of Legionella pneumophila and functional analysis of outer membrane vesicles. Infect Immun 76(5): 1825–1836

Ge J, Shao F (2011) Manipulation of host vesicular trafficking and innate immune defence by Legionella Dot/Icm effectors. Cell Microbiol 13(12):1870–1880

Gendrin C, Contreras-Martel C, Bouillot S, Elsen S, Lemaire D, Skoufias DA, Huber P, Attree I, Dessen A (2012) Structural basis of cytotoxicity mediated by the type III secretion toxin ExoU from Pseudomonas aeruginosa. PLoS Pathog 8(4):e1002637

Geng D, Baker DP, Foley SF, Zhou C, Stieglitz K, Roberts MF (1999) A 20-kDa domain is required for phosphatidic acid-induced allosteric activation of phospholipase D from Streptomyces chromofuscus. Biochimica et Biophysica Acta (BBA)-Protein Structure and Molecular Enzymology 1430 (2):234–244***

Geoffroy C, Raveneau J, Beretti J, Lecroisey A, Vazquez-Boland J, Alouf J, Berche P (1991) Purification and characterization of an extracellular 29-kilodalton phospholipase C from Listeria monocytogenes. Infect Immun 59(7):2382–2388

Glöckner G, Albert-Weissenberger C, Weinmann E, Jacobi S, Schunder E, Steinert M, Hacker J, Heuner K (2008) Identification and characterization of a new conjugation/type IVA secretion system (trb/tra) of Legionella pneumophila Corby localized on two mobile genomic islands. Int J Med Microbiol 298(5):411–428

Goldfine H, Knob C, Alford D, Bentz J (1995) Membrane permeabilization by Listeria monocytogenes phosphatidylinositol-specific phospholipase C is independent of phospholipid hydrolysis and cooperative with listeriolysin O. PNAS 92(7):2979–2983

Gomez-Valero L, Rusniok C, Cazalet C, Buchrieser C (2011) Comparative and functional genomics of Legionella identified eukaryotic like proteins as key players in host–pathogen interactions. Front Microbiol 2:208

González-Bulnes P, González-Roura A, Canals D, Delgado A, Casas J, Llebaria A (2010) 2-Aminohydroxamic acid derivatives as inhibitors of Bacillus cereus phosphatidylcholine preferred phospholipase C PC-PLC Bc. Bioorganic Med Chem 18(24):8549–8555

Hales LM, Shuman HA (1999) The Legionella pneumophila rpoS gene is required for growth within Acanthamoeba castellanii. J Bacteriol 181(16):4879–4889

Hammer BK, Tateda ES, Swanson MS (2002) A two-component regulator induces the transmission phenotype of stationary-phase Legionella pneumophila. Mol Microbiol 44(1):107–118

Hansen S, Hansen LK, Hough E (1993) The Crystal Structure of Tris-inhibited Phospholipase C from Bacillus cereus at 1· 9 Å Resolution: the nature of the metal ion in site 2. J Mol Biol 231(3):870–876

Hauser AR (2009) The type III secretion system of Pseudomonas aeruginosa: infection by injection. Nat Rev Microbiol 7(9):654–665

Heffernan BJ, Thomason B, Herring-Palmer A, Shaughnessy L, McDonald R, Fisher N, Huffnagle GB, Hanna P (2006) Bacillus anthracis phospholipases C facilitate macrophage-associated growth and contribute to virulence in a murine model of inhalation anthrax. Infect Immun 74(7):3756–3764

Heffernan BJ, Thomason B, Herring-Palmer A, Hanna P (2007) Bacillus anthracis anthrolysin O and three phospholipases C are functionally redundant in a murine model of inhalation anthrax. FEMS Microbiol Lett 271(1):98–105

Heidtman M, Chen EJ, Moy MY, Isberg RR (2009) Large-scale identification of Legionella pneumophila Dot/Icm substrates that modulate host cell vesicle trafficking pathways. Cell Microbiol 11(2):230–248

Hernandez F, Kirby B, Stanley T, Edelstein P (1980) Legionnaires' disease. Postmortem pathologic findings of 20 cases. Am J Clin Pathol 73(4):488

Hicklin M, Thomason B, Chandler F, Blackmon J (1980) Pathogenesis of acute Legionnaires' disease pneumonia. Am J Clin Pathol 73(4):480

Higgins J, Hitchin B, Low M (1989) Phosphatidylinositol-specific phospholipase C of Bacillus thuringiensis as a probe for the distribution of phosphatidylinositol in hepatocyte membranes. Biochem J 259(3):913

Hilbi H, Jarraud S, Hartland E, Buchrieser C (2010) Update on Legionnaires' disease: pathogenesis, epidemiology, detection and control. Mol Microbiol 76(1):1–11

Hirschberg HJ, Simons JW, Dekker N, Egmond MR (2001) Cloning, expression, purification and characterization of patatin, a novel phospholipase A. Eur J Biochem 268(19):5037–5044

Hodgson A, Bird P, Nisbet I (1990) Cloning, nucleotide sequence, and expression in Escherichia coli of the phospholipase D gene from Corynebacterium pseudotuberculosis. J Bacteriol 172(3):1256–1261

Hough E, Hansen LK, Birknes B, Jynge K, Hansen S, Hordvik A, Little C, Dodson E, Derewenda Z (1989) High-resolution (1.5 Å) crystal structure of phospholipase C from Bacillus cereus. Nature 338(6213):357–360

Huang F, Subbaiah PV, Holian O, Zhang J, Johnson A, Gertzberg N, Lum H (2005) Lysophosphatidylcholine increases endothelial permeability: role of PKCα and RhoA cross talk. Am J Physiol-Lung Cell Mol Physiol 289(2):L176–L185

Huang L, Boyd D, Amyot WM, Hempstead AD, Luo ZQ, O'Connor TJ, Chen C, Machner M, Montminy T, Isberg RR (2011) The E Block motif is associated with Legionella pneumophila translocated substrates. Cell Microbiol 13(2):227–245

Hubber A, Roy CR (2010) Modulation of host cell function by Legionella pneumophila type IV effectors. Ann Rev Cell Develop Biol 26:261–283

Ibarguren M, Bomans PH, Frederik PM, Stonehouse M, Vasil AI, Vasil ML, Alonso A, Goñi FM (2010) End-products diacylglycerol and ceramide modulate membrane fusion induced by a

phospholipase C/sphingomyelinase from Pseudomonas aeruginosa. Biochimica et Biophysica Acta (BBA)-Biomembranes 1798 (1):59–64
Isberg RR, O'Connor TJ, Heidtman M (2008) The Legionella pneumophila replication vacuole: making a cosy niche inside host cells. Nat Rev Microbiol 7(1):13–24
Istivan TS, Coloe PJ (2006) Phospholipase A in Gram-negative bacteria and its role in pathogenesis. Microbiology 152(5):1263–1274
Jacobi S, Heuner K (2003) Description of a putative type I secretion system in Legionella pneumophila. Int J Med Microbiol 293(5):349–358
Jendrossek D, Handrick R (2002) Microbial degradation of Polyhydroxyalkanoates*. Ann Rev Microbiol 56(1):403–432
Kato S, Yoshimura T, Hemmi H, Moriyama R (2010) Biochemical analysis of a novel lipolytic enzyme YvdO from Bacillus subtilis 168. Biosci Biotechnol Biochem 74(4):701–706
Kienesberger PC, Oberer M, Lass A, Zechner R (2009) Mammalian patatin domain containing proteins: a family with diverse lipolytic activities involved in multiple biological functions. J Lipid Res 50(Supplement):S63–S68
Kikkawa Y, Yoneda K, Smith F, Packard B, Suzuki K (1975) The type II epithelial cells of the lung. II. Chemical composition and phospholipid synthesis. Lab Investig; J Tech Methods Pathol 32(3):295
Koonin EV (1996) A duplicated catalytic motif in a new superfamily of phosphohydrolases and phospholipid synthases that includes poxvirus envelope proteins. Trends Biochem Sci 21(7):242–243
Kozak NA, Buss M, Lucas CE, Frace M, Govil D, Travis T, Olsen-Rasmussen M, Benson RF, Fields BS (2010) Virulence factors encoded by Legionella longbeachae identified on the basis of the genome sequence analysis of clinical isolate D-4968. J Bacteriol 192(4):1030–1044
Krug EL, Kent C (1984) Phospholipase C from Clostridium perfringens: preparation and characterization of homogeneous enzyme. Arch Biochem Biophys 231(2):400–410
Ku B, Lee K-H, Park WS, Yang C-S, Ge J, Lee S-G, Cha S-S, Shao F, Do Heo W, Jung JU (2012) VipD of Legionella pneumophila targets activated Rab5 and Rab22 to interfere with endosomal trafficking in macrophages. PLoS Pathog 8(12):e1003082
Laguna RK, Creasey EA, Li Z, Valtz N, Isberg RR (2006) A Legionella pneumophila-translocated substrate that is required for growth within macrophages and protection from host cell death. PNAS 103(49):18745–18750
Lang C, Flieger A (2011) Characterisation of Legionella pneumophila phospholipases and their impact on host cells. Eur J Cell Biol 90(11):903–912
Lang C, Rastew E, Hermes B, Siegbrecht E, Ahrends R, Banerji S, Flieger A (2012) Zinc metalloproteinase ProA directly activates Legionella pneumophila PlaC glycerophospholipid: cholesterol acyltransferase. J Biol Chem 287(28):23464–23478
Lin M-E, Herr DR, Chun J (2010) Lysophosphatidic acid (LPA) receptors: signaling properties and disease relevance. Prostaglandins Other Lipid Mediators 91(3):130–138
Linkous A, Yazlovitskaya E (2010) Cytosolic phospholipase A2 as a mediator of disease pathogenesis. Cell Microbiol 12(10):1369–1377
Lossi NS, Rolhion N, Magee AI, Boyle C, Holden DW (2008) The Salmonella SPI-2 effector SseJ exhibits eukaryotic activator-dependent phospholipase A and glycerophospholipid: cholesterol acyltransferase activity. Microbiology 154(9):2680–2688
Lucas E, Billington S, Carlson P, McGee D, Jost BH (2010) Phospholipase D promotes Arcanobacterium haemolyticum adhesion via lipid raft remodeling and host cell death following bacterial invasion. BMC Microbiol 10(1):270
Luo Z-Q, Isberg RR (2004) Multiple substrates of the Legionella pneumophila Dot/Icm system identified by interbacterial protein transfer. PNAS U S A 101(3):841–846
Lynch D, Fieser N, Glöggler K, Forsbach-Birk V, Marre R (2003) The response regulator LetA regulates the stationary-phase stress response in Legionella pneumophila and is required for efficient infection of Acanthamoeba castellanii. FEMS Microbiol Lett 219(2):241–248
Masamune A, Sakai Y, Satoh A, Fujita M, Yoshida M, Shimosegawa T (2001) Lysophosphatidylcholine induces apoptosis in AR42 J cells. Pancreas 22(1):75–83

Mason RJ, Williams MC (1980) Phospholipid composition and ultrastructure of A549 cells and other cultured pulmonary epithelial cells of presumed type II cell origin. Biochimica et Biophysica Acta (BBA)-Lipids and Lipid. Metabolism 617(1):36–50

McCoy-Simandle K, Stewart CR, Dao J, DebRoy S, Rossier O, Bryce PJ, Cianciotto NP (2011) Legionella pneumophila type II secretion dampens the cytokine response of infected macrophages and epithelia. Infect Immun 79(5):1984–1997

McMorran B, Town L, Costelloe E, Palmer J, Engel J, Hume D, Wainwright B (2003) Effector ExoU from the type III secretion system is an important modulator of gene expression in lung epithelial cells in response to Pseudomonas aeruginosa infection. Infect Immun 71(10):6035–6044

McNamara PJ, Bradley GA, Songer JG (1994) Targeted mutagenesis of the phospholipase D gene results in decreased virulence of Corynebacterium pseudotuberculosis. Mol Microbiol 12(6):921–930

McNamara PJ, Cuevas WA, Songer JG (1995) Toxic phospholipases D of Corynebacterium pseudotuberculosis, C. ulcerans and Arcanobacterium haemolyticum: cloning and sequence homology. Gene 156(1):113–118

Messaoudi A, Belguith H, Ghram I, Ben Hamida J (2010) Classification of EC 3.1. 1.3 bacterial true lipases using phylogenetic analysis. Afr J Biotechnol 9(48):8243–8247

Mølgaard A, Kauppinen S, Larsen S (2000) Rhamnogalacturonan acetylesterase elucidates the structure and function of a new family of hydrolases. Structure 8(4):373–383

Moliner C, Raoult D, Fournier P-E (2009) Evidence that the intra-amoebal Legionella drancourtii acquired a sterol reductase gene from eukaryotes. BMC Res Notes 2(1):51

Molofsky AB, Swanson MS (2004) Differentiate to thrive: lessons from the Legionella pneumophila life cycle. Mol Microbiol 53(1):29–40

Moolenaar WH, van Meeteren LA, Giepmans BN (2004) The ins and outs of lysophosphatidic acid signaling. Bioessays 26(8):870–881

Moreau H, Pieroni G, Jolivet-Reynaud C, Alouf J, Verger R (1988) A new kinetic approach for studying phospholipase C (Clostridium perfringens. alpha. toxin) activity on phospholipid monolayers. Biochemistry 27(7):2319–2323

Muder RR, Victor LY (2002) Infection due to Legionella species other than L. pneumophila. Clin Infect Dis 35(8):990–998

Murugesan G, Sandhya Rani M, Gerber CE, Mukhopadhyay C, Ransohoff RM, Chisolm GM, Kottke-Marchant K (2003) Lysophosphatidylcholine regulates human microvascular endothelial cell expression of chemokines. J Mol Cell Cardiol 35(11):1375–1384

Naylor CE, Eaton JT, Howells A, Justin N, Moss DS, Titball RW, Basak AK (1998) Structure of the key toxin in gas gangrene. Nat Struct & Mol Biol 5(8):738–746

Nelson DE, Crane DD, Taylor LD, Dorward DW, Goheen MM, Caldwell HD (2006) Inhibition of chlamydiae by primary alcohols correlates with the strain-specific complement of plasticity zone phospholipase D genes. Infect Immun 74(1):73–80

Newton AC (2010) Protein kinase C: poised to signal. Am J Physiol-Endocrinol Metab 298(3):E395–E402

Newton HJ, Ang DK, van Driel IR, Hartland EL (2010) Molecular pathogenesis of infections caused by Legionella pneumophila. Clin Microbiol Rev 23(2):274–298

Ninio S, Zuckman-Cholon DM, Cambronne ED, Roy CR (2005) The Legionella IcmS–IcmW protein complex is important for Dot/Icm-mediated protein translocation. Mol Microbiol 55(3):912–926

Ochi S, Miyawaki T, Matsuda H, Oda M, Nagahama M, Sakurai J (2002) Clostridium perfringens α-toxin induces rabbit neutrophil adhesion. Microbiology 148(1):237–245

Oliva JL, Oliva JL, Griner EM, Oliva JL, Griner EM, Kazanietz MG (2005) PKC isozymes and diacylglycerol-regulated proteins as effectors of growth factor receptors. Growth Factors 23(4):245–252

Ostroff R, Vasil A, Vasil M (1990) Molecular comparison of a nonhemolytic and a hemolytic phospholipase C from Pseudomonas aeruginosa. J Bacteriol 172(10):5915–5923

Otnaess AB, Little C, Sletten K, Wallin R, Johnsen S, Flengsrud R, Prydz H (1977) Some characteristics of phospholipase C from Bacillus cereus. Eur J Biochem 79(2):459–468
Papageorgiou AC, Hermawan S, Singh CB, Jendrossek D (2008) Structural Basis of Poly (3-Hydroxybutyrate) Hydrolysis by PhaZ7 Depolymerase from Paucimonas lemoignei. J Mol Biol 382(5):1184–1194
Pearce MM, Theodoropoulos N, Mandel MJ, Brown E, Reed KD, Cianciotto NP (2012) Legionella cardiaca sp. nov., isolated from a case of native valve endocarditis in a human heart. Int J Syst Evolution Microbiol 62 (Pt 12):2946–2954
Phillips RM, Six DA, Dennis EA, Ghosh P (2003) In vivo phospholipase activity of the Pseudomonas aeruginosa cytotoxin ExoU and protection of mammalian cells with phospholipase A2 inhibitors. J Biol Chem 278(42):41326–41332
Ponting C, Kerr I (1996) A novel family of phospholipase D homologues that includes phospholipid synthases and putative endonucleases: identification of duplicated repeats and potential active site residues. Prot Sci 5(5):914–922
Popoff MR, Bouvet P (2009) Clostridial toxins. Futur Microbiol 4(8):1021–1064
Preuss I, Kaiser I, Gehring U (2001) Molecular characterization of a phosphatidylcholine-hydrolyzing phospholipase C. Eur J Biochem 268(19):5081–5091
Prokazova N, Zvezdina N, Korotaeva A (1998) Review: effect of lysophosphatidylcholine on transmembrane signal transduction. Biochem (Mosc) 63:31–37
Qin T, Cui Y, Cen Z, Liang T, Ren H, Yang X, Zhao X, Liu Z, Xu L, Li D (2012) Draft genome sequences of two Legionella dumoffii strains, TEX-KL and NY-23. J Bacteriol 194(5): 1251–1252
Raederstorff D, Rohmer M (1985) Sterol biosynthesis de nova via cycloartenol by the soil amoeba Acanthamoeba polyphaga. Biochem J 231(3):609
Raynaud C, Guilhot C, Rauzier J, Bordat Y, Pelicic V, Manganelli R, Smith I, Gicquel B, Jackson M (2002) Phospholipases C are involved in the virulence of Mycobacterium tuberculosis. Mol Microbiol 45(1):203–217
Reilly TJ, Baron GS, Nano FE, Kuhlenschmidt MS (1996) Characterization and sequencing of a respiratory burst-inhibiting acid phosphatase from Francisella tularensis. J Biol Chem 271(18):10973–10983
Rooney S, Nardone L, Shapiro D, Motoyama E, Gobran L, Zaehringer N (1977) The phospholipids of rabbit type II alveolar epithelial cells: Comparison with lung lavage, lung tissue, alveolar macrophages, and a human alveolar tumor cell line. Lipids 12(5):438–442
Rossier O, Cianciotto NP (2001) Type II Protein Secretion Is a Subset of the PilD-Dependent Processes That Facilitate Intracellular Infection byLegionella pneumophila. Infect Immun 69(4):2092–2098
Rossier O, Cianciotto NP (2005) The Legionella pneumophila tatB gene facilitates secretion of phospholipase C, growth under iron-limiting conditions, and intracellular infection. Infect Immun 73(4):2020–2032
Rossier O, Starkenburg SR, Cianciotto NP (2004) Legionella pneumophila type II protein secretion promotes virulence in the A/J mouse model of Legionnaires' disease pneumonia. Infect Immun 72(1):310–321
Rossignol G, Merieau A, Guerillon J, Veron W, Lesouhaitier O, Feuilloley MG, Orange N (2008) Involvement of a phospholipase C in the hemolytic activity of a clinical strain of Pseudomonas fluorescens. BMC Microbiol 8(1):189
Rudolph AE, Stuckey JA, Zhao Y, Matthews HR, Patton WA, Moss J, Dixon JE (1999) Expression, Characterization, and Mutagenesis of theYersinia pestis Murine Toxin, a Phospholipase D Superfamily Member. J Biol Chem 274(17):11824–11831
Ruiz-Albert J, Yu XJ, Beuzón CR, Blakey AN, Galyov EE, Holden DW (2002) Complementary activities of SseJ and SifA regulate dynamics of the Salmonella typhimurium vacuolar membrane. Mol Microbiol 44(3):645–661
Rydel TJ, Williams JM, Krieger E, Moshiri F, Stallings WC, Brown SM, Pershing JC, Purcell JP, Alibhai MF (2003) The crystal structure, mutagenesis, and activity studies reveal that patatin is a lipid acyl hydrolase with a Ser-Asp catalytic dyad. Biochemistry 42(22):6696–6708

Sakurai J, Ochi S, Tanaka H (1993) Evidence for coupling of Clostridium perfringens alpha-toxin-induced hemolysis to stimulated phosphatidic acid formation in rabbit erythrocytes. Infect Immun 61(9):3711–3718

Sakurai J, Nagahama M, Oda M (2004) Clostridium perfringens alpha-toxin: characterization and mode of action. J Biochem 136(5):569–574

Salacha R, Kovačić F, Brochier-Armanet C, Wilhelm S, Tommassen J, Filloux A, Voulhoux R, Bleves S (2010) The Pseudomonas aeruginosa patatin-like protein PlpD is the archetype of a novel Type V secretion system. Environ Microbiol 12(6):1498–1512

Saliba A, Nascimento D, Silva M, Assis M, Gayer C, Raymond B, Coelho M, Marques E, Touqui L, Albano R (2005) Eicosanoid-mediated proinflammatory activity of Pseudomonas aeruginosa ExoU. Cell Microbiol 7(12):1811–1822

Saliba AM, De Assis M-C, Nishi R, Raymond B, Marques EdA, Lopes UG, Touqui L, Plotkowski M-C (2006) Implications of oxidative stress in the cytotoxicity of *Pseudomonas aeruginosa* ExoU. Microbes Infect 8(2):450–459

Sato H, Frank DW (2004) ExoU is a potent intracellular phospholipase. Mol Microbiol 53(5):1279–1290

Sato H, Frank DW, Hillard CJ, Feix JB, Pankhaniya RR, Moriyama K, Finck-Barbançon V, Buchaklian A, Lei M, Long RM (2003) The mechanism of action of the Pseudomonas aeruginosa-encoded type III cytotoxin. ExoU. EMBO J 22(12):2959–2969

Schmidt JA, Kalkofen DN, Donovan KW, Brown WJ (2010) A role for phospholipase A2 activity in membrane tubule formation and TGN trafficking. Traffic 11(12):1530–1536

Schmiel DH, Miller VL (1999) Bacterial phospholipases and pathogenesis. Microbes Infect 1(13):1103–1112

Schrag JD, Cygler M (1997) [4] Lipases and $\alpha\beta$ hydrolase fold. Methods Enzymol 284:85–107

Schroeder GN, Petty NK, Mousnier A, Harding CR, Vogrin AJ, Wee B, Fry NK, Harrison TG, Newton HJ, Thomson NR (2010) Legionella pneumophila strain 130b possesses a unique combination of type IV secretion systems and novel Dot/Icm secretion system effector proteins. J Bacteriol 192(22):6001–6016

Schunder E, Adam P, Higa F, Remer KA, Lorenz U, Bender J, Schulz T, Flieger A, Steinert M, Heuner K (2010) Phospholipase PlaB is a new virulence factor of Legionella pneumophila. Int J Med Microbiol 300(5):313–323

Segal G, Purcell M, Shuman HA (1998) Host cell killing and bacterial conjugation require overlapping sets of genes within a 22-kb region of the Legionella pneumophila genome. PNAS 95(4):1669–1674

Segal G, Russo JJ, Shuman HA (1999) Relationships between a new type IV secretion system and the icm/dot virulence system of Legionella pneumophila. Mol Microbiol 34(4):799–809

Selvy PE, Lavieri RR, Lindsley CW, Brown HA (2011) Phospholipase D-enzymology, functionality, and chemical modulation. Chem Rev 111(10):6064

Shewry PR (2003) Tuber storage proteins. Ann Botany 91(7):755–769

Shinoda S, Matsuoka H, Tsuchie T, Miyoshi S-I, Yamamoto S, Taniguchi H, Mizuguchi Y (1991) Purification and characterization of a lecithin-dependent haemolysin from Escherichia coli transformed by a Vibrio parahaemolyticus gene. J Gen Microbiol 137(12):2705–2711

Shohdy N, Efe JA, Emr SD, Shuman HA (2005) Pathogen effector protein screening in yeast identifies Legionella factors that interfere with membrane trafficking. PNAS U S A 102(13): 4866–4871

Singh G, Singh G, Jadeja D, Kaur J (2010) Lipid hydrolizing enzymes in virulence: Mycobacterium tuberculosis as a model system. Crit Rev Microbiol 36(3):259–269

Sitkiewicz I, Stockbauer KE, Musser JM (2007) Secreted bacterial phospholipase A_2 enzymes: better living through phospholipolysis. Trends Microbiol 15(2):63–69

Slepkov ER, Bitar AP, Marquis H (2010) Differentiation of propeptide residues regulating the compartmentalization, maturation and activity of the broad-range phospholipase C of Listeria monocytogenes. Biochem J 432(3):557

Smith GA, Marquis H, Jones S, Johnston NC, Portnoy DA, Goldfine H (1995) The two distinct phospholipases C of Listeria monocytogenes have overlapping roles in escape from a vacuole and cell-to-cell spread. Infect Immun 63(11):4231–4237

Songer JG (1997) Bacterial phospholipases and their role in virulence. Trends Microbiol 5(4):156–161

Sreenivas A, Patton-Vogt JL, Bruno V, Griac P, Henry SA (1998) A role for phospholipase D (Pld1p) in growth, secretion, and regulation of membrane lipid synthesis in yeast. J Biol Chem 273(27):16635–16638

Stonehouse MJ, Cota-Gomez A, Parker SK, Martin WE, Hankin JA, Murphy RC, Chen W, Lim KB, Hackett M, Vasil AI (2002) A novel class of microbial phosphocholine-specific phospholipases C. Mol Microbiol 46(3):661–676

Sutterwala FS, Mijares LA, Li L, Ogura Y, Kazmierczak BI, Flavell RA (2007) Immune recognition of Pseudomonas aeruginosa mediated by the IPAF/NLRC4 inflammasome. J Exp Med 204(13):3235–3245

Takahashi M, Okazaki H, Ogata Y, Takeuchi K, Ikeda U, Shimada K (2002) Lysophosphatidylcholine induces apoptosis in human endothelial cells through a p38-mitogen-activated protein kinase-dependent mechanism. Atherosclerosis 161(2):387–394

Takeshita S, Inoue N, Gao D, Rikitake Y, Kawashima S, Tawa R, Sakurai H, Yokoyama M (2000) Lysophosphatidylcholine enhances superoxide anions production via endothelial NADH/NADPH oxidase. J Atheroscler Thrombosis 7(4):238

Tamura M, Ajayi T, Allmond LR, Moriyama K, Wiener-Kronish JP, Sawa T (2004) Lysophospholipase A activity of Pseudomonas aeruginosa type III secretory toxin ExoU. Biochem Biophys Res Commun 316(2):323–331

Titball RW (1993) Bacterial phospholipases C. Microbiol Rev 57(2):347–366

Titball R (1998) Bacterial phospholipases. J Appl Microbiol 84(S1):127S–137S

Titball RW, Rubidge T (1990) The role of histidine residues in the alpha toxin of Clostridium perfringens. FEMS Microbiol Lett 68(3):261–265

Titball RW, Naylor CE, Basak AK (1999) The Clostridium perfringens α-toxin. Anaerobe 5(2):51–64

Tyson JY, Pearce MM, Vargas P, Bagchi S, Mulhern BJ, Cianciotto NP (2013) Multiple Legionella pneumophila type II secretion substrates, including a novel protein, contribute to differential infection of amoebae Acanthamoeba castellanii, Hartmannella vermiformis, and Naegleria lovaniensis. Infect Immun 81(5):1399–1410

Upton C, Buckley JT (1995) A new family of lipolytic enzymes? Trends Biochem Sci 20(5):178–179

van der Meer-Janssen YP, van Galen J, Batenburg JJ, Helms JB (2010) Lipids in host–pathogen interactions: pathogens exploit the complexity of the host cell lipidome. Prog Lipid Res 49(1):1–26

Van Dijk M, Postma F, Hilkmann H, Jalink K, van Blitterswijk WJ, Moolenaar WH (1998) Exogenous phospholipase D generates lysophosphatidic acid and activates Ras, Rho and Ca2 + signaling pathways. Curr Biol 8(7):386–392

Van Kampen M, Simons J-W, Dekker N, Egmond M, Verheij H (1998) The phospholipase activity of Staphylococcus hyicus lipase strongly depends on a single Ser to Val mutation. Chem Phys Lipids 93(1):39–45

van Kampen MD, Rosenstein R, Götz F, Egmond MR (2001) Cloning, purification and characterisation of Staphylococcus warneri lipase 2. Biochimica et Biophysica Acta (BBA)-Protein Structure and Molecular Enzymology 1544 (1):229–241

van Leeuwen FN, Olivo C, Grivell S, Giepmans BN, Collard JG, Moolenaar WH (2003) Rac Activation by Lysophosphatidic Acid LPA1Receptors through the Guanine Nucleotide Exchange Factor Tiam1. J Biol Chem 278(1):400–406

van Meeteren LA, Frederiks F, Giepmans BN, Pedrosa MFF, Billington SJ, Jost BH, Tambourgi DV, Moolenaar WH (2004) Spider and bacterial sphingomyelinases D target cellular lysophosphatidic acid receptors by hydrolyzing lysophosphatidylcholine. J Biol Chem 279(12):10833–10836

Van Oort MG, Deveer AM, Dijkman R, Tjeenk ML, Verheij HM, De Haas GH, Wenzig E, Goetz F (1989) Purification and substrate specificity of Staphylococcus hyicus lipase. Biochemistry 28(24):9278–9285

VanRheenen SM, Luo Z-Q, O'Connor T, Isberg RR (2006) Members of a Legionella pneumophila family of proteins with ExoU (phospholipase A) active sites are translocated to target cells. Infect Immun 74(6):3597–3606

Vazquez-Boland J, Kocks C, Dramsi S, Ohayon H, Geoffroy C, Mengaud J, Cossart P (1992) Nucleotide sequence of the lecithinase operon of Listeria monocytogenes and possible role of lecithinase in cell-to-cell spread. Infect Immun 60(1):219–230

Viner R, Chetrit D, Ehrlich M, Segal G (2012) Identification of Two Legionella pneumophila Effectors that Manipulate Host Phospholipids Biosynthesis. PLoS Pathog 8(11):e1002988

Virtanen JA, Cheng KH, Somerharju P (1998) Phospholipid composition of the mammalian red cell membrane can be rationalized by a superlattice model. PNAS 95(9):4964–4969

Vogel JP, Andrews HL, Wong SK, Isberg RR (1998) Conjugative transfer by the virulence system of Legionella pneumophila. Science 279(5352):873–876

Volkman J (2003) Sterols in microorganisms. Appl Microbiol Biotechnol 60(5):495–506

Wei Y, Schottel JL, Derewenda U, Swenson L, Patkar S, Derewenda ZS (1995) A novel variant of the catalytic triad in the Streptomyces scabies esterase. Nat Struct & Mol Biol 2(3):218–223

Weissenmayer BA, Prendergast JG, Lohan AJ, Loftus BJ (2011) Sequencing illustrates the transcriptional response of Legionella pneumophila during infection and identifies seventy novel small non-coding RNAs. PloS One 6(3):e17570

Weltzien HU (1979) Cytolytic and membrane-perturbing properties of lysophosphatidylcholine. Biochimica et biophysica acta 559(2–3):259

Wilderman PJ, Vasil AI, Johnson Z, Vasil ML (2001) Genetic and biochemical analyses of a eukaryotic-like phospholipase D of Pseudomonas aeruginosa suggest horizontal acquisition and a role for persistence in a chronic pulmonary infection model. Mol Microbiol 39(2):291–304

Winn WC Jr, Myerowitz RL (1981) The pathology of the Legionella pneumonias: a review of 74 cases and the literature. Human Pathol 12(5):401–422

Zambonelli C, Casali M, Roberts MF (2003) Mutagenesis of putative catalytic and regulatory residues of Streptomyces chromofuscus phospholipase D differentially modifies phosphatase and phosphodiesterase activities. J Biol Chem 278(52):52282–52289

Zhao Y, Natarajan V (2012) Lysophosphatidic acid (LPA) and its Receptors: Role in Airway Inflammation and Remodeling. Biochimica et Biophysica Acta (BBA)-Molecular and Cell Biology of Lipids

Zhao Y, Stuckey JA, Lohse DL, Dixon JE (1997) Expression, characterization, and crystallization of a member of the novel phospholipase D family of phosphodiesterases. Prot Sci 6(12):2655–2658

Zhu W, Banga S, Tan Y, Zheng C, Stephenson R, Gately J, Luo Z-Q (2011) Comprehensive identification of protein substrates of the Dot/Icm type IV transporter of Legionella pneumophila. PloS One 6(3):e17638

Cytotoxic Glucosyltransferases of *Legionella pneumophila*

Yury Belyi, Thomas Jank and Klaus Aktories

Abstract *Legionella* is a gram-negative bacterium and the causative pathogen of legionellosis—a severe pneumonia in humans. A large number of *Legionella* effectors interfere with numerous host cell functions, including intracellular vacuole trafficking and maturation, phospholipid metabolism, protein ubiquitination, pro-/anti-apoptotic balances or inflammatory responses. Moreover, eukaryotic protein synthesis is affected by *L. pneumophila* glucosyltransferases Lgt1, Lgt2, and Lgt3. Structurally, these enzymes are similar to large clostridial cytotoxins, use UDP-glucose as a co-substrate and modify a conserved serine residue (Ser-53) in elongation factor 1A (eEF1A). The ternary complex consisting of eEF1A, GTP, and aminoacylated-tRNA seems to be the substrate for Lgts. Studies with *Saccharomyces cerevisiae* corroborated that eEF1A is the major target responsible for Lgt-induced cytotoxic activity. In addition to Lgt proteins, *Legionella* produces other effector glycosyltransferase, including the modularly composed protein SetA, which displays tropism for early endosomal compartments, subverts host cell vesicle trafficking and demonstrates toxic activities toward yeast and mammalian cells. Here, our current knowledge about both groups of *L. pneumophila* glycosylating effectors is reviewed.

Y. Belyi
Gamaleya Research Institute, Moscow, Russia 123098

T. Jank · K. Aktories (✉)
Institute for Experimental and Clinical Pharmacology and Toxicology,
Albert-Ludwigs University of Freiburg, Freiburg, Germany,
e-mail: klaus.aktories@pharmakol.uni-freiburg.de

Current Topics in Microbiology and Immunology (2013) 376: 211–226
DOI: 10.1007/82_2013_338

Published Online: 31 July 2013

Contents

1 *Legionella* Infection: A Plethora of Effectors and Mechanisms 212
2 Lgt-Family of Glucosylating Enzymes.. 213
3 Structure of Lgt1 215
4 Substrates of Lgt1 217
5 *Legionella* Effector SetA 219
6 Eukaryotic Protein Glycosylation and Intracellular Life Cycle of *L. pneumophila*....... 221
References.. 223

1 *Legionella* Infection: A Plethora of Effectors and Mechanisms

Legionella is a gram-negative bacterium and an infectious agent of legionellosis, which is characterized by a severe pneumonia in humans. Among multiple species of *Legionella* the most significant human pathogen is *L. pneumophila*. Its strains account for more than 90 % of cases of the infection (Diederen 2008). Therefore, it is not surprising that the majority of studies dedicated to molecular pathogenesis of legionellosis are centered on this single species, while virulence mechanisms of other pathogens (e.g., *L. longbeachae*, *L. micdadei*, *L. bozemanii*) are obviously neglected.

The infection life cycle of *L. pneumophila* is dependent on the intracellular proliferation of the bacterium. The natural habitats of *Legionellae* are free-living unicellular organisms (amoebae and ciliated protozoa) where bacteria multiply intracellularly. During the infection process in humans, bacteria invade predominantly macrophages, monocytes, and lung epithelial cells (Richards et al. 2013). After uptake by host cells, *Legionellae* multiply within a specialized phagosome-derived organelle, the *Legionella*-containing vacuole (LCV), which interferes with vesicles of the secretory pathway and avoids lysosome fusion and degradation (Isberg et al. 2009). Such transformation is dependent upon a type IVb secretion system (T4bSS), encoded by *dot*/*icm* genes. It translocates bacterial proteins (i.e., effectors) into the target cell. General consensus is that *L. pneumophila* can produce about 300 different effectors (Segal 2013; Zhu et al. 2011). Their highly specialized activities are prerequisites for the conversion of the *Legionella*-containing phagosome into a "replicative vacuole" and successful proliferation of the bacteria (Xu and Luo 2013).

Several *Legionella* effectors target small eukaryotic GTPases which are involved in vesicular trafficking of the host cell by switching back and forth from a GDP- to GTP-bound state regulated in a GTPase cycle. *Legionella* effectors are able to specifically turn on or turn off these GTP-binding proteins. For example, the mammalian protein Arf1, which is involved in vesicle formation in the Golgi, is activated by *Legionella* protein RalF. RalF acts as a GDP/GTP exchange factor

(GEF) for this small GTPase (Nagai et al. 2002). Another example is the GTP-binding protein Rab1, which regulates various steps of vesicle trafficking in eukaryotic cells. Rab1 is manipulated by several *Legionella* effectors. SidM/DrrA stimulates the activation of Rab1 and recruits the GTPase to the membrane of *Legionella*-containing vacuole by means of its GEF and adenylylation (AMPylation) activity (Ingmundson et al. 2007; Machner and Isberg 2007; Murata et al. 2006). *Legionella* LidA binds the GTP-bound form of Rab with extraordinary high affinity making LidA a "supereffector" (Schoebel et al. 2011). However, when the replication vacuole is formed, Rab1 is deactivated by the effector protein SidD through de-AMPylation (Neunuebel et al. 2011; Tan and Luo 2011). Afterward the active GTP-bound state of Rab1 is terminated by the *Legionella*-produced LepB, a GTPase-activating protein (GAP) (Ingmundson et al. 2007). Another *Legionella* effector AnkX is able to phosphocholinate Rab1/35 (Mukherjee et al. 2011; Tan et al. 2011), while Lem3 reverses this modification (Goody et al. 2012; Tan et al. 2011). The process of phosphocholination is thought to represent an alternative to GDP/GTP exchange, stabilizing modified Rab molecules in membranes in the GDP-bound form, whereas de-phosphocholination apparently alleviates such an effect.

Besides vesicular trafficking, many other cellular processes are affected during *Legionella* replication in the host cell. For example, the T4bSS effector SidK targets vacuolar ATPase by interacting with one of the key components of the vesicular proton pump (VatA) and thereby inhibiting ATP hydrolysis, proton translocation, and vacuole acidification (Xu et al. 2010). Several *Legionella* effector proteins participate in controlling pro- and anti-apoptotic balances either directly or indirectly (Abu-Zant et al. 2007; Laguna et al. 2006; Banga et al. 2007). Modulation of NF-κB activities by several *L. pneumophila* proteins might contribute to manipulation of death pathways by the bacterium (Bartfeld et al. 2009; Ge et al. 2009; Hsu et al. 2012; Losick et al. 2010). Recent data demonstrate that *L. pneumophila* utilizes effectors to manipulate important steps in phospholipid metabolism (Viner et al. 2012), uses phosphoinositide lipid-binding domains to anchor effectors to specific vesicular membranes (Hilbi et al. 2011), or influences the phosphorylation status of different phospholipids (Haenssler and Isberg 2011).

2 Lgt-Family of Glucosylating Enzymes

The first glucosyltransferase isolated from *L. pneumophila* Philadelphia-1 cultures was termed "Lgt1" (*Legionella* glucosyltransferase 1). Lgt1 has a molecular mass of 59.7 kDa, consists of 525 amino acid residues and has an isoelectric point of 7.0 (Belyi et al. 2003).

The primary amino acid sequence of Lgt1 shares little homology with known proteins. Only the central region of Lgt1 demonstrates limited but significant similarity to the enzymatic domain of large clostridial glucosylating toxins (Fig. 1). Here, several conserved amino acid residues of a catalytic core can be

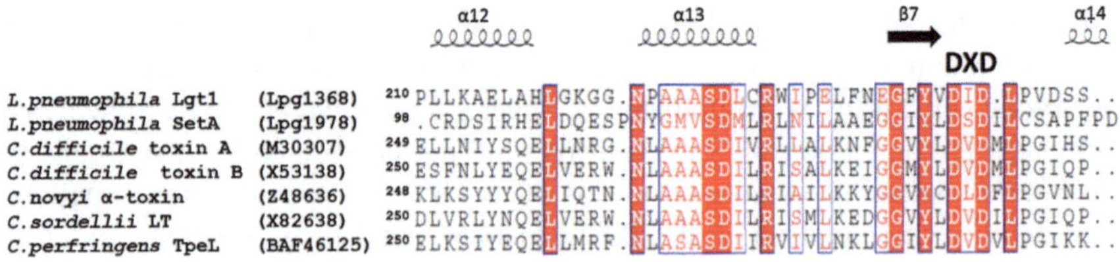

Fig. 1 Partial amino acid sequence alignment of *Legionella* glucosyltransferases and clostridial glucosylating toxins. Lgt1, SetA from *L. pneumophila* Philadelphia-1, toxins A and B from *C. difficile*, α-toxin from *C. novyi*, lethal toxin from *C. sordellii*, and TpeL from *C. perfringens*. Gene bank accession numbers of the corresponding coding sequences are shown in *brackets*. Glycosyltransferase specific DXD-motif is marked. Structure-based sequence alignment was based on the structure of Lgt1 and created with Espript (http://espript.ibcp.fr/Espript/Espript). Conserved residues are in *red* and strictly conserved residues are *highlighted red*

identified, including N_{224}, D_{230}, R_{233} and the two aspartic amino acid residues D_{246}, and D_{248}, representing the DXD motif of many glycosyltransferases. Also similar to *Clostridium difficile* toxins A and B, Lgt1 uses UDP-glucose but no other sugars as a donor in glucosylation reactions (Belyi et al. 2006; Just et al. 1995).

Database searches for similar sequences in the genomes of six *L. pneumophila* strains (Philadelphia-1, Corby, Lens, Paris, 2300/99 Alcoy, and 130b) disclosed altogether 13 open-reading frames exhibited significant sequence homology to Lgt1 (Belyi et al. 2011). The gene products demonstrate considerable amino acid sequence variations and are grouped into three subfamilies: Lgt1, Lgt2, and Lgt3 (Belyi et al. 2008). Sequence homology within each subfamily reaches 90 %, whereas similarity between different Lgt groups is in the range 15–30 %. *L. pneumophila* Philadelphia-1 strain contains the coding sequences of *lgt1*, *lgt2*, and *lgt3*, whereas the other strains possess only *lgt1* and *lgt3*. In a separate study it was shown that clinical isolates of *L. pneumophila* contained more often the full set of Lgt enzymes, while environmental isolates in many instances have only *lgt1* and *lgt3* coding sequences (Fig. 2) (Sadretdinova et al. 2012). Lgt1, Lgt2, and Lgt3 are serologically distinct and do not display antigenic cross-reactivity with the monospecific sera. Purified recombinant proteins Lgt2 and Lgt3 from different strains demonstrate glucosylation activities identical to that of Lgt1 (Belyi et al. 2008).

All Lgt glucosyltransferases are substrates of T4bSS. This was shown with chimeras, containing adenylyl cyclase or *β*-lactamase reporter domains (de Felipe et al. 2005, 2008; Hurtado-Guerrero et al. 2010). *Legionella* effectors, which are transported by the Dot/Icm secretion machinery, are mainly produced during the stationary phase of bacterial growth, when bacterial cells become remarkably more virulent and display a transmission phenotype (Bruggemann et al. 2006; Byrne and Swanson 1998; Zusman et al. 2007). Interestingly, production of Lgt glucosyltransferases in *Legionella* cultures is also tightly regulated. Thus, the level of Lgt1 and Lgt2 is strongly increased at the stationary phase of bacterial growth in broth medium, while Lgt3 is detectable in the pre-logarithmic phase of in vitro

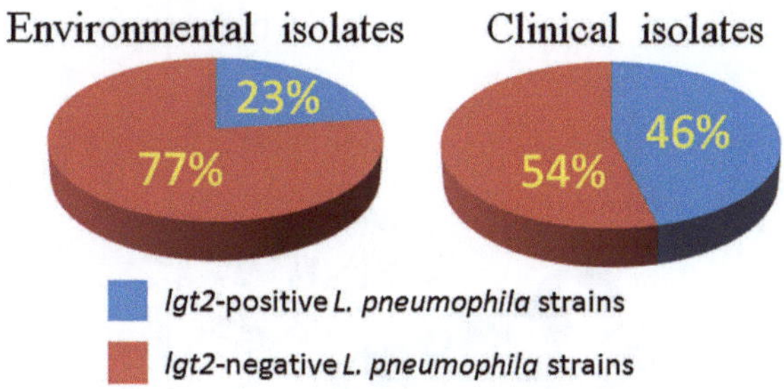

Fig. 2 Frequency of *lgt2*-positive strains in *L. pneumophila*. Fifty-three strains of *L. pneumophila* (26 clinical isolates and 27 environmental strains) were tested by PCR for the presence of *lgt2* gene (Sadretdinova et al. 2012)

cultivation. Similar results were obtained during in vivo experiments using the protozoan *Acanthamoeba castellanii* as a host for *L. pneumophila*. Levels of mRNA coding for Lgt1 were maximal at late time points of bacteria-amoeba co-infection, while *lgt3* was expressed mainly at the initial stage of *Legionella—A. castellanii* interaction (Belyi et al. 2008). These experiments suggest differential regulation of glucosyltransferase activity in *L. pneumophila*, which, in turn, indicates toward specific roles of each enzyme in bacterial virulence. Speculatively, Lgt3 can be important for initiation of infection cycle, while Lgt1/Lgt2 might be necessary for egress of *Legionella* from the host cell.

3 Structure of Lgt1

Resolving the crystal structure of Lgt1 considerably improved our understanding in mechanistic features of this glucosyltransferase (Hurtado-Guerrero et al. 2010; Lu et al. 2010). Lgt1 is composed of three structural domains and is classified into the GT-A type retaining glucosyltransferase family GT88 in the carbohydrate modifying enzymes database (http://www.cazy.org/GT88.html). N-terminal Domain I consists of seven α-helices with yet unknown function (Fig. 3). Domain II constitutes the typical glycosyltransferase GT-A core assembly with a central β-sheet surrounded by α-helices presenting a double Rossmann fold-like structure. This nucleotide binding domain harbors the donor substrate-binding site and contains conserved amino acid residues known to be important for glycosyltransfer reaction (Jank et al. 2007). C-terminal domain III is a predominantly α-helical, suggested being involved in acceptor substrate binding (Hurtado-Guerrero et al. 2010). Lgt1 also possesses a flexible loop at the very C-terminus, which seems to be important for proper arrangement of the substrate-binding site, accommodation of UDP-glucose in catalytic center within domain II and the release of the reaction products after catalysis.

The DXD-motif (Asp-246 and Asp-248) is the remarkable feature of glucosyltransferases of the GT-A type and is crucial for divalent cation binding (Busch et al. 1998; Wiggins and Munro 1998). In Lgt1 the Mn^{2+}/Mg^{2+} is coordinated within an octahedral complex where two valences are occupied by the α- and

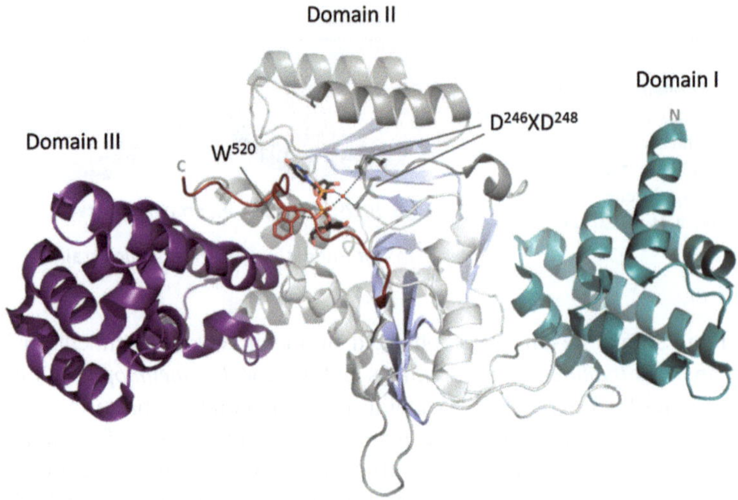

Fig. 3 Crystal structure of Lgt1 in complex with UDP–glucose and Mg^{2+} (pdb code 3JSZ). The N-terminal domain (Domain I) is depicted in green, the central domain (Domain II) in gray and the protrusion domain (Domain III) in purple. The central beta sheet is shaded in *light blue*. UDP-glucose is shown in sticks and Mg^{2+} as a *red sphere*. The C-terminal flexible loop harboring tryptophane-520 is *highlighted in dark red*. The canonical DXD-motif of GT-A type glucosyltransferase coordinating Mn^{2+} is *highlighted*

β-phosphates of UDP. Only the second aspartic acid (Asp-248) of the DXD-motif is involved in direct cation coordination, while the first residue (Asp-246) coordinates the cation through a water molecule. The role of the divalent metal in Lgt1 functioning seems to be manifold. On one hand, binding of the metal ion in conjunction with the donor UDP-glucose substrate is a prerequisite for the induction of a conformational change in the C-terminal flexible loop region. On the other hand, the ion is necessary for the stabilization of the transition state during catalysis by compensating the negative charge of the β-phosphate of the nucleotide and facilitating the release of the leaving group (Qasba et al. 2005; Ziegler et al. 2008).

The reaction mechanism of Lgt1 is suggested to follow a S_Ni-like mechanism (Lairson et al. 2008). The glycosyltransfer reaction may start with the binding of the divalent cation to Asp-248 of the DXD-motif and the association/binding of the donor substrate UDP-glucose into the open cleft of the enzyme. Subsequently, the C-terminal flexible loop rearranges to the closed conformation where Trp-520 flips into the catalytic pocket and interacts with the nucleotide β-phosphate. In this state the catalytically competent conformation and the substrate-binding site is arranged. Upon cleavage of UDP-glucose the positive charged oxocarbenium glucosyl-intermediate is then attacked by serine-53 residue of the acceptor substrate (eEF1A) leading to the products UDP, Mn^{2+} and glucosyl-eEF1A. After dissociation of modified eEF1A the flexible loop reorganizes to release UDP and the cation and a new reaction cycle starts.

4 Substrates of Lgt1

In pioneering experiments it was shown that glucosyltransferase Lgt1 modifies an ~50 kDa component in the cytoplasmic fraction of eukaryotic cells (Belyi et al. 2003). Later this component was identified as eukaryotic elongation factor 1A (eEF1A). In subsequent experiments it was demonstrated that other enzymes (i.e., Lgt2 and Lgt3) also glucosylated this factor and the site of modification by all three glucosyltransferases was a single serine residue in position 53 (Belyi et al. 2006).

Elongation factor eEF1A plays a pivotal role in protein synthesis (Ramakrishnan 2002). It is required for the GTP-dependent delivery of aminoacylated tRNAs to the A-site of mRNA-charged ribosomes. As shown for yeast elongation factor 1A, the molecule consists of three structural domains (Fig. 4) (Andersen et al. 2000). Domain 1 harbors a typical GTP-binding "G domain." Key feature of this domain is binding and hydrolysis of GTP in a canonical GTPase cycle. Domains 2 and 3 are involved in interaction with different functional targets like ribosomes, aminoacyl-tRNA, guanine nucleotide exchange factor (elongation factor eEF1Bα), and probably some others (Sasikumar et al. 2012).

Serine-53 of eEF1A, which is modified by Lgt, is located in the G domain near the switch-1 region of the GTPase on a protruding loop between the two helices A* and A′. Switch-1 region of the prokaryotic analog EF-Tu is known to be subjected to conformational changes, depending on the nucleotide bound (GDP or GTP) (Abel et al. 1996). Therefore, it is suggested that modification of serine-53 in such an important region of eEF1A alters structural rearrangements of the molecule. However, prokaryotic elongation factor EF-Tu does not contain a helix-loop-helix region with a serine residue in the proper position for modification by Lgt, limiting conclusions drawn from EF-Tu.

Initial experiments with purified recombinant elongation factor gave puzzling results in respect to substrate specificity and the efficiency of Lgt-induced glycosylation. Systematic truncation analysis revealed that a considerable portion of the molecule was dispensable for recognition of the elongation factor by *Legionella* enzyme. Eventually, the length of the whole protein was reduced down to a decapeptide with residues 50-GKGSFKYAWV-59. Such peptide was modified much more efficiently than the full length eEF1A (Belyi et al. 2009). Altogether these findings suggested that the substrate properties of eEF1A depend on a specific conformation of the full length protein, which allows its modification by the *Legionella* enzymes.

The answer to this puzzle was obtained recently. Not eEF1A alone but the elongation-competent ternary complex, consisting of eEF1A, GTP, and aminoacyl-tRNA is the most efficient substrate for modification by Lgt. Glucosylation of the ternary complex by Lgts in vitro exceeds substrate properties of the elongation factor in the apo form by several orders of magnitude and might reflect high efficiency of Lgt-induced glucosylation inside of host macrophages (Tzivelekidis et al. 2011).

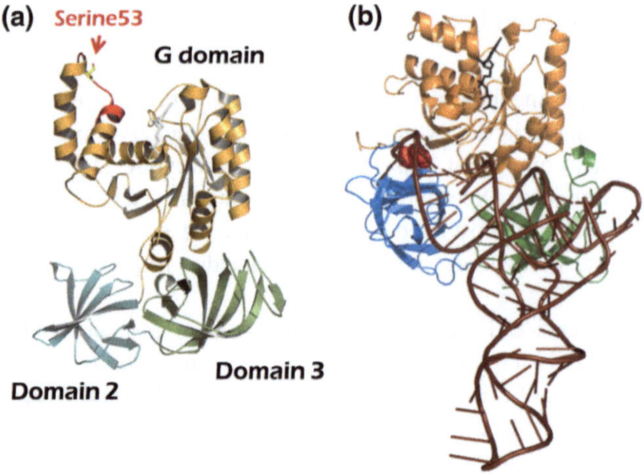

Fig. 4 Structure of the eukaryotic elongation factor 1A (eEF1A) in comparison to prokaryotic elongation factor (EF-TU) ternary complex. **a** Structural view of yeast elongation factor eEF1A (adapted from pdb 1IJF). Elongation factor eEF1A consists of three main structural parts: domain 1 (G-domain), domain 2, and domain 3. The decapeptide (GKGSFKYAWV), which is a sufficient substrate for glucosylation by Lgt, is shown in *red*. Serine-53, which is modified by glucosyltransferases Lgt, is shown in *yellow*. The complexed fragment of eEF1Bα molecule, which is present in the original structure, is omitted. **b** Crystal structure of the ternary complex of prokaryotic elongation factor EF Tu (pdb 1TTT) consisting of EF Tu (coloring as in **a**), GTP (*black sticks*), Mg^{2+}, and Phe-tRNA^{Phe} (*brown*, phenylalanine is shown as *red spheres*). Note, the helix-loop-helix region is missing in EF Tu

The decapeptide 50-GKGSFKYAWV-59, which is the minimal substrate of Lgts, is part of the first turn of helix A′ of the helix-loop-helix region formed by helices A* and A′ of eEF1A. Substitution of Phe-54, Tyr-56, or Trp-58 with alanine prevented or strongly decreased glucosylation. In contrast, amino acid residues Gly-50, Lys-51, Gly-52, Lys-55, or Val-59 seemed to be dispensable for substrate activity of the decapeptide since their replacement by alanine residues failed to affect glucosylation efficiency of the target molecule (Belyi et al. 2009). Such low specificity of the recognition fragment raised the question of the existence of other substrate proteins, containing similar amino acid sequences.

Indeed, BLAST screenings with the minimal peptide sequence as a query retrieved Hbs1 protein (Hsp70 subfamily B suppressor 1 (Nelson et al. 1992)) as another possible substrate for Lgt1. Hbs1 is a conserved protein and can be found in diverse eukaryotic organisms ranging from yeast to humans. Both yeast and human Hbs1 contain the decapeptides 210-GKSSFKFAWI-219 and 311-GKASFAYAWV-320, respectively, which are similar to the corresponding sequence in eEF1A. Recombinant yeast Hbs1 as well as the decapeptide fused to glutathione-S-transferase (GST) were modified by Lgt in vitro (Belyi et al. 2009).

Cellular functions of Hbs1 have been scrutinized mainly using *S. cerevisiae* as a model. These studies suggest importance of the protein as a component of eukaryotic RNA quality control system. Ribosomal complexes, stalled in elongation due to inhibitory secondary structures or defects of translated mRNA (e.g., hairpin loops, rare codons, chemical damage, nonsense mutations, etc.), are rescued by Hbs1 complexed with the release factor (RF) 1-related protein Dom34. According to reported data, Hbs1/Dom34 similarly to the corresponding pair of release factors (RF3/RF1) directly binds to the A-site of the ribosome in a codon-independent manner and destabilizes the mRNA:ribosome complex. Thereby, Hbs1/Dom34 promotes recycling of ribosomal components and endonucleolytic cleavage of defective mRNA (Shoemaker and Green 2012). So far, however, it is not known if modification of Hbs1 by Lgt influences its functions. The question to what extent alterations of recovery processes in stalled ribosome are beneficial to *Legionella* also remains absolutely enigmatic.

Eukaryotic substrates, modified by glucosyltransferases Lgt (eEF1A and Hbs1) include crucial components of translational machinery. It was not surprising, therefore, that the addition of purified Lgt1, Lgt2, or Lgt3 to in vitro translational extracts resulted in a dose-dependent inhibition of protein synthesis. Furthermore, delivery of the proteins into mammalian cells by electroporation or expression of the corresponding genes in *S. cerevisiae* results in eEF1A modification, protein synthesis inhibition, and cell death (Belyi et al. 2006, 2008, 2012; Heidtman et al. 2009). However, bearing in mind low structural stringency of substrate recognition by Lgt one can imply existence of other yet unidentified targets. Using *S. cerevisiae* as a model for studies on functional consequences of Lgt actions, some of these questions were addressed recently.

S. cerevisiae strains were engineered, containing an eEF1A S53A variant insensitive to glucosylation, lacking Hbs1, and combining both features (Belyi et al. 2012). Transformation of these strains with fully active Lgt1 revealed that yeast, possessing eEF1A S53A as the only elongation factor present in a cell, were efficiently protected from the toxic activity of the enzyme in spite of high production of the protein in yeast cultures. In contrast, deletion of the gene coding for Hbs1 neither in a wild type nor in *eef1a S53A* genetic backgrounds contributed to resistance of yeast cells to the toxic action of *L. pneumophila* glucosyltransferase. These experiments indicate that elongation factor eEF1A represents the major target of a toxic phenotype accomplished by the *Legionella* enzyme.

5 *Legionella* Effector SetA

In order to identify further *Legionella* effectors, Heidtman et al. applied yeast in a lethal phenotype screen and identified 79 putative effectors. Among those, the Lgt family members Lgt1, 2, and 3 were found to exhibit very strong toxicity. Another protein Lpg1978/SetA was found to impact yeast growth and interfered with the trafficking of secretory proteins (Heidtman et al. 2009). Following sorting of the secretory proteins caboxypeptidase Y and alkaline phosphatase in yeast revealed moderate but significant effects of SetA upon protein and vesicle maturation from the ER to the yeast vacuole, which is the equivalent of the mammalian lysosome. These results indicated that SetA might modulate host cell vesicle trafficking during infection. Therefore, the former protein Lpg1978 was named into Subversion of eukaryotic traffic A (SetA). SetA also provoked dramatic yeast cell deformations seen in scanning electron microscopy (Jank et al. 2012).

SetA exhibits considerable sequence similarity to Lgts and clostridial glucosylating toxins and possesses the canonical glycosyltransferase DxD-motif (Heidtman et al. 2009) (Fig. 1). The gene of SetA is present in all pathogenic *L. peumophila* strains (*L.p.* Corby Lpc1464, *L.p.* Paris Lpp1961, and *L.p.* Lens Lpl1955) but is not present in *L. longbeachae* (Cazalet et al., Cazalet 2010; Kozak et al., 2010). On the chromosome of *L. pneumophila*, *setA* is located in a T4SS effector rich region (Franco et al. 2009). The gene product SetA itself is a translocated effector and is secreted by *L. pneumophila* in a Dot/Icm-dependent manner (Heidtman et al. 2009; Huang et al. 2011). Notably, a deletion strain of *L. pneumophila* lacking SetA exhibited no reduced replication in murine bone marrow-derived macrophages or *Dictyostelium discoideum* (Heidtman et al. 2009). This seems to be a general phenomenon attributed to other functional redundant Dot/Icm substrates secreted by *L. pneumophila*.

SetA effects were not only recognized in yeast, also microinjection of SetA into mammalian cells caused dramatic cell morphological alterations, which finally resulted in cell death (Jank et al. 2012). Yeast and mammalian toxicity and also vesicle trafficking defects were depending on a functional glycosyltransferase domain, while a mutant of the essential DxD-motif (SetA NxN or AxA) did not result in cellular effects.

Analyses with purified recombinant protein showed that SetA possesses enzymatic activity and revealed UDP-glucose as specific nucleotide sugar donor for glycosyltransferase activity, hydrolase activity, and autoglucosylation. Owing the lack of the eukaryotic substrate, basic histones proteins were identified as artificial substrates for SetA and allowed to follow enzyme catalysis and glucosyltransferase activity. Accordingly, SetA enzyme activity depended on an intact DxD-motif (Jank et al. 2012). Mass spectrometric analyses revealed that histones H4 and H3.1 were mono-*O*-glycosylated at distinct serine and threonine residues. Thus, SetA is suggested as glucose specific mono-*O*-glycosyltransferase, modifying serine, or threonine residues.

Truncation analysis revealed that SetA consists of at least two domains. The glucosyltransferase domain, which is restricted to the N-terminus (aa1-300), and a localization domain in the C-terminus (aa401-644), which is important to guide SetA to the outer leaflet of pleiomorphic vesicular compartments. In order to identify the target vesicles of SetA, colocalization experiments in HeLa cells were performed with several endocytic marker proteins. SetA was recruited by Rab5/EEA1 positive early endosomes and only to a minor extend by late endosomal or lysosomal compartments (Rab7/LAMP1-positive vesicles) (Jank et al. 2012) but not by the ER, Golgi, or the autophagosome (Heidtman et al. 2009).

Consistent with this finding, SetA was found to interact specifically with phosphatidylinositol-3-phosphate [PtdIns(3)P], a phospholipid species tightly regulated and found almost exclusively on early endosomes and multi-vesicular bodies of the endocytic pathway (Gillooly et al. 2003). Surface plasmon resonance spectroscopy was used to measure the affinity of SetA to biotin-PtdIns(3)P, which is comparable to the eukaryotic p40phox PX domain of NADPH-oxidase. As phosphopinositides serve as binding platforms for distinct eukaryotic as well as prokaryotic protein domains, it was shown that mCherry-labeled SetA is recruited to the outer leaflet of the *Legionella*-containing vacuole in *L. pneumophila* infected RAW-macrophages (Jank et al. 2012). This corroborates the assumption that SetA localizes to distinct endocytic compartments. Interactions with PtdInsP were also found in other *Legionella* effectors (SidC, SidM/DrrA, SdcA) but here recruitment is mediated mainly by PtdIns(4)P (Brombacher et al. 2009; Ragaz et al. 2008; Schoebel et al. 2010; Weber et al. 2009). Next to SetA, the LpnE was reported to bind PtdIns(3)P (Weber et al. 2009). Moreover, it was shown that PtdIns(3)P is essential for *Legionella* intracellular replication and participates in the modulation of the LCV, but seems not to be essential for phagocytosis (Weber et al. 2006). It needs to be determined, which functional role SetA or LpnE plays in the context of *Legionella* infection. The C-terminal PtdIns(3)P-binding domain of SetA is unique in amino acid sequence and not observed in other proteins in nature. Structural information would be valuable to elucidate the mechanism underlying the specific binding of SetA to PtdIns(3)P and will certainly promote *Legionella*-localization-domains as tools in cell biology.

Interestingly, both domains of SetA, the glycosyltransferase domain and the localization domain were shown to be important for the toxic effect in yeast. Therefore, we propose that after secretion into the host cytosol, SetA is attracted by LCV-containing PtdIns(3)P and, thereby, guided to its substrate for glucosylation, which might be involved in the initial steps of LCV maturation.

Ectopic expression showed that SetA is polyubiquitinated in both yeast and mammalian cells (Heidtman et al. 2009). Unfortunately, it was not possible to detect ubiquitination of SetA during *L. pneumophila* infection of host cells. Therefore, the connection of SetA glycosyltransferase to the ubiquitination machinery remains unclear.

6 Eukaryotic Protein Glycosylation and Intracellular Life Cycle of *L. pneumophila*

The importance of protein synthesis inhibition for the intracellular biology of *L. pneumophila* is not clear. One hypothesis, assuming a *direct* role of eEF1A-targeting glucosyltransferases, implies that the action of Lgts results in an impaired defense mechanism of a host. On the other hand, at final stages of the intracellular life cycle, *Legionella* tends to kill the eukaryotic cell and Lgt can participate in such a task as a cytotoxic virulence factor (Belyi and Aktories 2010; Belyi et al. 2011). Furthermore, *Legionella* intracellular replication is dependent on the provision of nutrients, e.g., amino acids. Lgt might induce the shutdown of host protein synthesis in order to exploit the remaining amino acid or amino-acyl-tRNA pool of the cell.

According to another hypothesis, which acknowledges a *smart* role of Lgt, suppression of translation can trigger different unrelated processes in a cell being for one or another reason advantageous for the infection agent. Targeting elongation factor eEF1A by Lgt can lead to pleiotropic outcomes and the translation arrest observed in our experiments can be a side effect of some other pro-bacterial consequence of eEF1A glucosylation (Ensminger and Isberg 2009). For instance, *Legionella* effector SidI was demonstrated both to inhibit protein synthesis and to participate in activation of stress response mechanisms (Shen et al. 2009). Stress response is triggered in eukaryotic cell suffering under unfavorable conditions and results in production of a panel of heat-shock proteins (Sarge et al. 1991). How this process can be related to *L. pneumophila* virulence is not clear. However, one speculation is that such production of stress-induced heat-shock proteins could be important for correct folding of *Legionella* effector proteins.

Experimental data illustrating such *smart* roles of glucosyltransferases Lgt in *L. pneumophila* biology have just started to evolve. Systematic study with *Drosophila* cells, treated by interfering RNA to deplete certain eukaryotic factors, disclosed pairs of seemingly unrelated T4bSS effectors that perform compensatory functions during intracellular infection of *Legionella* (O'Connor et al. 2012). In particular, although deletion even of all three Lgt genes did not produce any growth disadvantage for intracellular *Legionella* (Ivanov and Roy 2009), attenuation of *L. pneumophila* was observed when the gene coding for glucosyltransferase Lgt2 was deleted in combination with the coding sequence of *mavP* (*lpg2884*). In contrast, combined mutations in *wipB* (*lpg0642*) and *lgt2* resulted in gain of intracellular growth of the engineered *L. pneumophila* strain. Unfortunately, products of both *mavP* and *wipB* sequences remain so far uncharacterized and carry no defined functions in *Legionella* virulence. However, the information presented suggests the existence of a network of functional interactions between different *L. pneumophila* effectors and can be helpful in elucidating roles of different *L. pneumophila* T4bSS substrates in intracellular biology of the host and their importance for the pathogen.

In silico analysis of available genomes of *Legionella* can be helpful in detection of novel enzymes with glycosylation activities. Thus, BLAST search for proteins similar to SetA in the *L. pneumophila* Philadelphia-1 genome revealed a ~59 kDa protein, representing a product of *lpg1961* gene, which was shown to be toxic for *S. cerevisiae* and caused a mild defect in alkaline phosphatase sorting in yeast (Heidtman et al. 2009). This protein possessed a canonical DXD-motif, which pointed toward possible glycosyltransferase activity in this *L. pneumophila* product. Similarly, genome analyses of two non-*pneumophila* species (*L. drancourtii* (Moliner et al. 2010) and *L. longbeachae* (Cazalet et al. 2010; Kozak et al. 2010)) disclosed several proteins showing identity of around 15 % with Lgt1 (Belyi et al. 2011). All these proteins possess a DXD-containing region resembling that of typical glycosyltransferases.

On the other hand, it should be noted, that some known effectors of *L. pneumophila*, shown to repress translation of eukaryotic cells, do not display any substantial structural similarities to Lgt-s. These are SidI and SidL proteins (Fontana et al. 2011; Shen et al. 2009). Such proteins could contribute to *Legionella* pathogenicity in a redundant manner enriching array of host-targeting reactions accomplished by glucosylating factors.

Taken together, data available suggest important roles of *L. pneumophila* enzymes, glycosylating eukaryotic proteins. Such *Legionella* effectors can modify different host targets in a highly specific manner, producing extensive alterations in the cellular signaling and/or metabolism of the host cell, which contribute to a successful colonization of the invading pathogen.

References

Abel K et al (1996) An alpha to beta conformational switch in EF-Tu. Structure 4(10):1153–1159

Abu-Zant A et al (2007) Anti-apoptotic signalling by the Dot/Icm secretion system of *L. pneumophila* Cell. Microbiol 9(1):246–264

Andersen GR et al (2000) Structural basis for nucleotide exchange and competition with tRNA in the yeast elongation factor complex eEF1A:eEF1Balpha. Mol Cell 6(5):1261–1266

Banga S et al (2007) *Legionella pneumophila* inhibits macrophage apoptosis by targeting pro-death members of the Bcl_2 protein family. Proc Natl Acad Sci USA 104(12):5121–5126

Bartfeld S et al (2009) Temporal resolution of two-tracked NF-kappaB activation by *Legionella pneumophila*. Cell Microbiol 11(11):1638–1651

Belyi I, Popoff MR, Cianciotto NP (2003) Purification and characterization of a UDP-glucosyltransferase produced by *Legionella pneumophila*. Infect Immun 71(1):181–186

Belyi Y, Aktories K (2010) Bacterial toxin and effector glycosyltransferases. Biochim Biophys Acta 1800(2):134–143

Belyi Y, Jank T, Aktories K (2011) Effector glycosyltransferases in *Legionella*. Front Microbiol 2:76

Belyi Y et al (2006) *Legionella pneumophila* glucosyltransferase inhibits host elongation factor 1A. Proc Natl Acad Sci U S A 103(45):16953–16958

Belyi Y et al (2009) Region of elongation factor 1A1 involved in substrate recognition by *Legionella pneumophila* glucosyltransferase Lgtl: identification of Lgtl as a retaining glucosyltransferase. J Biol Chem 284(30):20167–20174

Belyi Y et al (2008) Lgt: a family of cytotoxic glucosyltransferases produced by *Legionella pneumophila*. J Bacteriol 190(8):3026–3035

Belyi Y et al (2012) Elongation factor 1A is the target of growth inhibition in yeast caused by *Legionella pneumophila* glucosyltransferase Lgt1. J Biol Chem 287(31):26029–26037

Brombacher E et al (2009) Rab1 guanine nucleotide exchange factor SidM is a major phosphatidylinositol 4-phosphate-binding effector protein of *Legionella pneumophila*. J Biol Chem 284(8):4846–4856

Bruggemann H et al (2006) Virulence strategies for infecting phagocytes deduced from the in vivo transcriptional program of *Legionella pneumophila*. Cell Microbiol 8(8):1228–1240

Busch C et al (1998) A common motif of eukaryotic glycosyltransferases is essential for the enzyme activity of large clostridial cytotoxins. J Biol Chem 273(31):19566–19572

Byrne B, Swanson MS (1998) Expression of *Legionella pneumophila* virulence traits in response to growth conditions. Infect Immun 66(7):3029–3034

Cazalet C et al (2010) Analysis of the *Legionella longbeachae* genome and transcriptome uncovers unique strategies to cause Legionnaires' disease. PLoS Genet 6(2):e1000851

de Felipe KS et al (2008) *Legionella* eukaryotic-like type IV substrates interfere with organelle trafficking. PLoS Pathog 4(8):e1000117

de Felipe KS et al (2005) Evidence for acquisition of *Legionella* type IV secretion substrates via interdomain horizontal gene transfer. J Bacteriol 187(22):7716–7726

Diederen BM (2008) *Legionella* spp. and Legionnaires' disease. J Infect 56(1):1–12

Ensminger AW, Isberg RR (2009) *Legionella pneumophila* Dot/Icm translocated substrates: a sum of parts. Curr Opin Microbiol 12(1):67–73

Fontana MF et al (2011) Secreted bacterial effectors that inhibit host protein synthesis are critical for induction of the innate immune response to virulent *Legionella pneumophila.* PLoS Pathog 7(2):e1001289

Franco IS, Shuman HA, Charpentier X (2009) The perplexing functions and surprising origins of *Legionella pneumophila* type IV secretion effectors. Cell Microbiol 11(10):1435–1443

Ge J et al (2009) A *Legionella* type IV effector activates the NF-kappaB pathway by phosphorylating the IkappaB family of inhibitors. Proc Natl Acad Sci U S A 106(33):13725–13730

Gillooly DJ, Raiborg C, Stenmark H (2003) Phosphatidylinositol 3-phosphate is found in microdomains of early endosomes. Histochem Cell Biol 120(6):445–453

Goody PR et al (2012) Reversible phosphocholination of Rab proteins by *Legionella pneumophila* effector proteins. EMBO J 31(7):1774–1784

Haenssler E, Isberg RR (2011) Control of host cell phosphorylation by *legionella pneumophila.* Front Microbiol 2:64

Heidtman M et al (2009) Large-scale identification of *Legionella pneumophila* Dot/Icm substrates that modulate host cell vesicle trafficking pathways. Cell Microbiol 11(2):230–248

Hilbi H, Weber S, Finsel I (2011) Anchors for effectors: subversion of phosphoinositide lipids by *Legionella.* Front Microbiol 2:91

Hsu F et al (2012) Structural basis for substrate recognition by a unique *Legionella* phosphoinositide phosphatase. Proc Natl Acad Sci U S A 109(34):13567–13572

Huang L et al (2011) The E block motif is associated with *Legionella pneumophila* translocated substrates. Cell Microbiol 13(2):227–245

Hurtado-Guerrero R et al (2010) Molecular mechanism of elongation factor 1A inhibition by a *Legionella pneumophila* glycosyltransferase. Biochem J 426(3):281–292

Ingmundson A et al (2007) *Legionella pneumophila* proteins that regulate Rab1 membrane cycling. Nature 450(7168):365–369

Isberg RR, O'Connor TJ, Heidtman M (2009) The *Legionella pneumophila* replication vacuole: making a cosy niche inside host cells. Nat Rev Microbiol 7(1):13–24

Ivanov SS, Roy CR (2009) Modulation of ubiquitin dynamics and suppression of DALIS formation by the *Legionella pneumophila* Dot/Icm system. Cell Microbiol 11(2):261–278

Jank T et al (2012) Domain organization of *Legionella* effector SetA. Cell Microbiol 14(6):852–868

Jank T, Giesemann T, Aktories K (2007) Rho-glucosylating Clostridium difficile toxins A and B: new insights into structure and function. Glycobiology 17(4):15R–22R

Just I et al (1995) Glucosylation of Rho proteins by *Clostridium difficile* toxin B. Nature 375(6531):500–503

Kozak NA et al (2010) Virulence factors encoded by *Legionella longbeachae* identified on the basis of the genome sequence analysis of clinical isolate D-4968. J Bacteriol 192(4):1030–1044

Laguna RK et al (2006) A Legionella pneumophila-translocated substrate that is required for growth within macrophages and protection from host cell death. Proc Natl Acad Sci USA 103(49):18745–18750

Lairson LL et al (2008) Glycosyltransferases: structures, functions, and mechanisms. Annu Rev Biochem 77:521–555

Losick VP et al (2010) LnaB: a *Legionella pneumophila* activator of NF-kappaB. Cell Microbiol 12(8):1083–1097

Lu W et al (2010) Structural basis of the action of glucosyltransferase Lgt1 from *Legionella pneumophila*. J Mol Biol 396(2):321–331
Machner MP, Isberg RR (2007) A bifunctional bacterial protein links GDI displacement to Rab1 activation. Science 318(5852):974–977
Moliner C, Fournier PE, Raoult D (2010) Genome analysis of microorganisms living in amoebae reveals a melting pot of evolution. FEMS Microbiol Rev 34(3):281–294
Mukherjee S et al (2011) Modulation of Rab GTPase function by a protein phosphocholine transferase. Nature 477(7362):103–106
Murata T et al (2006) The *Legionella pneumophila* effector protein DrrA is a Rab1 guanine nucleotide-exchange factor. Nat Cell Biol 8(9):971–977
Nagai H et al (2002) A bacterial guanine nucleotide exchange factor activates ARF on *Legionella* phagosomes. Science 295(5555):679–682
Nelson RJ et al (1992) The translation machinery and 70 kd heat shock protein cooperate in protein synthesis. Cell 71(1):97–105
Neunuebel MR et al (2011) De-AMPylation of the small GTPase Rab1 by the pathogen *Legionella pneumophila*. Science 333(6041):453–456
O'Connor TJ et al (2012) Aggravating genetic interactions allow a solution to redundancy in a bacterial pathogen. Science 338(6113):1440–1444
Qasba PK, Ramakrishnan B, Boeggeman E (2005) Substrate-induced conformational changes in glycosyltransferases. Trends Biochem Sci 30(1):53–62
Ragaz C et al (2008) The *Legionella pneumophila* phosphatidylinositol-4 phosphate-binding type IV substrate SidC recruits endoplasmic reticulum vesicles to a replication-permissive vacuole. Cell Microbiol 10(12):2416–2433
Ramakrishnan V (2002) Ribosome structure and the mechanism of translation. Cell 108(4):557–572
Richards AM et al (2013) Cellular microbiology and molecular ecology of Legionella-amoeba interaction. Virulence 4(4):307–314
Sadretdinova OV et al (2012) Prevalence of glucosyl transferase Lgt among *Legionella pneumophila* strains isolated from various sources. Zh Mikrobiol Epidemiol Immunobiol 3:8–12
Sarge KD et al (1991) Cloning and characterization of two mouse heat shock factors with distinct inducible and constitutive DNA-binding ability. Genes Dev 5(10):1902–1911
Sasikumar AN, Perez WB, Kinzy TG (2012) The many roles of the eukaryotic elongation factor 1 complex. Wiley Interdiscip Rev RNA 3(4):543–555
Schoebel S et al (2010) High-affinity binding of phosphatidylinositol 4-phosphate by *Legionella pneumophila* DrrA. EMBO Rep 11(8):598–604
Schoebel S et al (2011) Protein LidA from *Legionella* is a Rab GTPase supereffector. Proc Natl Acad Sci U S A 108(44):17945–17950
Segal G (2013) Identification of *Legionella* effectors using bioinformatic approaches. Methods Mol Biol 954:595–602
Shen X et al (2009) Targeting eEF1A by a *Legionella pneumophila* effector leads to inhibition of protein synthesis and induction of host stress response. Cell Microbiol 11(6):911–926
Shoemaker CJ, Green R (2012) Translation drives mRNA quality control. Nat Struct Mol Biol 19(6):594–601
Tan Y, Arnold RJ, Luo ZQ (2011) *Legionella pneumophila* regulates the small GTPase Rab1 activity by reversible phosphorylcholination. Proc Natl Acad Sci U S A 108(52):21212–21217
Tan Y, Luo ZQ (2011) *Legionella pneumophila* SidD is a deAMPylase that modifies Rab1. Nature 475(7357):506–509
Tzivelekidis T et al (2011) Aminoacyl-tRNA-charged eukaryotic elongation factor 1A is the bona fide substrate for *Legionella pneumophila* effector glucosyltransferases. PLoS ONE 6(12):e29525
Viner R et al (2012) Identification of two *Legionella pneumophila* effectors that manipulate host phospholipids biosynthesis. PLoS Pathog 8(11):e1002988

Weber SS, Ragaz C, Hilbi H (2009) Pathogen trafficking pathways and host phosphoinositide metabolism. Mol Microbiol 71(6):1341–1352

Weber SS et al (2006) *Legionella pneumophila* exploits PI(4)P to anchor secreted effector proteins to the replicative vacuole. PLoS Pathog 2(5):e46

Wiggins CA, Munro S (1998) Activity of the yeast MNN1 alpha-1,3-mannosyltransferase requires a motif conserved in many other families of glycosyltransferases. Proc Natl Acad Sci U S A 95(14):7945–7950

Xu L, Luo ZQ (2013) Cell biology of infection by *Legionella pneumophila*. Microbes Infect 15(2):157–167

Xu L et al (2010) Inhibition of host vacuolar H+-ATPase activity by a *Legionella pneumophila* effector. PLoS Pathog 6(3):e1000822

Zhu W et al (2011) Comprehensive identification of protein substrates of the Dot/Icm type IV transporter of *Legionella pneumophila*. PLoS ONE 6(3):e17638

Ziegler MO et al (2008) Conformational changes and reaction of clostridial glycosylating toxins. J Mol Biol 377(5):1346–1356

Zusman T et al (2007) The response regulator PmrA is a major regulator of the icm/dot type IV secretion system in *Legionella pneumophila* and *Coxiella burnetii*. Mol Microbiol 63(5):1508–1523

Modulation of the Ubiquitination Machinery by *Legionella*

Andree Hubber, Tomoko Kubori and Hiroki Nagai

Abstract The bacterial pathogen *Legionella pneumophila* manipulates its intracellular fate by co-opting host processes. Using bacterial proteins translocated into host cells, *L. pneumophila* targets pathways shared by unicellular protozoa and higher eukaryotes. In eukaryotes, an important mechanism that regulates numerous cellular processes, including those designed to kill invading microorganisms, is ubiquitination. Post-translational modification of proteins with ubiquitin is a highly regulated process that either targets proteins for degradation or modifies their activity. It is emerging that *L. pneumophila* possesses functional mimics of eukaryotic E3 ubiquitin ligases that function with the host ubiquitination machinery to select and modify substrates for polyubiquitination. *L. pneumophila* proteins have been identified that ubiquitinate both host and bacterial proteins, and ubiquitination of the bacterial protein SidH results in its degradation by the host proteasome. This pathway allows *L. pneumophila* to temporally regulate effector function inside host cells, and facilitates optimal *L. pneumophila* replication by undefined mechanisms. This review will focus on our current knowledge of the proteins used by *L. pneumophila* to co-opt the host ubiquitination machinery, and current progress toward understanding the ubiquitin-mediated processes manipulated by *L. pneumophila* to facilitate intracellular survival and propagation.

Contents

1 Introduction 228
  1.1 Bacterial Pathogens and Ubiquitination 228
  1.2 Proteins that Modulate Host Ubiquitin Pathways 229
2 *Legionella* Recruits Ubiquitin to the *Legionella*-Containing Vacuole and Requires the Proteasome for Optimal Growth 231

A. Hubber · T. Kubori · H. Nagai (✉)
Research Institute for Microbial Diseases, Osaka University,
Yamadaoka 3-1, Suita, Osaka 565-0871, Japan
e-mail: hnagai@biken.osaka-u.ac.jp

Current Topics in Microbiology and Immunology (2013) 376: 227–247
DOI: 10.1007/82_2013_343

Published Online: 6 August 2013

3 F-Box- and U-Box-Type E3 Ubiquitin Ligases of *L. pneumophila* ... 233
3.1 Identification of Putative L. pneumophila E3 Ligases ... 233
3.2 Biochemical Analysis of Putative E3 Ligases of L. pneumophila ... 235
4 Substrates of *L. pneumophila* E3 Ubiquitin Ligases ... 235
4.1 A Host Kinase Clk1 ... 235
4.2 Another Effector Protein SidH ... 237
4.3 Hsp70 Co-chaperone BAT3 ... 237
4.4 Unknown Substrate of LegAU13/AnkB ... 238
5 The Role of Ubiquitin in Selective Autophagy and *L. pneumophila* ... 240
6 Concluding Remarks ... 241
References ... 242

1 Introduction

1.1 Bacterial Pathogens and Ubiquitination

Ubiquitin is a 76 amino acid polypeptide that can be covalently attached to amino groups of mostly lysine residues of target proteins in a process known as ubiquitination, ubiquitylation, or ubiquitinylation. This post-translational modification is one of the most conserved in eukaryotic cells (Hershko and Ciechanover 1998). Ubiquitination requires E1, E2, and E3 proteins that are dedicated to the selection and modification of target proteins. The consequences of ubiquitin-modification vary. Modification often results in degradation by either the 26S proteasome or in lysosomal compartments, yet other modified proteins are not degraded but rather their functions are altered by the addition of ubiquitin (Clague and Urbe 2010; Ikeda et al. 2010; Pickart and Fushman 2004; Xu et al. 2009). Proteins targeted for ubiquitination are often key regulators in many cellular pathways including cell cycle, ribosomal function, DNA repair, signal transduction, and vesicular trafficking (reviewed in Koepp et al. (1999), Hicke and Dunn (2003), Haglund and Dikic (2005, 2012), Mocciaro and Rape (2012), Ulrich and Walden (2010) and Vandenabeele and Bertrand (2012)). It is emerging that several human pathogens manipulate ubiquitin-regulated host cellular pathways including *Burkholderia pseudomallei, Listeria monocytogenes, Salmonella* sp., and *Shigella flexneri* (reviewed in Jiang and Chen (2012)). Targeting of ubiquitin modification processes by bacteria is not restricted to human pathogens as numerous plant pathogens and symbionts, including *Agrobacterium tumefaciens*, *Xanthomonas pampestris*, *Ralstonia solanacearum*, and *Rhizobium* sp., also co-opt the host ubiquitination machinery (Magori and Citovsky 2011; Price and Kwaik 2010; Xin et al. 2012). Many of these bacteria manipulate host processes using bacterial proteins, known as effectors, that are translocated into the cytosol of host cells. Identifying bacterial effectors that co-opt the host ubiquitination machinery is paramount in the quest to understand the pathways manipulated by these bacteria. Thus interference with the host ubiquitination machinery is a common strategy used by eukaryote-interacting prokaryotes to promote growth and survival during host association.

1.2 Proteins that Modulate Host Ubiquitin Pathways

In eukaryotes, research focus has been on examination of how target proteins are selected, how proteins are modified, and what determines the fate of modified proteins. We will briefly summarize key aspects of host ubiquitination, as the molecules mediating these processes are likely targets for mimicry or manipulation by bacterial effector proteins. Key examples of how other bacteria utilize the host ubiquitin-modification system will also be provided to highlight the various strategies used by bacteria to survive in concert with eukaryotic hosts.

Modification of proteins with ubiquitin is dynamic and tightly regulated by host cells. In eukaryotes, an enzymatic cascade is required to covalently attach free ubiquitin to substrate proteins. The three proteins E1, E2, and E3 successively participate in the attachment process (Hershko et al. 1983). The E1 and E2 proteins are colloquially known as the ubiquitin activating and conjugating enzymes, respectively. They function to prepare and position the ubiquitin for transfer by the E3 enzyme to the target protein. In this three-step reaction much focus had been placed on E3 proteins because of their role in substrate selection (Ardley and Robinson 2005).

Two crucial aspects of E3 enzymes are catalysis of the attachment and specificity of the reaction. Two major eukaryotic domains HECT (homologous to the E6AP carboxyl terminus) and RING (really interesting new gene) as well as several less common domains achieve catalysis through bringing E2-bound ubiquitin close to the attachment site on substrates, whereas at least 20 different substrate binding domains contribute to the specificity of the reaction. The importance of E3 ligases for reaction specificity is demonstrated by the hierarchical nature of the cascade. In humans there are hundreds of E3, compared to over 40 E2s and only two E1 enzymes (Li et al. 2008; Hutchins et al. 2013; van Wijk and Timmers 2010). E3 ligases with HECT domains are single molecules with a modular domain structure; the N-terminal selection domain followed by the HECT enzymatic domain. The *Salmonella enterica* protein SopA is considered a functional mimic of HECT E3 ligases and catalytic residues conserved in all HECT domains are found in SopA (Diao et al. 2008). The RING finger domain and RING finger-like U-box domain consist of a three strand and a single helix structure that are variably stabilized by zinc ions (RING-finger) or salt bridges and hydrogen bonds (U-box). This scaffold is thought to provide a framework for the transfer of ubiquitin to substrate proteins. The RING-type E3 ligases have two architectures of single subunit and multisubunit. An important group of multisubunit RING-type E3 ligases is the SCF (Skp1-Cullin-F-box protein) complex. The F-box domain recognizes the SCF complex and additional domains within the F-box-containing protein recognize the substrate (Kipreos and Pagano 2000). In vitro ubiquitin modification can be achieved using an F-box containing protein, together with Skp, a RING-finger containing protein, Cullin, and an E2 enzyme. In recent years numerous bacterial pathogens and symbionts, which share the ability to interact with eukaryotes, have been found to encode E3 ligases with structural similarity (reviewed in Hicks and Galan (2010)). The IpaH/SspH E3 ubiquitin ligases from *Shigella flexneri* and *Salmonella enterica* are the archetypal members of this new NEL (Novel E3 Ligase)-type family of E3 ligases (Quezada et al. 2009; Singer et al. 2008; Zhu et al. 2008). Despite the lack of structural similarity between the NEL domain and RING-finger or HECT domains, the bacterial NEL-type ligases are functional mimics of eukaryotic E3-ligases.

The fate of ubiquitinated proteins relies on the combination of E2 and E3 proteins and the nature of the ubiquitin modification. Ubiquitination is an extremely versatile modification, as proteins can either become mono or poly ubiquitinated in diverse chain conformations (reviewed in Ikeda et al. (2010)). Linear chains form by linkage of the carboxy terminus of one ubiquitin to the amino-terminal methionine of the next, and nonlinear chains form by linkage at one of the seven lysines found within ubiquitin (K6, K11, K27, K29, K33, K48, and K63). Different E2 enzymes, which share a highly conserved ubiquitin-conjugating catalytic fold (van Wijk and Timmers 2010), appear to be responsible for the formation of different chain topologies (reviewed in Windheim et al. (2008), Ye and Rape (2009) and David et al. (2010)). In polyubiquitinated proteins, the two chain linkages K48 and K63 have been associated with proteins destined for proteasomal degradation and nonproteasomal processing, respectively. Recently it was reported that the reason K63-polyubiquitin chains do not associate with the proteasome in vivo, despite being an excellent substrate for degradation with purified 26S proteasome, is that accessory soluble proteins bind to K63 conjugates and block their binding to the proteasome (Nathan et al. 2013). Conversely, other proteins preferentially bind K48-ubiquitinated proteins and aid binding to the 26S complex (Nathan et al. 2013). Thus, E2 enzymes and accessory proteins, which recognize specific modifications, play a key role in controlling the fate of modified substrates.

Numerous mechanisms regulate the ubiquitination machinery itself and these may be targets for bacterial manipulation. Spatial control of several E2 enzymes has been described and several E2's are themselves regulated by ubiquitination (regulation of E2s reviewed in van Wijk and Timmers (2010)). Post-translational modifications, including phosphorylation, ubiquitination, modification by ubiquitin-like peptides such as SUMO (small ubiquitin-like modifier) and NEDD8 (neural precursor cell expressed, developmentally down-regulated 8) also regulate E3 enzymes (reviewed in de Bie and Ciechanover (2011)). These modifications regulate E3s by both proteolytic and nonproteolytic mechanisms. The subcellular localization of ubiquitinated proteins is also thought to influence their fate and function. Ubiquitination of the *Salmonella enterica* serovar Typhimurium translocated effector SopB mediates recruitment of SopB from the plasma membrane to the *Salmonella*-containing vacuole (Patel et al. 2009). Thus, ubiquitination temporally regulates the site at which this protein acts during different stages of infection. Therefore, bacterial manipulation of the host ubiquitination machinery may involve co-opting the localization or post-translational modification status of host E2 and E3 enzymes, and bacterial effectors may themselves be regulated inside host cells by ubiquitination.

Finally, ubiquitination is a reversible process. A final group of enzymes regulating protein ubiquitination in host cells are deubiquitinating enzymes, abbreviated to the commonly used term DUBs. These proteins counter-regulate ubiquitin modifications with over 100 known DUBs in humans specifically recognizing different ubiquitin chain modifications (reviewed in Hutchins et al. (2013), Clague et al. (2012) and Reyes-Turcu et al. (2009)). Briefly, the DUBs are proteases that can be classified into two classes: Cysteine proteases and metalloproteases. Bacterial DUBs have been identified in a number of pathogens including *Yersinia*, *Salmonella,* and *Chlamydia* (Mukherjee et al. 2006; Rytkonen et al. 2007; Mesquita et al. 2012; Misaghi et al. 2006; Ye et al. 2007; Zhou et al. 2005), and the targets for some of these bacterial DUBs are immune signaling proteins (Zhou et al. 2005). The DUBs encoded by these pathogens all belong to the CE clan of cysteine proteases. Recently, the genome of the intracellular amoeba symbiont *Candidatus Amoebophilus asiaticus* revealed two putative bacterial DUBs with similarity the ubiquitin proteases of CA clan family C19 (Schmitz-Esser et al. 2010), and one further bacterial DUB from the clan CA was described in the equine pathogen *Burkholderia mallei* (Shanks et al. 2009). Eukaryotic DUBs have recently been implicated in modulating innate immune responses (reviewed in (Harhaj and Dixit (2012)). Thus, another strategy to modulate host responses to infection is likely bacterial manipulation of host DUBs. Indeed, *Helicobacter pylori* has been proposed to influence the activity and expression of at least one eukaryotic DUB (Coombs et al. 2011).

2 *Legionella* Recruits Ubiquitin to the *Legionella*-Containing Vacuole and Requires the Proteasome for Optimal Growth

The bacterial pathogen *Legionella pneumophila* normally resides in the environment where it replicates inside unicellular protozoa such as amoeba. Its reputation as a human pathogen is due to an illness, known as Legionnaires disease, caused by incidental inhalation of contaminated water particles and subsequent replication of *L. pneumophila* within lung macrophages (McDade et al. 1977). Defense mechanisms utilized to clear invading organisms that are conserved between unicellular protozoa and humans are those likely to be co-opted by *L. pneumophila* to facilitate its intracellular survival. The key virulence mechanism of *L. pneumophila*, which is required to manipulate host pathways, is the Dot/Icm system

(Berger and Isberg 1993; Horwitz 1987; Marra et al. 1992). This system acts as a macromolecular machine, classified as a type IV secretion system, transporting bacterial proteins into host cells (Nagai et al. 2002; Segal et al. 1998; Segal and Shuman 1999; Vogel et al. 1998). *L. pneumophila* transports at least 275 effector proteins into host cells. Once inside the host these effectors modulate numerous processes including phagosomal trafficking, vesicular trafficking, host translation, and defense responses (reviewed in Hubber and Roy (2010)). Mutants lacking a functional Dot/Icm system are avirulent and fail to replicate inside host cells (Marra et al. 1992). Direct evidence for manipulation of innate immune responses by *L. pneumophila* via manipulation of ubiquitin-modification has not yet been reported. However, both the ubiquitin-modification system and ubiquitin-dependent autophagy pathway are utilized by both protozoa and higher eukaryotes (Hutchins et al. 2013; Calvo-Garrido et al. 2010), and examination of how *L. pneumophila* co-opts these processes is a growing field of research. Here we review the phenotypic, bioinformatic, and biochemical data that support co-option of the host ubiquitin system by *L. pneumophila.*

A key observation first reported by Dorer et al. (2006) was that shortly after infection *L. pneumophila* recruits polyubiquitin conjugates around the vacuolar membrane in a Dot/Icm-dependent manner, based on staining with the polyubiquitin-specific antibody FK1. In the same study MG132, a chemical inhibitor of the host proteasome, was shown to inhibit intracellular *L. pneumophila* replication in a replicative vacuole assay. Also, the Cdc48/p97 complex was identified as a host factor required for optimal growth of *L. pneumophila* in *Drosophila* cells. Cdc48/p97 belongs to AAA (ATPases associated with various cellular activities) ATPase family and is required for proteasomal degradation of many polyubiquitinated proteins including those extracted from the endoplasmic reticulum (ER) (Ye et al. 2001; Jarosch et al. 2002). RNA silencing of factors associated with ER-associated degradation (ERAD) (for review, see Meusser et al. (2005)), namely Npl4, Ufd1, Ufd3, Dsk2, Pac10, and CG32566, also decreased *L. pneumophila* replication. Further, Cdc48/p97 localized to *Legionella*-Containing Vacuoles (LCVs) in a Dot/Icm-dependent manner and removes at least some effector proteins including LidA and polyubiquitinated conjugates from LCVs. Although the effector SidC was reported to behave similar to LidA, it is not clear if this clearance mechanism is important for removing other effector proteins from the LCV and if these effectors are themselves ubiquitinated at the relevant stage of infection to facilitate their removal from the LCV. Taken together, these data suggested that the ERAD pathway contributes to optimal intracellular growth of *L. pneumophila.* This seminal paper provided the first evidence that ubiquitination of proteins during *L. pneumophila* infection is actively directed by the pathogen for optimum survival in host cells. However, the identity of the proteins polyubiquitinated at LCVs and the molecular mechanisms behind the co-option of Cdc48/p97 awaits further clarification.

3 F-Box- and U-Box-Type E3 Ubiquitin Ligases of *L. pneumophila*

3.1 Identification of Putative L. pneumophila E3 Ligases

As early as 2005, it was reported that *L. pneumophila* encodes proteins with regions of similarity to F-box and U-box domains (de Felipe et al. 2005). Currently, six *L. pneumophila* proteins LegU1, LicA, Lpg1975/Lpp1959, LegAU13/AnkB, Lpg2525, and Lpp2486 have been found to possess regions with similarity to F-box domains and LubX/LegU2 and Lpg2455 possess regions with similarity to U-box domains (Angot et al. 2007; Bruggemann et al. 2006; Habyarimana et al. 2008; Kubori et al. 2008; Zeng et al. 2008) (Table 1). Indicative of a role within host cells, expression of LegAU13/AnkB in yeast caused a slow growth phenotype, and LubX was found to cause a mild defect in vesicular trafficking of alkaline phosphatase (Heidtman et al. 2009). Indeed, the majority of the predicted putative F-box and U-box-containing proteins have been confirmed as substrates for Dot/Icm-mediated translocation into host cells (Kubori et al. 2008; Al-Khodor et al. 2008; de Felipe et al. 2008; Lomma et al. 2010; Zhu et al. 2011). Translocation of Lpg2455 has not yet been shown, but it was identified as a likely candidate based on the presence of the so-called 'E-block' motif that is associated with translocated substrates (Huang et al. 2011). Lpg2455 has orthologs in all of *L. pneumophila* strains sequenced so far. This is also true for LicA and LegAU13/AnkB, whereas LegU1, Lpp2486, Lpg2525, and LubX/LegU2 appear to be more strain specific (Table 1). Whether this distribution represents a host-specific requirement for E3-ligase activities with differing specificities or whether some ubiquitin modulating proteins are more essential than others for *L. pneumophila* pathogenesis is currently unknown. The finding that non-pneumophila strains of *Legionella* also possess proteins with F-box and U-box domains supports an important role for ubiquitin-modification in *Legionella's* intracellular survival strategy (Cazalet et al. 2010; Kozak et al. 2010).

Clues to the stage of infection that requires the activity of a given effector may be provided by analysis of when a given effector is produced by the bacterium. Comprehensive analysis of all predicted F-box containing effectors in the Philadelphia-1 strain failed to find significant transcription of *legU1*, *licA*, *or legAU13/ankB* when grown in broth culture, suggesting that they are likely upregulated during growth in vivo (Ensminger and Isberg 2010). In strain AA100/130b, *ankB* was reported to be expressed during post-exponential phase in broth cultures (Al-Khodor et al. 2008). Detailed analysis of LubX production by *L. pneumophila* grown in broth cultures and after infection showed this protein was only detectable by Western-blot analysis after 8 h of infection, suggesting this protein is maximally produced during the mid to late phase of infection (Kubori et al. 2008). This provided the first evidence for an E3 ligase-effector that functions during a specific phase of the infection process.

Table 1 F-box and U-box containing proteins in *Legionella pneumophila*[1]

Strains								Pfam hits			Aliases
Philadelphia 1[2]	ATCC 43290	Paris	HL06041035	Lorraine	Lens	Alcoy	Corby	E-values[3]		Other hits	
								F-box	F-box-like		
F-box containing proteins											
lpg0171	lp12_0173	lpp0233	LPV_0254	LPO_0202	lpl0234	–	–	6.5E – 07	2.7E – 05	U3snoRNP10	legU1
lpg1408	lp12_1346	lpp1363	LPV_1525	LPO_1404	lpl1359	lpa_02071	LPC_0824	3.4E – 03	9.3E – 04	Choline_kinase APH	licA
lpg1975-1976[4]	lp12_1915-1916[4]	lpp1959	LPV_2275	LPO_2073	lpl1953	lpa_02889	LPC_1462	–	3.9E – 03*	RCC1	(lpg1976 = legG1)
lpg2144	lp12_2136	lpp2082	LPV_2392	LPO_2207	lpl2072	lpa_03071	LPC_1593	5.4E – 04	1.3E – 06	Ankylin repeats	legAU13 ankB
–	–	lpp2486	–	–	–	–	–	1.0E – 05*	7.8E – 04*		
lpg2525	lp12_2518	–	–	–	–	–	–	1.7E – 02	1.2E – 11		
U-box containing proteins								U-box	zf-RING_LisH		
lpg2455	lp12_2447	lpp2521	LPV_2780	LPO_2646	lpl2374	lpa_03581	LPC_2022	6.8E – 03	–		
lpg2830	lp12_2820	lpp2887	LPV_3185	LPO_3124	–	–	–	5.0E – 26	2.3E – 10	Rtf2 zf-Nse DUF1076	legU2 lubX

[1] PFAM protein family database (Pfam26.0) was scanned by HMMER3 program with all protein sequences extracted from complete genome sequences of various *Legionella pneumophila* strains available on March, 2013 (accessions AE017354, NC_016811, NC_006368, FQ958211, FQ958210, NC_006369, NC_014125, NC_009494). Locus tags of proteins hit with e-values less than inclusion threshold (0.01) and their orthologs are shown. At the time of writing, the entries of the complete genome sequence (accession FR687201) as well as most of protein sequences from *L. pneumophila* strain 130b were removed from database at the submitter's request; therefore proteins from this strain are not included in this table

[2] We were not able to confirm the presence of F-box in lpg2224/ppgA previously described by Angot et al. (2007) in this search

[3] E-values of relevant search with proteins from strain Philadelphia 1 or Paris (denoted by asterisks following e-values) are shown

[4] In these strains, orthologous proteins of lpp1959 are split into two adjacent proteins: one contains F-box, and the other contains RCC1 domain

3.2 Biochemical Analysis of Putative E3 Ligases of L. pneumophila

In order to examine whether the putative *L. pneumophila* F-box-containing proteins possessed conventional activities, researchers examined the ability of these proteins to interact with SCF components. Several of the proposed F-box domain containing proteins, namely LegU1, LegAU13/AnkB, and LicA, were shown to interact with the SCF component Skp1 (Lomma et al. 2010; Ensminger and Isberg 2010; Al-Quadan and Kwaik 2011; Price et al. 2009). In addition, Ensminger and Isberg (2010) found interaction of LegU1 and LegAU13/AnkB with Cullin 1, and demonstrated that the F-box domain was required for pull-down of SCF components, thereby confirming these proteins possess functional F-box domains. These data support formation of a canonical SCF-complex containing these effector proteins inside host cells. In pull-down experiments, Lpg2525 failed to show any association with Skp1 (Ensminger and Isberg 2010). The lack of evidence for F-box mediated binding of Lpg2525 to SCF suggests this protein may not function as canonical F-box protein.

The ultimate test of E3 ligase activity is provided by in vitro E3 ligase assays. In vitro ligase activity has been demonstrated for the effectors LegU1, LegAU13/AnkB, and LubX (Kubori et al. 2008; Ensminger and Isberg 2010; Kubori and Nagai 2011). Many E3 ligases have the ability to autoubiquitinate themselves and this feature may be used for in vitro assays of putative E3 ligases with unknown target proteins. E1, a panel of E2 enzymes and LubX were screened for formation of LubX-dependent polyubiquitin chains. In the case of LubX, the E2 enzymes UbcH5a and UbcH5c were functional. This assay was also utilized to test which of the two U-boxes found in LubX acts as a canonical U-box. U-box 1 was found to be required for formation of polyubiquitin chains supporting a canonical function in binding to E2 ubiquitin conjugation enzymes. E2 enzymes were also screened to assess autoubiquitination of LegU1 and LegAU13/AnkB in in vitro assays performed with ubiquitin, E1 and the F-box effectors (Ensminger and Isberg 2010). These assays were performed slightly differently from the assay used for LubX. In this case, FLAG-tagged F-box proteins and SCF components were expressed in HEK-293T cells. After pull-down of complexes containing F-box proteins, polyubiquitin formation was assessed by incubation with ubiquitin, E1, and E2 enzymes in a suitable buffer. Once again the E2 proteins UbcH5a and UbcH5c were able to form functional complexes with *L. pneumophila* E3 enzymes.

4 Substrates of *L. pneumophila* E3 Ubiquitin Ligases

4.1 A Host Kinase Clk1

Yeast two-hybrid analysis identified the host protein Clk1 (Cdc2-like kinase 1) as an interaction partner for LubX (Kubori et al. 2008) (Fig. 1). This result was further confirmed by cotransfection and coimmunoprecipitation assays. In vitro ubiquitination assays further confirmed that Clk1 could act as a specific substrate for ubiquitination by LubX, however the consequence of the ubiquitination remains unclear. Direct protein–protein binding experiments support a role for U-box 2 in binding Clk1 suggesting this domain is required for substrate binding. The Clk1 kinase, which is involved in the regulation of alternative splicing (Prasad et al. 1999; Hartmann et al. 2001; Schwertz et al. 2006), may play a role in facilitating the optimum conditions for *L. pneumophila* survival as a chemical inhibitor of Clk family kinases reduced *L. pneumophila* replication in mouse macrophages (Kubori et al. 2008).

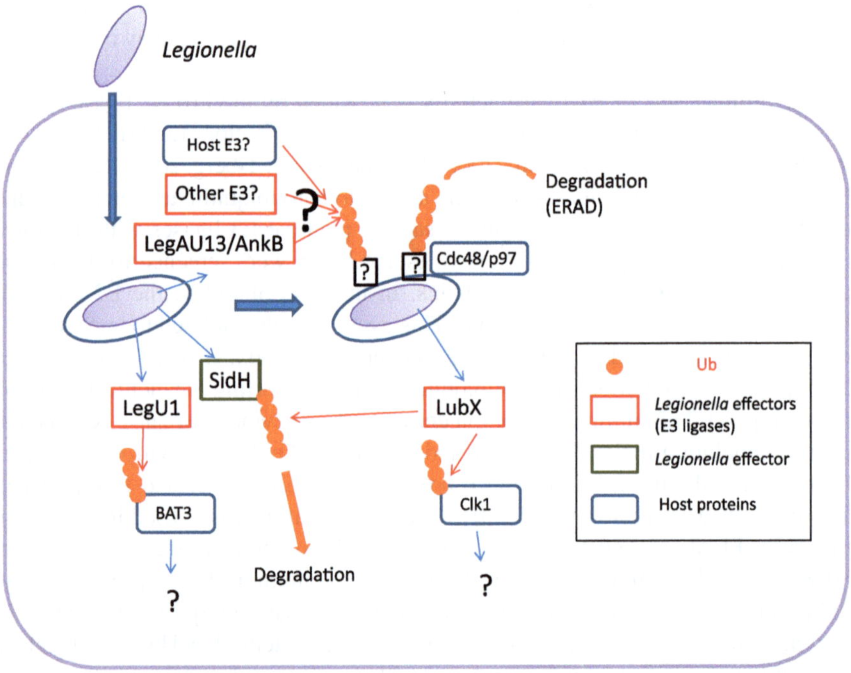

Fig. 1 Manipulation of host ubiquitin system by *L. pneumophila*, *L. pneumophila* translocates at least three (LegU1, LegAU13/AnkB, LubX) E3 ubiquitin ligases into the host cytoplasm. LubX, a U-box-type E3 ligase, polyubiquitinates a host kinase Clk1, however the consequence of this remains unclear. In addition, LubX leads another effector SidH to proteasomal degradation at later stages of infection. Thus LubX functions as a meta-effector, which is an effector protein that targets and regulates another effector protein in host cells. SCF complex containing LegU1 targets and polyubiquitinates host protein BAT3, the consequence of this modification is currently unknown. Several reports suggested that LegAU13/AnkB plays a role in ubiquitin recruitment to LCVs. Ubiquitin conjugates around LCVs are targeted to ERAD in a manner depending on Cdc48/p97 AAA-ATPase

4.2 Another Effector Protein SidH

U-box 2 of LubX is also sufficient for binding of an effector protein SidH (Kubori et al. 2010) (Fig. 1). In addition, modification of SidH with polyubiquitin was observed both in vitro and in vivo. Ubiquitination of SidH depended on the catalytic activity of LubX, as LubX with the amino acids mutation I39A (U-box 1) failed to ubiquitinate SidH. Unlike LubX that is not detected in host cells at early time points post-infection, SidH could be detected as early as 15 min post-infection then declined to undetectable levels by 8 h post-infection. Two lines of evidence support temporal regulation of SidH function by LubX: delayed delivery of LubX translocation correlates with the kinetics of SidH degradation and inhibition of the 26S proteasome with MG132 prevents LubX-dependent degradation of SidH. The term 'metaeffector' was coined to describe LubX. It represents the first member of a new group of effectors that regulate the function of other effectors inside host cells. To date other *L. pneumophila* effectors with such functions have not yet been described, although it has been reported that *Agrobacterium tumefaciens* effector protein VirD5 binds to another effector VirF to regulate its stability in plant cells (Magori and Citovsky 2011).

4.3 Hsp70 Co-chaperone BAT3

A host protein HLA-B-associated transcript 3 (BAT3) was found to specifically bind to LegU1 (Ensminger and Isberg 2010) (Fig. 1). As expected for a substrate protein of SCF complex containing LegU1 (SCF^{LegU1}), the association between LegU1 and BAT3 did not require the F-box domain of LegU1. In vitro ubiquitination assays demonstrated that BAT3 is a specific target of SCF^{LegU1} E3 ligase activity, and the F-box domain of LegU1 was required for ubiquitin modification of BAT3. Another effector protein Lpg2160 was also found to bind LegU1 and BAT3 in mammalian cells, but SCF^{LegU1} does not direct polyubiquitination of Lpg2160. The role of Lpg2160 in regards to SCF^{LegU1} function remains unknown.

The BAT3 protein has been observed to play a role in many host processes, and it is now becoming clear that this is due to its regulatory role in protein quality control systems targeting substrates for ERAD and/or proteasome-mediated degradation (reviewed in (Kawahara et al. 2013)): depletion of BAT3 from cells was found to inhibit ERAD, BAT3 interacts with MG132-induced polyubiquitin substrates, and coimmunoprecipitates with an E3 ligase and cytosolic ER-derived misfolded proteins (Ernst et al. 2011; Wang et al. 2011; Minami et al. 2010). BAT3 has an N-terminal ubiquitin-like domain that is essential for recruiting ubiquitination machinery to BAT3 substrates (Wang et al. 2011; Hessa et al. 2011). Of note BAT3 interacts with puromycin-induced aggresome-like structures known as ALIS (Szeto et al. 2006), and is proposed to control the in vivo formation of these structures (Minami et al. 2010). Dendritic cell aggresome-like induced structures (DALIS) are structures related to ALIS (Canadien et al. 2005). These aggregates of ubiquitinated proteins are thought to provide a pool of antigens for antigen presentation after immune cell migration to sites of presentation (reviewed in (Pierre 2005). DALIS induction in response to TLR (Toll-like receptor) stimulation by LPS released by *L. pneumophila* infection has been observed (Ivanov and Roy 2009). However, wild-type *L. pneumophila* suppresses DALIS formation in infected cells in a process that requires bacterial protein synthesis and the Dot/Icm system, which suggests an effector protein may be involved. Although single mutants deficient for LegAU13/AnkB, Lpg2525, LicA and LubX, as well as a quadruple mutant lacking all of these, suppressed DALIS similar to wild-type, it was not reported whether *legU1* mutant strains suppress DALIS (Ivanov and Roy 2009). Considering the Dot/Icm dependent inhibition of DALIS formation, the interaction between LegU1 and BAT3, and the link between BAT3 and ALIS, further analysis of this relationship is warranted.

4.4 Unknown Substrate of LegAU13/AnkB

Currently, the substrates targeted by LegAU13/AnkB for ubiquitin modification during infection are unknown (Fig. 1). One report found that Lpp2082, the Paris strain ortholog of the Philadelphia-1 protein LegAU13/AnkB, interacts with the host protein Parvin *β*/ParvB, which is normally associated with cell shape and motility (Sepulveda and Wu 2006). However, this protein was endogenously ubiquitin modified in host cells and Lpp2082 did not increase ubiquitination levels (Lomma et al. 2010). Rather overexpression of Lpp2082 in host cells and infection with an *lpp2082*-deficient strain were reported to show increased levels of ubiquitin-modified ParvB compared to control transfections or wild-type infections, respectively. It was postulated that reduced levels of ubiquitinated ParvB may reduce its pro-apoptotic effects. Indeed *lpp2082* deletion strains showed reduced caspase-3 activity and nuclear DNA fragmentation compared to wild-type strains at late stages of infection, and overexpression of Lpp2082 in A549 cells resulted in enhanced caspase-3 activity. The relationship between apoptotic signal induction and Lpp2082 awaits further clarification. In particular it would be of interest to determine whether the F-box domain of Lpp2082/LegAU13/AnkB is required for the increased caspase-3 activity observed upon expression of the effector in host cells.

Dot/Icm-dependent association of ubiquitin with the LCV (Sect. 2) promoted the hypothesis that effectors that possess domains for manipulation of the host ubiquitination machinery may be responsible for this phenotype. Deletion mutants lacking the F-box effector LegAU13/AnkB were reported to have no effect in ubiquitin recruitment to LCV compared to the wild-type *L. pneumophila* Philadelphia-1 (Ensminger and Isberg 2010; Ivanov and Roy 2009). In contrast, two other groups have reported that mutants of LegAU13/AnkB showed defects in ubiquitin recruitment to LCVs containing *L. pneumophila* strains Paris and

AA100/130b. Lomma et al. (2010) reported moderate (~50 %) reduction of ubiquitin-positive LCVs containing deletion mutants lacking *legAU13/ankB (lpp2082* in *L. pneumophila* strain Paris). Price et al. (2010) reported that a kanamycin-cassette insertion mutant of *legAU13/ankB* in *L. pneumophila* strain AA100/130b showed severe defects in ubiquitin accumulation on LCVs. These observations raise the possibility that $SCF^{LegAU13/AnkB}$ plays some roles in formation of ubiquitin-conjugates around LCVs containing these *L. pneumophila* strains. However, it is not yet clear if there is a mechanistic basis for the differing requirements for LegAU13/AnkB in *L. pneumophila* strains Philadelphia-1, Paris, and AA100/130b.

Classical genetic experiments, whereby the contribution of individual genes towards the virulence mechanism of an organism is assessed by gene disruption and growth analysis, have failed to detect significant growth effects for the majority of *L. pneumophila* effectors analyzed. It is commonly believed the large number of *L. pneumophila* effectors and numerous host pathways targeted by the organism builds redundancy into *Legionella's* survival strategy masking the effects of individual mutations. This view was recently given support by a genetic screen called insertional mutagenesis and depletion (iMAD) that combined bacterial mutagenesis and RNA interference to identify effectors that contribute to redundant pathways (O'Connor et al. 2012). In addition, some effectors may function in a host-specific manner. Therefore, it is not surprising that the majority of research groups have not found significant growth defects for *L. pneumophila* strains lacking one or more genes encoding F-box or U-box-domain containing proteins. One notable exception has been reported: the kanamycin-cassette insertion mutant of *legAU13/ankB* in *L. pneumophila* strain AA100/130b showed severe defects in intracellular growth within mammalian macrophages and unicellular protozoa *Acanthamoeba* (Al-Khodor et al. 2008). The observed defects were restored by supplying wild-type gene *in trans*, suggesting the replication defect is due to the mutation of *legAU13/ankB*. Growth defects due to the presence of proteasome inhibitors or the *legAU13/ankB* mutation are restored by supplementation of amino acids including cysteine to the culture medium, suggesting a role for LegAU13/AnkB and the host proteasome in supplying nutrients for robust intracellular growth (Price et al. 2011). In contrast, deletion mutants of *legAU13/ankB* of *L. pneumophila* strains Philadelphia-1 and Paris are able to replicate within amoeba to a similar extent as isogenic wild-type strains. The *legAU13/ankB* deletion mutant in strain Paris showed slight defects in survival in the human monocytic cell line THP-1 and the lung epithelial cell line A549.

The drastic disagreements mentioned above regarding the behavior of *legAU13/ankB* mutants in various *L. pneumophila* strains may be accounted for by the difference in genetic background of strains used and/or the nature of mutant construction employed. Future studies toward understanding the molecular basis of $SCF^{LegAU13/AnkB}$ function will shed light on the causes of this controversial situation.

5 The Role of Ubiquitin in Selective Autophagy and *L. pneumophila*

Autophagy is an important mechanism to restrict bacterial invaders (Jiang and Chen 2012; Fujita and Yoshimori 2011; Ligeon et al. 2011; Shahnazari and Brumell 2011). The process of autophagy is intimately linked to any discussion of ubiquitin-modification in host cells because ubiquitin has emerged as a factor for selective autophagy of bacteria, termed xenophagy (Johansen and Lamark 2011; Kirkin et al. 2009a). Although the role of autophagy in *Legionella's* intracellular survival is of general interest, we will limit our discussion to ubiquitin-related aspects of autophagy during *L. pneumophila* infection. Recently a specialized review on the role autophagy in *L. pneumophila* infection was published (Joshi and Swanson 2011). During autophagy, a double membrane structure called the autophagosome is formed and this compartment engulfs particles that are later degraded by fusion of autophagosomes with lysosomes. In yeast there are 15 core autophagy related (ATG) genes, and the ubiquitin-like modifier Atg8 is commonly used as an autophagy marker; specifically the LC3 protein, which is one of seven Atg8 proteins in mammals. Ubiquitin-mediated identification of autophagy targets is mediated by adaptor proteins that recognize both ubiquitin modified cargo and autophagosomes. The autophagic adaptors p62/SQSTM1 (sequestosome 1) and NBR1 (neighbor of Brca1 gene) are two such adaptors (Kirkin et al. 2009b; Pankiv et al. 2007; Ichimura et al. 2008), and p62/SQSTM1 is involved in targeting invading bacteria to the autophagy pathway (Zheng et al. 2009). The bacterial pathogen *Listeria monocytogenes* avoids autophagosomal destruction using a host protein disguise that prevents association of LC3 and the adaptor p62/SQSTM1 (Yoshikawa et al. 2009). It is of interest to explore the question of how *L. pneumophila*, which resides in an ubiquitin-modified compartment, evades degradation by autophagy. Autophagic components are not required for *L. pneumophila* replication in the natural host *Dictyostelium discoideum* (Otto et al. 2004), and autophagy likely functions to restrict *L. pneumophila* replication (Khweek et al. 2013; Matsuda et al. 2009). A protective role for autophagy in *L. pneumophila* infections is further supported by the finding that *L. pneumophila* replication was enhanced in *Dictyostelium discoideum* Atg9 knockouts (Tung et al. 2010). The transcriptional response to *L. pneumophila* infection has also been linked to increased transcription of a number of autophagy proteins (Farbrother et al. 2006). Despite potential targeting by autophagy during infection, the majority of *L. pneumophila* survive and replicate within human macrophage-like cell lines and amoeba. In light of this, it is not surprising that recently *L. pneumophila* was found to possess at least one effector RavZ that has been shown to actively interfere with autophagy (Choy et al. 2012). This is the first bacterial effector protein reported to directly interfere with LC3/Atg8. RavZ acts as a deconjugating enzyme to uncouple LC3/Atg8 from phosphatidylethanolamine (PE) on early autophagosomes like the host deconjugating enzyme Atg4. Importantly, RavZ cleaves a peptide bond that is distinct from that is cleaved by Atg4, which results in conversion of LC3 into the form not able to be reconjugated to PE by host autophagic enzymes. Because the Δ*ravZ* strain retains the ability to prevent LC3 recruitment to LCVs, it is likely that multiple effectors disrupt recognition by the autophagy pathway. Indeed, a key question becomes whether *L. pneumophila* possesses an effector that targets an adaptor? Whether p62/SQSTM1 associates with LCVs in amoeba or human cell lines has not been reported, but in macrophages, obtained from non-permissive C57BL/6 mice, p62/SQSTM1 was reported to associate with LCVs (Khweek et al. 2013). *Salmonella* possesses a DUB, SseL that has the ability to deubiquitinate p62/SQSTM1-bound proteins and reduces the recruitment of LC3 to ubiquitinated structures (Mesquita et al. 2012). Notably, a feature shared between *Legionella* and *Salmonella* is inhibition of 'ALIS' formation (Sect. 4.3) (Thomas et al. 2012). Due to the action of SseL and the shared inhibition of 'ALIS' formation, Thomas et al. (2012) proposed that a *L. pneumophila* DUB might exist with a similar function.

6 Concluding Remarks

Much is left undiscovered regarding the mechanisms by which *L. pneumophila* manipulates the host ubiquitination machinery. From the wide array of host proteins involved in the regulated modification of proteins with ubiquitin, it is surprising that so far the only *L. pneumophila* proteins implicated are those possessing domains with similarity to E3 ligases. Additional effectors may act as mimics of the ubiquitination machinery, function as DUBs, or modify the activities of host ubiquitination-components through post-translational modifications. Due to the limited amount of effector proteins delivered into host cells, it is thought the majority of effectors possess enzymatic functions. Supporting this notion, modulation of the host GTPase Rab1 during infection involves by the post-translational addition of phosphatidylcholine and adenosine monophosphosphate by effector proteins (Goody et al. 2012; Muller et al. 2010; Mukherjee et al. 2011; Neunuebel et al. 2011; Tan et al. 2011; Tan and Luo 2011). Due to the multiple post-translational mechanisms that regulate host-ubiquitination it seems likely that effectors targeting the host ubiquitination machinery will be identified. Perhaps *L. pneumophila* possesses the ability to modify proteins with SUMO or NEDD8. It is as yet unclear which host E2 enzymes are used by *L. pneumophila* during infection, and due to the role of E2s in directing the types of ubiquitin chains formed it will be of interest to examine this. Most important will be the discovery of additional targets that are directed for ubiquitination by *L. pneumophila*, the types of modifications, and the consequence of these modifications for the protein. Because the association of polyubiquitin with LCVs is associated with successful vacuolar remodeling, protein turnover or modulation of protein activities at this membrane is expected to be important for this process. It will be of interest to determine the identities of effectors, arguably required in addition to AnkB, responsible for vacuolar ubiquitin. Also, it seems probable given the requirement to modulate the localization and function of so many effectors, that effectors other than LubX act to temporally or spatially regulate effectors by ubiquitin-modifications. Ultimately, the goal will be to discover the pathways co-opted by *L. pneumophila's* ubiquitin-modification network. Given the importance of avoiding host innate immune responses and the growing body of evidence for ubiquitin-modifications in modulating these responses, we anticipate that *L. pneumophila* may possess effectors directed against these ubiquitin-modulated responses. Bacterial strategies to evade host autophagic clearance are becoming one of the major topics in the field of bacterial pathogenesis. It remains an interesting question whether host adaptor proteins that recognize ubiquitin-conjugated substrates for selective autophagy are manipulated by *L. pneumophila*. Identification of the ubiquitin-modified proteins at the LCV membrane may help to unravel the issue of how the ubiquitin-modified LCV avoids destruction by xenophagy.

Acknowledgments We thank Drs Xuan Bui Thanh and Masafumi Koike for critical reading of the manuscript. Research in the Nagai laboratory was supported by Grants-in-Aid for Scientific Research (23117002, 23390105, 24659198) from Ministry of Education, Culture, Sports, Science and Technology, Japan. Andree Hubber is supported by a postdoctoral fellowship for foreign researchers awarded by the Japanese Society for the Promotion of Science (JSPS).

References

Al-Khodor S, Price CT, Habyarimana F, Kalia A, Abu Kwaik Y (2008) A Dot/Icm-translocated ankyrin protein of *Legionella pneumophila* is required for intracellular proliferation within human macrophages and protozoa. Mol Microbiol 70:908–923

Al-Quadan T, Kwaik YA (2011) Molecular characterization of exploitation of the polyubiquitination and farnesylation machineries of *Dictyostelium discoideum* by the AnkB F-box effector of *Legionella pneumophila*. Front Microbiol 2:23

Angot A, Vergunst A, Genin S, Peeters N (2007) Exploitation of eukaryotic ubiquitin signaling pathways by effectors translocated by bacterial type III and type IV secretion systems. PLoS Pathog 3:e3

Ardley HC, Robinson PA (2005) E3 ubiquitin ligases. Essays Biochem 41:15–30

Berger KH, Isberg RR (1993) Two distinct defects in intracellular growth complemented by a single genetic locus in *Legionella pneumophila*. Mol Microbiol 7:7–19

Bruggemann H, Hagman A, Jules M, Sismeiro O, Dillies MA et al (2006) Virulence strategies for infecting phagocytes deduced from the in vivo transcriptional program of *Legionella pneumophila*. Cell Microbiol 8:1228–1240

Calvo-Garrido J, Carilla-Latorre S, Kubohara Y, Santos-Rodrigo N, Mesquita A et al (2010) Autophagy in *Dictyostelium*: genes and pathways, cell death and infection. Autophagy 6:686–701

Canadien V, Tan T, Zilber R, Szeto J, Perrin AJ et al (2005) Cutting edge: microbial products elicit formation of dendritic cell aggresome-like induced structures in macrophages. J Immunol 174:2471–2475

Cazalet C, Gomez-Valero L, Rusniok C, Lomma M, Dervins-Ravault D et al (2010) Analysis of the *Legionella longbeachae* genome and transcriptome uncovers unique strategies to cause Legionnaires' disease. PLoS Genet 6:e1000851

Choy A, Dancourt J, Mugo B, O'Connor TJ, Isberg RR et al (2012) The *Legionella* effector RavZ inhibits host autophagy through irreversible Atg8 deconjugation. Science 338:1072–1076

Clague MJ, Urbe S (2010) Ubiquitin: same molecule, different degradation pathways. Cell 143:682–685

Clague MJ, Coulson JM, Urbe S (2012) Cellular functions of the DUBs. J Cell Sci 125:277–286

Coombs N, Sompallae R, Olbermann P, Gastaldello S, Goppel D et al (2011) Helicobacter pylori affects the cellular deubiquitinase USP7 and ubiquitin-regulated components TRAF6 and the tumour suppressor p53. Int J Med Microbiol 301:213–224

David Y, Ziv T, Admon A, Navon A (2010) The E2 ubiquitin-conjugating enzymes direct polyubiquitination to preferred lysines. J Biol Chem 285:8595–8604

de Bie P, Ciechanover A (2011) Ubiquitination of E3 ligases: self-regulation of the ubiquitin system via proteolytic and non-proteolytic mechanisms. Cell Death Differ 18:1393–1402

de Felipe KS, Pampou S, Jovanovic OS, Pericone CD, Ye SF et al (2005) Evidence for acquisition of *Legionella* type IV secretion substrates via interdomain horizontal gene transfer. J Bacteriol 187:7716–7726

de Felipe KS, Glover RT, Charpentier X, Anderson OR, Reyes M et al (2008) *Legionella* eukaryotic-like type IV substrates interfere with organelle trafficking. PLoS Pathog 4:e1000117

Diao J, Zhang Y, Huibregtse JM, Zhou D, Chen J (2008) Crystal structure of SopA, a *Salmonella* effector protein mimicking a eukaryotic ubiquitin ligase. Nat Struct Mol Biol 15:65–70

Dorer MS, Kirton D, Bader JS, Isberg RR (2006) RNA interference analysis of *Legionella* in *Drosophila* cells: exploitation of early secretory apparatus dynamics. PLoS Pathog 2:315–327

Ensminger AW, Isberg RR (2010) E3 ubiquitin ligase activity and targeting of BAT3 by multiple *Legionella pneumophila* translocated substrates. Infect Immun 78:3905–3919

Ernst R, Claessen JH, Mueller B, Sanyal S, Spooner E et al (2011) Enzymatic blockade of the ubiquitin-proteasome pathway. PLoS Biol 8:e1000605

Farbrother P, Wagner C, Na J, Tunggal B, Morio T et al (2006) *Dictyostelium* transcriptional host cell response upon infection with *Legionella*. Cell Microbiol 8:438–456

Fujita N, Yoshimori T (2011) Ubiquitination-mediated autophagy against invading bacteria. Curr Opin Cell Biol 23:492–497

Goody PR, Heller K, Oesterlin LK, Muller MP, Itzen A et al (2012) Reversible phosphocholination of Rab proteins by *Legionella pneumophila* effector proteins. EMBO J 31:1774–1784

Habyarimana F, Al-Khodor S, Kalia A, Graham JE, Price CT et al (2008) Role for the Ankyrin eukaryotic-like genes of *Legionella pneumophila* in parasitism of protozoan hosts and human macrophages. Environ Microbiol 10:1460–1474

Haglund K, Dikic I (2005) Ubiquitylation and cell signaling. EMBO J 24:3353–3359

Haglund K, Dikic I (2012) The role of ubiquitylation in receptor endocytosis and endosomal sorting. J Cell Sci 125:265–275

Harhaj EW, Dixit VM (2012) Regulation of NF-kappaB by deubiquitinases. Immunol Rev 246:107–124

Hartmann AM, Rujescu D, Giannakouros T, Nikolakaki E, Goedert M et al (2001) Regulation of alternative splicing of human tau exon 10 by phosphorylation of splicing factors. Mol Cell Neurosci 18:80–90

Heidtman M, Chen EJ, Moy MY, Isberg RR (2009) Large-scale identification of *Legionella pneumophila* Dot/Icm substrates that modulate host cell vesicle trafficking pathways. Cell Microbiol 11:230–248

Hershko A, Ciechanover A (1998) The ubiquitin system. Annu Rev Biochem 67:425–479

Hershko A, Heller H, Elias S, Ciechanover A (1983) Components of ubiquitin-protein ligase system. Resolution, affinity purification, and role in protein breakdown. J Biol Chem 258:8206–8214

Hessa T, Sharma A, Mariappan M, Eshleman HD, Gutierrez E et al (2011) Protein targeting and degradation are coupled for elimination of mislocalized proteins. Nature 475:394–397

Hicke L, Dunn R (2003) Regulation of membrane protein transport by ubiquitin and ubiquitin-binding proteins. Annu Rev Cell Dev Biol 19:141–172

Hicks SW, Galan JE (2010) Hijacking the host ubiquitin pathway: structural strategies of bacterial E3 ubiquitin ligases. Curr Opin Microbiol 13:41–46

Horwitz MA (1987) Characterization of avirulent mutant *Legionella pneumophila* that survive but do not multiply within human monocytes. J Exp Med 166:1310–1328

Huang L, Boyd D, Amyot WM, Hempstead AD, Luo ZQ et al (2011) The E Block motif is associated with *Legionella pneumophila* translocated substrates. Cell Microbiol 13:227–245

Hubber A, Roy CR (2010) Modulation of host cell function by *Legionella pneumophila* type IV effectors. Annu Rev Cell Dev Biol 26:261–283

Hutchins AP, Liu S, Diez D, Miranda-Saavedra D (2013) The repertoires of ubiquitinating and deubiquitinating enzymes in eukaryotic genomes. Mol Biol Evol 30:1172–1187

Ichimura Y, Kumanomidou T, Sou YS, Mizushima T, Ezaki J et al (2008) Structural basis for sorting mechanism of p62 in selective autophagy. J Biol Chem 283:22847–22857

Ikeda F, Crosetto N, Dikic I (2010) What determines the specificity and outcomes of ubiquitin signaling? Cell 143:677–681

Ivanov SS, Roy CR (2009) Modulation of ubiquitin dynamics and suppression of DALIS formation by the *Legionella pneumophila* Dot/Icm system. Cell Microbiol 11:261–278

Jarosch E, Taxis C, Volkwein C, Bordallo J, Finley D et al (2002) Protein dislocation from the ER requires polyubiquitination and the AAA-ATPase Cdc48. Nat Cell Biol 4:134–139

Jiang X, Chen ZJ (2012) The role of ubiquitylation in immune defence and pathogen evasion. Nat Rev Immunol 12:35–48

Johansen T, Lamark T (2011) Selective autophagy mediated by autophagic adapter proteins. Autophagy 7:279–296

Joshi AD, Swanson MS (2011) Secrets of a successful pathogen: legionella resistance to progression along the autophagic pathway. Front Microbiol 2:138

Kawahara H, Minami R, Yokota N (2013) BAG6/BAT3: emerging roles in quality control for nascent polypeptides. J Biochem 153:147–160

Khweek AA, Caution K, Akhter A, Abdulrahman BA, Tazi M, et al (2013) A bacterial protein promotes the recognition of the *Legionella pneumophila* vacuole by autophagy. Eur J Immunol 43:1333–1344

Kipreos ET, Pagano M (2000) The F-box protein family. Genome Biol 1:REVIEWS3002

Kirkin V, McEwan DG, Novak I, Dikic I (2009a) A role for ubiquitin in selective autophagy. Mol Cell 34:259–269

Kirkin V, Lamark T, Sou YS, Bjorkoy G, Nunn JL et al (2009b) A role for NBR1 in autophagosomal degradation of ubiquitinated substrates. Mol Cell 33:505–516

Koepp DM, Harper JW, Elledge SJ (1999) How the cyclin became a cyclin: regulated proteolysis in the cell cycle. Cell 97:431–434

Kozak NA, Buss M, Lucas CE, Frace M, Govil D et al (2010) Virulence factors encoded by *Legionella longbeachae* identified on the basis of the genome sequence analysis of clinical isolate D-4968. J Bacteriol 192:1030–1044

Kubori T, Nagai H (2011) Bacterial effector-involved temporal and spatial regulation by hijack of the host ubiquitin pathway. Front Microbiol 2:145

Kubori T, Hyakutake A, Nagai H (2008) *Legionella* translocates an E3 ubiquitin ligase that has multiple U-boxes with distinct functions. Mol Microbiol 67:1307–1319

Kubori T, Shinzawa N, Kanuka H, Nagai H (2010) *Legionella* metaeffector exploits host proteasome to temporally regulate cognate effector. PLoS Pathog 6:e1001216

Li W, Bengtson MH, Ulbrich A, Matsuda A, Reddy VA et al (2008) Genome-wide and functional annotation of human E3 ubiquitin ligases identifies MULAN, a mitochondrial E3 that regulates the organelle's dynamics and signaling. PLoS ONE 3:e1487

Ligeon LA, Temime-Smaali N, Lafont F (2011) Ubiquitylation and autophagy in the control of bacterial infections and related inflammatory responses. Cell Microbiol 13:1303–1311

Lomma M, Dervins-Ravault D, Rolando M, Nora T, Newton HJ et al (2010) The *Legionella pneumophila* F-box protein Lpp 2082 (AnkB) modulates ubiquitination of the host protein parvin B and promotes intracellular replication. Cell Microbiol 12:1272–1291

Magori S, Citovsky V (2011a) Hijacking of the host SCF ubiquitin ligase machinery by plant pathogens. Front Plant Sci 2:87

Magori S, Citovsky V (2011b) Agrobacterium counteracts host-induced degradation of its effector F-box protein. Sci Signal 4:ra69

Marra A, Blander SJ, Horwitz MA, Shuman HA (1992) Identification of a *Legionella pneumophila* locus required for intracellular multiplication in human macrophages. Proc Natl Acad Sci U S A 89:9607–9611

Matsuda F, Fujii J, Yoshida S (2009) Autophagy induced by 2-deoxy-D-glucose suppresses intracellular multiplication of *Legionella pneumophila* in A/J mouse macrophages. Autophagy 5:484–493

McDade JE, Shepard CC, Fraser DW, Tsai TR, Redus MA et al (1977) Legionnaires' disease: isolation of a bacterium and demonstration of its role in other respiratory disease. N Engl J Med 297:1197–1203

Mesquita FS, Thomas M, Sachse M, Santos AJ, Figueira R et al (2012) The *Salmonella* deubiquitinase SseL inhibits selective autophagy of cytosolic aggregates. PLoS Pathog 8:e1002743

Meusser B, Hirsch C, Jarosch E, Sommer T (2005) ERAD: the long road to destruction. Nat Cell Biol 7:766–772

Minami R, Hayakawa A, Kagawa H, Yanagi Y, Yokosawa H et al (2010) BAG-6 is essential for selective elimination of defective proteasomal substrates. J Cell Biol 190:637–650

Misaghi S, Balsara ZR, Catic A, Spooner E, Ploegh HL et al (2006) Chlamydia trachomatis-derived deubiquitinating enzymes in mammalian cells during infection. Mol Microbiol 61:142–150

Mocciaro A, Rape M (2012) Emerging regulatory mechanisms in ubiquitin-dependent cell cycle control. J Cell Sci 125:255–263

Mukherjee S, Keitany G, Li Y, Wang Y, Ball HL et al (2006) Yersinia YopJ acetylates and inhibits kinase activation by blocking phosphorylation. Science 312:1211–1214

Mukherjee S, Liu X, Arasaki K, McDonough J, Galan JE et al (2011) Modulation of Rab GTPase function by a protein phosphocholine transferase. Nature 477:103–106

Muller MP, Peters H, Blumer J, Blankenfeldt W, Goody RS et al (2010) The *Legionella* effector protein DrrA AMPylates the membrane traffic regulator Rab1b. Science 329:946–949

Nagai H, Kagan JC, Zhu X, Kahn RA, Roy CR (2002) A bacterial guanine nucleotide exchange factor activates ARF on *Legionella* phagosomes. Science 295:679–682

Nathan JA, Tae Kim H, Ting L, Gygi SP, Goldberg AL (2013) Why do cellular proteins linked to K63-polyubiquitin chains not associate with proteasomes? EMBO J 32:552–565

Neunuebel MR, Chen Y, Gaspar AH, Backlund PS Jr, Yergey A et al (2011) De-AMPylation of the small GTPase Rab1 by the pathogen *Legionella pneumophila*. Science 333:453–456

O'Connor TJ, Boyd D, Dorer MS, Isberg RR (2012) Aggravating genetic interactions allow a solution to redundancy in a bacterial pathogen. Science 338:1440–1444

Otto GP, Wu MY, Clarke M, Lu H, Anderson OR et al (2004) Macroautophagy is dispensable for intracellular replication of *Legionella pneumophila* in *Dictyostelium discoideum*. Mol Microbiol 51:63–72

Pankiv S, Clausen TH, Lamark T, Brech A, Bruun JA et al (2007) p62/SQSTM1 binds directly to Atg8/LC3 to facilitate degradation of ubiquitinated protein aggregates by autophagy. J Biol Chem 282:24131–24145

Patel JC, Hueffer K, Lam TT, Galan JE (2009) Diversification of a *Salmonella* virulence protein function by ubiquitin-dependent differential localization. Cell 137:283–294

Pickart CM, Fushman D (2004) Polyubiquitin chains: polymeric protein signals. Curr Opin Chem Biol 8:610–616

Pierre P (2005) Dendritic cells, DRiPs, and DALIS in the control of antigen processing. Immunol Rev 207:184–190

Prasad J, Colwill K, Pawson T, Manley JL (1999) The protein kinase Clk/Sty directly modulates SR protein activity: both hyper- and hypophosphorylation inhibit splicing. Mol Cell Biol 19:6991–7000

Price CT, Kwaik YA (2010) Exploitation of host polyubiquitination machinery through molecular mimicry by eukaryotic-like bacterial F-box effectors. Front Microbiol 1:122

Price CT, Al-Khodor S, Al-Quadan T, Santic M, Habyarimana F et al (2009) Molecular mimicry by an F-box effector of *Legionella pneumophila* hijacks a conserved polyubiquitination machinery within macrophages and protozoa. PLoS Pathog 5:e1000704

Price CT, Al-Khodor S, Al-Quadan T, Abu Kwaik Y (2010) Indispensable role for the eukaryotic-like ankyrin domains of the ankyrin B effector of *Legionella pneumophila* within macrophages and amoebae. Infect Immun 78:2079–2088

Price CT, Al-Quadan T, Santic M, Rosenshine I, Abu Kwaik Y (2011) Host proteasomal degradation generates amino acids essential for intracellular bacterial growth. Science 334:1553–1557

Quezada CM, Hicks SW, Galan JE, Stebbins CE (2009) A family of *Salmonella* virulence factors functions as a distinct class of autoregulated E3 ubiquitin ligases. Proc Natl Acad Sci U S A 106:4864–4869

Reyes-Turcu FE, Ventii KH, Wilkinson KD (2009) Regulation and cellular roles of ubiquitin-specific deubiquitinating enzymes. Annu Rev Biochem 78:363–397

Rytkonen A, Poh J, Garmendia J, Boyle C, Thompson A et al (2007) SseL, a *Salmonella* deubiquitinase required for macrophage killing and virulence. Proc Natl Acad Sci U S A 104:3502–3507

Schmitz-Esser S, Tischler P, Arnold R, Montanaro J, Wagner M et al (2010) The genome of the amoeba symbiont "*Candidatus Amoebophilus asiaticus*" reveals common mechanisms for host cell interaction among amoeba-associated bacteria. J Bacteriol 192:1045–1057

Schwertz H, Tolley ND, Foulks JM, Denis MM, Risenmay BW et al (2006) Signal-dependent splicing of tissue factor pre-mRNA modulates the thrombogenicity of human platelets. J Exp Med 203:2433–2440

Segal G, Shuman HA (1999) Possible origin of the *Legionella pneumophila* virulence genes and their relation to *Coxiella burnetii*. Mol Microbiol 33:669–670

Segal G, Purcell M, Shuman HA (1998) Host cell killing and bacterial conjugation require overlapping sets of genes within a 22-kb region of the *Legionella pneumophila* genome. Proc Natl Acad Sci U S A 95:1669–1674

Sepulveda JL, Wu C (2006) The parvins. Cell Mol Life Sci 63:25–35

Shahnazari S, Brumell JH (2011) Mechanisms and consequences of bacterial targeting by the autophagy pathway. Curr Opin Microbiol 14:68–75

Shanks J, Burtnick MN, Brett PJ, Waag DM, Spurgers KB et al (2009) Burkholderia mallei tssM encodes a putative deubiquitinase that is secreted and expressed inside infected RAW 264.7 murine macrophages. Infect Immun 77:1636–1648

Singer AU, Rohde JR, Lam R, Skarina T, Kagan O et al (2008) Structure of the Shigella T3SS effector IpaH defines a new class of E3 ubiquitin ligases. Nat Struct Mol Biol 15:1293–1301

Szeto J, Kaniuk NA, Canadien V, Nisman R, Mizushima N et al (2006) ALIS are stress-induced protein storage compartments for substrates of the proteasome and autophagy. Autophagy 2:189–199

Tan Y, Luo ZQ (2011) *Legionella pneumophila* SidD is a deAMPylase that modifies Rab1. Nature 475:506–509

Tan Y, Arnold RJ, Luo ZQ (2011) *Legionella pneumophila* regulates the small GTPase Rab1 activity by reversible phosphorylcholination. Proc Natl Acad Sci U S A 108:21212–21217

Thomas M, Mesquita FS, Holden DW (2012) The DUB-ious lack of ALIS in *Salmonella* infection: a *Salmonella* deubiquitinase regulates the autophagy of protein aggregates. Autophagy 8:1824–1826

Tung SM, Unal C, Ley A, Pena C, Tunggal B et al (2010) Loss of *Dictyostelium* ATG9 results in a pleiotropic phenotype affecting growth, development, phagocytosis and clearance and replication of *Legionella pneumophila*. Cell Microbiol 12:765–780

Ulrich HD, Walden H (2010) Ubiquitin signalling in DNA replication and repair. Nat Rev Mol Cell Biol 11:479–489 van Wijk SJ, Timmers HT (2010) The family of ubiquitin-conjugating enzymes (E2s): deciding between life and death of proteins. FASEB J 24:981–993

Vandenabeele P, Bertrand MJ (2012) The role of the IAP E3 ubiquitin ligases in regulating pattern-recognition receptor signalling. Nat Rev Immunol 12:833–844

Vogel JP, Andrews HL, Wong SK, Isberg RR (1998) Conjugative transfer by the virulence system of *Legionella pneumophila*. Science 279:873–876

Wang Q, Liu Y, Soetandyo N, Baek K, Hegde R et al (2011) A ubiquitin ligase-associated chaperone holdase maintains polypeptides in soluble states for proteasome degradation. Mol Cell 42:758–770

Windheim M, Peggie M, Cohen P (2008) Two different classes of E2 ubiquitin-conjugating enzymes are required for the mono-ubiquitination of proteins and elongation by polyubiquitin chains with a specific topology. Biochem J 409:723–729

Xin DW, Liao S, Xie ZP, Hann DR, Steinle L et al (2012) Functional analysis of NopM, a novel E3 ubiquitin ligase (NEL) domain effector of *Rhizobium* sp. strain NGR234. PLoS Pathog 8:e1002707

Xu P, Duong DM, Seyfried NT, Cheng D, Xie Y et al (2009) Quantitative proteomics reveals the function of unconventional ubiquitin chains in proteasomal degradation. Cell 137:133–145

Ye Y, Rape M (2009) Building ubiquitin chains: E2 enzymes at work. Nat Rev Mol Cell Biol 10:755–764

Ye Y, Meyer HH, Rapoport TA (2001) The AAA ATPase Cdc48/p97 and its partners transport proteins from the ER into the cytosol. Nature 414:652–656

Ye Z, Petrof EO, Boone D, Claud EC, Sun J (2007) *Salmonella* effector AvrA regulation of colonic epithelial cell inflammation by deubiquitination. Am J Pathol 171:882–892

Yoshikawa Y, Ogawa M, Hain T, Yoshida M, Fukumatsu M et al (2009) Listeria monocytogenes ActA-mediated escape from autophagic recognition. Nat Cell Biol 11:1233–1240

Zeng LR, Park CH, Venu RC, Gough J, Wang GL (2008) Classification, expression pattern, and E3 ligase activity assay of rice U-box-containing proteins. Mol Plant 1:800–815

Zheng YT, Shahnazari S, Brech A, Lamark T, Johansen T et al (2009) The adaptor protein p62/SQSTM1 targets invading bacteria to the autophagy pathway. J Immunol 183:5909–5916

Zhou H, Monack DM, Kayagaki N, Wertz I, Yin J et al (2005) Yersinia virulence factor YopJ acts as a deubiquitinase to inhibit NF-kappa B activation. J Exp Med 202:1327–1332

Zhu Y, Li H, Hu L, Wang J, Zhou Y et al (2008) Structure of a Shigella effector reveals a new class of ubiquitin ligases. Nat Struct Mol Biol 15:1302–1308

Zhu W, Banga S, Tan Y, Zheng C, Stephenson R et al (2011) Comprehensive identification of protein substrates of the Dot/Icm type IV transporter of *Legionella pneumophila*. PLoS ONE 6:e17638

Host Signal Transduction and Protein Kinases Implicated in *Legionella* Infection

Andrew D. Hempstead and Ralph R. Isberg

Abstract Modulation of the phosphorylation status of proteins by both kinases and phosphatases plays an important role in cellular signal transduction. Challenge of host cells by *Legionella pneumophila* manipulates the phosphorylation state of multiple host factors. These changes play roles in bacterial uptake, vacuole modification, cellular survival, and the immune response. In addition to modification by host cell kinases in response to the bacterium, *L. pneumophila* translocates bacterial kinases into the host cell that may contribute to further signaling modifications. Proper regulation of host cell signaling by *L. pneumophila* is necessary for its ability to replicate intracellulary, while avoiding host defenses.

Abbreviations

BMDMs	Bone marrow-derived macrophages
CAPE	Caffeic acid phenethyl ester
CR3	Complement receptor 3
DAG	Diacylglycerol
Dusp	Dual-specificity protein phosphatase
hBD-2	Human β-defensin-2
IDTS	Icm/Dot translocated substrates
IκB	Inhibitor of κB

R. R. Isberg (✉)
Howard Hughes Medical Institute, Tufts University School of Medicine,
Boston, MA 02111, USA
e-mail: ralph.isberg@tufts.edu

A. D. Hempstead · R. R. Isberg
Department of Molecular Biology and Microbiology,
Tufts University School of Medicine, Boston, MA 02111, USA

A. D. Hempstead
Graduate Program in Molecular Microbiology,
Sackler School of Graduate Biomedical Science,
Tufts University School of Medicine, Boston, MA 02111, USA

Current Topics in Microbiology and Immunology (2013) 376: 249–269
DOI: 10.1007/82_2013_342

Published Online: 6 August 2013

Icm/Dot	Intracellular multiplication/defect in organelle trafficking
IKK	IκB kinase
LCV	*Legionella*-containing vacuole
LF	Lethal factor
MAPKK	MAPK kinase
MAPKKK	MAPK kinase kinase
MAPK	Mitogen-activated protein kinase
NF-κB	Nuclear factor kappa-light-chain-enhancer of activated B cells
NLR	Nucleotide-binding oligomerization domain like receptor
PAMP	Pathogen-associated molecular pattern
PRR	Pattern recognition receptor
PI3K	Phosphoinositide 3-kinase
PGE_2	Prostaglandin E_2
PKB/Akt	Protein kinase B
PKC	Protein kinase C
TLR	Toll-like receptor
T4SS	Type IV secretion system
TPK	Tyrosine protein kinase

Contents

1 Introduction 251
2 Modulation of Protein Phosphorylation During Binding and Uptake by Host Cells 252
2.1 Protein Phosphorylation During Uptake by Mammalian Cells 253
2.2 Modulation of Amoebal Protein Tyrosine Phosphorylation 254
3 MAPK Signaling Pathways 255
3.1 Pathogen-Associated Molecular Pattern and Icm/Dot Dependent Induction of MAPK Activation 256
3.2 Effects of MAPK Activation During Infection 258
3.3 Amoebal MAPK Activation 258
3.4 Role of Dusps in the MAPK Response 259
4 The Role of Protein Kinase C in the Response to *L. pneumophila* 259
4.1 Involvement of PKCs During Uptake 260
4.2 Innate Immune Signaling Mediated by PKC 260
5 The NF-κB Response to *L. pneumophila* Challenge 261
5.1 The Biphasic Activation of NF-κB 261
5.2 Translocated Substrates Implicated in NF-κB Activity 263
5.3 NF-κB and Host Cell Survival 263
6 *L. pneumophila* T4SS Translocated Kinases 264
7 Conclusions 265
References 265

1 Introduction

Cellular signaling pathways allow cells to sense changing environmental conditions by responding to both extracellular and intracellular stimuli. One mechanism of signal transduction involves the reversible phosphorylation of protein substrates. Phosphorylation and dephosphorylation reactions are performed by cellular kinases and phosphatases, respectively. Signaling through phosphorylation allows for rapid cellular responses to varying stimuli at distinct cellular localizations, allowing for signal specificity.

Signaling through phosphorylation is an evolutionarily conserved mechanism found in all domains of life (Manning et al. 2002). In eukaryotes, the common sites of phosphorylation by kinases on target proteins are either serine or threonine residues, or less commonly, tyrosines (Blom et al. 1999). The kinases that target these residues can either be specific serine/threonine or tyrosine kinases, or may be more promiscuous dual-specificity kinases which can target multiple residues (Ubersax and Ferrell 2007).

Signal transduction through phosphorylation plays an important role in many cellular processes. Responses such as innate immune signaling, cell cycle control, metabolism, cytoskeletal modification, response to cellular stress, and recognition of extracellular ligands are all controlled by phosphorylation-mediated signaling cascades (Manning et al. 2002). Substrate phosphorylation induces changes important for signaling and function including modulation of enzymatic activity, substrate stability, and interactions with other factors (Zhang et al. 2002). Defects in kinase-mediated signaling are implicated in multiple disease states including cancer, diabetes, severe combined immunodeficiency, and rheumatoid arthritis (Cohen 2001).

Because of the importance of kinase-mediated signaling pathways in cellular physiology and the immune response, pathogens have developed mechanisms to subvert these pathways for their own benefit. One of the earliest identified examples of this was a *Yersinia* effector protein, YopH, which is translocated into the host cytosol where it dephosphorylates tyrosine residues on multiple substrates (Guan and Dixon 1990). CagA is a *Helicobacter pylori* virulence factor which, once translocated into the host, is phosphorylated by Abl and Src family kinases, resulting in its binding to host cell proteins through their Src homology 2 domains, leading to host cell cytoskeletal modifications (Backert et al. 2010). Direct inactivation of host kinases is a mechanism of action of the lethal factor (LF) component of the *Bacillus anthracis* multi-subunit anthrax toxin. LF is a metalloprotease that cleaves host kinases, inhibiting their activity by limiting their ability to interact with substrates (Duesbery et al. 1998; Vitale et al. 1998). Lastly, *Shigella* encodes a phosphothreonine lyase, OspF, which irreversibly removes a phosphate by cleavage of the carbon–oxygen bond of target phosphothreonine residues (Li et al. 2007).

The intracellular pathogen *Legionella pneumophila* regulates host cell function in order to develop a niche permissive for replication. Much of the ability of

L. pneumophila to accomplish this is dependent on its type IV secretion system (T4SS), termed Icm/Dot (intracellular multiplication/defect in organelle trafficking) (Marra et al. 1992; Berger and Isberg 1993). This system translocates $\sim$300 proteins into the host cell after contact with the bacterium (Burstein et al. 2009; Huang et al. 2011; Zhu et al. 2011). These Icm/Dot translocated substrates (IDTS) have been shown to play roles in modulating host cell processes such as translation, cell survival, membrane trafficking, ubiquitination, and cytoskeletal dynamics (Nagai et al. 2002; Laguna et al. 2006; Kubori et al. 2008; Fontana et al. 2011; Franco et al. 2012). Although the absence of a single IDTS rarely results in an intracellular growth defect, likely due to functional redundancy among these substrates, the host cell factors and processes that they target are often required for high levels of replication (Dorer et al. 2006; O'Connor et al. 2012).

Host cell signaling through modulation of the phosphorylation states of proteins plays an important role in *L. pneumophila* intracellular replication. These signaling pathways are activated in response to *L. pneumophila* challenge and are further altered by the pathogen for its own benefit. During *Legionella* host cell binding and uptake, the phosphorylation status of multiple proteins is modulated (Venkataraman et al. 1997; Coxon et al. 1998; Tachado et al. 2008). Mitogen-activated protein kinase (MAPK) pathways are also activated during challenge (Welsh et al. 2004; Shin et al. 2008; Fontana et al. 2012). These pathways are altered by *Legionella* in an Icm/Dot dependent and independent manner to regulate the host response. Also activated is the transcription factor NF-κB (nuclear factor kappa-light-chain-enhancer of activated B cells) which is important for maintaining host cell viability during intracellular replication (Losick and Isberg 2006; Abu-Zant et al. 2007; Shin et al. 2008; Bartfeld et al. 2009; Fontana et al. 2011). Both MAPK and NF-κB signaling are mediated by Protein kinase C (PKC) which has also been shown to activate innate immune pathways in response to *L. pneumophila* challenge (N'Guessan et al. 2007; Vardarova et al. 2009). Lastly, *Legionella* translocates protein kinases into the host which may further modulate these pathways (de Felipe et al. 2008; Ge et al. 2009; Hervet et al. 2011).

2 Modulation of Protein Phosphorylation During Binding and Uptake by Host Cells

Initial studies of signal transduction following the interaction of host cells with *L. pneumophila* showed the global modification of the phosphorylation status of multiple proteins. Large-scale changes in protein phosphorylation have been linked to bacterial binding, uptake, and intracellular replication (Venkataraman et al. 1997; Coxon et al. 1998; Susa and Marre 1999). The change in the phosphorylation status of these proteins is mediated by both host kinases and phosphatases and there appears to be cell type specificity to these modifications (Venkataraman et al. 1997; Coxon et al. 1998; Tachado et al. 2008; Charpentier et al. 2009).

2.1 Protein Phosphorylation During Uptake by Mammalian Cells

Opsonization of *L. pneumophila* with complement C3b and C3bi induces binding to the host cell complement receptor 3 (CR3). This results in one of the most efficient mechanisms of *L. pneumophila* uptake by monocytes. CR3-mediated uptake of *L. pneumophila* is inhibited by the tyrosine protein kinase (TPK) inhibitors genistein and tyrphostin (Coxon et al. 1998). The inhibition of uptake by TPK inhibitors correlates with a decrease in actin polymerization in response to *L. pneumophila* as cells treated with these inhibitors did not show the marked increase in actin polymerization that was seen in untreated infected cells (Coxon et al. 1998).

Due to the role of tyrosine kinases in complement-mediated uptake of *L. pneumophila*, the global change in protein phosphorylation in response to *L. pneumophila* challenge was analyzed. CR3-mediated uptake of *L. pneumophila* by monocytes induces the tyrosine phosphorylation of multiple protein targets (Coxon et al. 1998). This phosphorylation is not dependent on the ability of *L. pneumophila* to replicate intracellularly, as an avirulent mutant, as well as *Escherichia coli*, also induce similar patterns of tyrosine phosphorylation (Coxon et al. 1998).

Challenge of MRC-5 lung epithelial cells also induces changes in the phosphorylation state of multiple proteins (Susa and Marre 1999). Both bacterial and eukaryotic proteins are believed to be phosphorylated as pretreatment with cycloheximide did not inhibit synthesis of all proteins that are phosphorylated during infection. Interestingly, little to no change in serine/threonine phosphorylation was observed while tyrosine phosphorylation was seen for many proteins (Susa and Marre 1999). Tyrosine phosphorylation is believed to be required for later stages of replication as an intracellular growth defect was observed for genistein treated cells at 24, but not four, hours post-infection (Susa and Marre 1999).

Studies utilizing chemical inhibitors have further revealed the importance of tyrosine phosphorylation for *L. pneumophila* uptake. A screen designed to identify host factors important for translocation of IDTS showed that the compound RWJ-60475, which has been shown to target the receptor protein tyrosine phosphate phosphatase, CD45, inhibited translocation, as well as bacterial uptake (Charpentier et al. 2009). Further studies revealed that bone marrow-derived macrophages (BMDMs) from CD45/CD148 CD148 double knockout mice were inhibited for uptake, while BMDMs from CD45 knockout mice were not. CD148 is also a receptor protein tyrosine phosphate phosphatase which is believed to function redundantly to (Zhu et al. 2008) and it, as well as other tyrosine phosphatases, may be an off target substrate of RWJ-60475. While the CD45/CD148-deficient BMDMs where defective in *L. pneumophila* uptake, they were able to efficiently phagocytose *E. coli* (Charpentier et al. 2009). This may point to a *L. pneumophila* specific mechanism of uptake by which it is taken up into a replication-competent niche within the host.

Though large-scale global changes in serine/threonine phosphorylation are not seen during host cell challenge with *L. pneumophila*, there is evidence that specific serine/threonine kinases may play a role in the early stages of *L. pneumophila* infection. Protein kinase B (PKB/Akt) Protein kinase B (PKB/Akt) is a serine/threonine kinase which functions in signaling pathways downstream phosphoinositide 3-kinase (PI3K) of phosphoinositide 3-kinase (PI3K) (Franke et al. 1995). Though there appears to be cell type differences in the requirement of PI3K activation for *Legionella* uptake (Khelef et al. 2001), J774A.1 macrophages are inhibited for *L. pneumophila* uptake when treated with chemical PI3K inhibitors (Tachado et al. 2008; Charpentier et al. 2009). Consistent with these results, Akt is phosphorylated within 15 min of J774A.1 challenge, but this activation is not seen when PI3K is inhibited (Tachado et al. 2008). Akt signaling may also be involved in *L. pneumophila* induced apoptosis of T-cells, when challenged at a high MOI, as it is dephosphorylated under these conditions (Takamatsu et al. 2010).

2.2 Modulation of Amoebal Protein Tyrosine Phosphorylation

Hartmannella vermiformis is a protozoan host within which *L. pneumophila* is found in the environment (Fields et al. 1990; Fields 1996). Invasion of *L. pneumophila* into this host is mediated by the host cell Gal/GalNAc lectin receptor, a homolog of the mammalian β_2 transmembrane receptors (Adams et al. 1993; Venkataraman et al. 1997). Uptake of *L. pneumophila* can be blocked by the addition of Gal or GalNAc, as well as by monoclonal antibodies targeting this receptor (Venkataraman et al. 1997). Challenge of *H. vermiformis* by *L. pneumophila* induces the tyrosine dephosphorylation of the Gal/GalNAc lectin receptor and this dephosphorylation is inhibited by the addition of Gal or GalNac, consistent with a requirement for *L. pneumophila* receptor binding (Venkataraman et al. 1997).

In addition to the dephosphorylation of the Gal/GalNAc lectin receptor, *L. pneumophila* induces the dephosphorylation of multiple other tyrosine phosphorylated proteins, including those associated with this lectin (Venkataraman et al. 1998; Venkataraman and Kwaik 2000). Protein dephosphorylation is mediated by the activation of protein tyrosine phosphatases, rather than inactivation of a kinase, and this activation appears to be unique to *L. pneumophila* as *E. coli* does not induce this dephosphorylation (Venkataraman et al. 1997; Venkataraman et al. 1998). Host protein dephosphorylation is associated with *L. pneumophila* binding, but not uptake, as it is seen in *H. vermiformis* pretreated with methylamine, which inhibits uptake, or infection with invasion-defective *L. pneumophila* mutants (Venkataraman et al. 1998). The dephosphorylation is also reversible as washing away extracellular bacteria results in the tyrosine phosphorylation of these protein substrates (Venkataraman et al. 1998).

The proteins which are dephosphorylated in response to *Legionella* challenge include cytoskeletal proteins involved in actin rearrangement such as paxillin, vinculin, and $pp125^{FAK}$. Dephosphorylation of these actin-associated proteins may result in cytoskeletal disassembly which could be responsible for the unique mechanisms of uptake of *L. pneumophila* into amoebal hosts (Venkataraman et al. 1998). While the tyrosine phosphorylation status of host proteins appears to be important for *L. pneumophila* uptake by a variety of cell types, it appears that dephosphorylation of protein targets, rather than phosphorylation seen in mammalian systems, mediates uptake into *H. vermiformis*.

3 MAPK Signaling Pathways

Mitogen-activated protein kinases are serine/threonine kinases that are activated by an evolutionally conserved signal transduction pathway, allowing eukaryotic cells to respond to environmental conditions through a kinase-mediated signaling cascade (Caffrey et al. 1999). MAPK signaling has been shown to play roles in cell cycle progression, metabolism, cytoskeletal dynamics, apoptosis, and the inflammatory response (Johnson and Lapadat 2002).

MAPK signaling proceeds through a phosphorelay system beginning with MAPK kinase kinase (MAPKKK) activation in response to a stimulus. Activation of MAPKKKs occurs by the phosphorylation of specific tyrosine and threonine residues. As the signaling cascade continues, MAPKKKs transfer a phosphate to a MAPK kinase (MAPKK), activating it and allowing it to phosphorylate a specific MAPK (Ray and Sturgill 1988; Johnson and Lapadat 2002). MAPK activation, by phosphorylation of a Thr-X-Tyr motif, induces cellular changes through the phosphorylation of transcription factors, kinases, and cytoskeletal proteins (Cargnello and Roux 2011).

There are four well-characterized conventional MAPK families found in multicellular eukaryotes: ERK1/2, SAPK/JNK, p38, and ERK5. These families are activated by cellular stresses, growth factors, protein synthesis inhibition, and cytokines (Cargnello and Roux 2011). Detection of pathogens, by pattern recognition receptors (PRRs), is also an important activator of MAPK signaling as Toll-like receptor (TLR) and nucleotide-binding oligomerization domain like receptor (NLR) detection of pathogen-associated molecular patterns (PAMPs) is linked to the MAPK response (Weinstein et al. 1992; Swantek et al. 2000; Girardin et al. 2001).

Modulation of MAPK pathways appears to be a common theme of host cell subversion by pathogens. SAPK/JNK and p38 pathways are activated following pathogen detection by TLRs or NLRs, leading to an enhanced immune response (Kobayashi et al. 2005; Huang et al. 2009). Inhibition of MAPK signaling has been shown for multiple pathogens including *B. anthracis*, by the activity of its LF toxin, and *Vibrio parahaemolyticus*, through the action of a type III secretion system effector protein, VopA (Duesbery et al. 1998; Trosky et al. 2004).

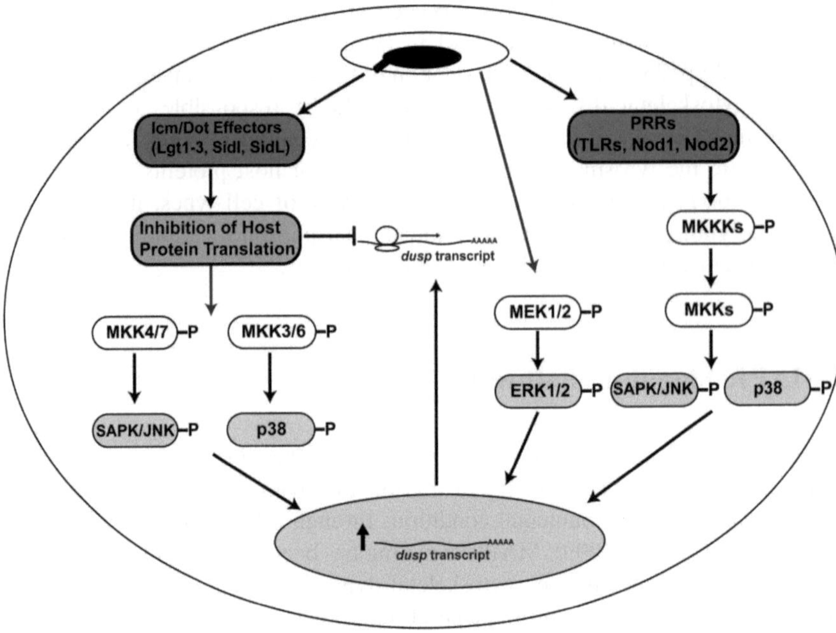

Fig. 1 MAPK signaling during *L. pneumophila* infection. *L. pneumophila* induced MAPK activation occurs by both Icm/Dot dependent and independent pathways. ERK1/2, SAPK/JNK, and p38 pathways are activated in an Icm/Dot independent manner. Activation of SAPK/JNK and p38 occurs by pathogen recognition receptor (PRR) signaling while induction of ERK1/2 occurs by an unknown mechanism (Shin et al. 2008). Icm/Dot dependent MAPK activation is induced by the inhibition of protein synthesis by five translocated effectors (Lgt1-3, SidI, and SidL) which, through an unknown mechanism, activate MAPKKs that phosphorylate p38 and SAPK/JNK (Fontana et al. 2012). Signaling through MAPKs results in enhanced transcription of target genes, including those encoding dual-specificity protein phosphatases (Dusps), which dephosphorylate MAPKs in a feedback response (Losick and Isberg 2006; Shin et al. 2008; Li et al. 2009). Dusp translation is inhibited by the action of protein synthesis inhibitors; preventing this response (Isberg Lab, unpublished results)

3.1 Pathogen-Associated Molecular Pattern and Icm/Dot Dependent Induction of MAPK Activation

MAPK activation in response to *L. pneumophila* challenge has been detected in numerous mammalian cell types, including primary macrophages and epithelial cells (Fig. 1) (Welsh et al. 2004; N'Guessan et al. 2007; Shin et al. 2008). In monocyte-derived macrophages, phosphorylation of SAPK/JNK, ERK1/2, and p38 is seen within 15 min of bacterial challenge (Welsh et al. 2004). This early MAPK response is independent of the translocation of substrates by the T4SS as an *icm/dot*$^{-}$ mutant (GL10) also induces this host cell response. Interestingly, when later time points are observed, p38 and JNK phosphorylation has returned to basal levels in the *icm/dot*$^{-}$ mutant-challenged cells while activation is maintained in cells infected with wild-type *L. pneumophila* (Welsh et al. 2004). Activation at late time points is also dependent on the translocation of specific effectors as a Δ*icmS* mutant, which retains T4SS dependent pore formation but is defective in the translocation of many effectors, did not induce SAPK/JNK or p38 phosphorylation (Zuckman et al. 1999; Coers et al. 2000; Bardill et al. 2005; Ninio et al. 2005; Cambronne and Roy 2007; Shin et al. 2008). ERK phosphorylation does not show this pattern of activation as its activation is not maintained at 1 h post-challenge with either an *icm/dot*$^-$ mutant or wild-type strain (Welsh et al. 2004; Shin et al. 2008).

Signaling both upstream and downstream of MAPK activation is seen during *L. pneumophila* challenge. MAPKKs in the p38, SAPK/JNK, and ERK pathway are activated with kinetics consistent with their ability to induce MAPK phosphorylation (Fig. 1) (Welsh et al. 2004; Shin et al. 2008). MAPK activation during challenge is also productive as c-Jun, a transcription factor that is modulated through phosphorylation by SAPK/JNK, is activated (Shin et al. 2008; Scharf et al. 2010). Lysates from challenged cells have also been shown to have activity toward ELK and ATF, ERK, and p38 substrates, respectively (Welsh et al. 2004).

Maximal induction of MAPK signaling by *Legionella* is dependent on extracellular and cytosolic sensing of bacterial PAMPs. *Myd88*$^{-/-}$*Trif*$^{-/-}$ BMDMs, which are defective in TLR dependent pathogen recognition, challenged with *L. pneumophila*, do not show p38 or SAPK/JNK phosphorylation until 1 h post-challenge. This pattern of activation is also seen in *Myd88*$^{-/-}$*Rip2*$^{-/-}$ BMDMs, which are defective in TLR, as well as Nod1 and Nod2 signaling. In these cells, no activation, at any time point, of either SAPK/JNK or p38 was detected when challenged with an *icm/dot*$^-$ *L. pneumophila* mutant (Shin et al. 2008). In contrast to SAPK/JNK and p38, ERK activation is independent of both TLR and Nod signaling (Fig. 1) (Shin et al. 2008).

The differential responses to wild-type and Icm/Dot deficient strains pointed to an early MAPK response that is Myd88/Nod dependent, Icm/Dot independent, and a later response that is Myd88/Nod independent, Icm/Dot dependent. This Icm/Dot dependent response appears to be upstream of MAPKK activation as MKK4 and MKK3/6 are both phosphorylated in *Myd88*$^{-/-}$*Rip2*$^{-/-}$ BMDMs when challenged with wild type, but not *icm/dot*$^-$, *L. pneumophila* (Fig. 1) (Shin et al. 2008).

Five IDTS, which inhibit host protein synthesis, have been shown to be the T4SS factors which result in the late activation of SAPK/JNK and p38 (Fontana et al. 2011, 2012). A strain lacking these five effectors (Lgt1, Lgt2, Lgt3, SidI, and SidL) is incapable of eliciting a SAPK/JNK or p38 response in *Myd88*$^{-/-}$-*Nod1*$^{-/-}$*Nod2*$^{-/-}$ BMDMs (Fig. 1). Complementation of this strain with a single effector (Lgt3), but not one with an inactivating point mutation, restores MAPK signaling, indicating that the cellular response to inhibition of protein biosynthesis, rather than a single effector, elicits this phenotype (Fontana et al. 2012).

3.2 Effects of MAPK Activation During Infection

The implications of MAPK signaling in response to *L. pneumophila* challenge have been analyzed using chemical inhibitors of MAPK pathways. Macrophages challenged with *L. pneumophila*, and treated with p38 or JNK inhibitors, are defective in the transcriptional induction of IL-1α and IL-1β (Shin et al. 2008; Fontana et al. 2012). Challenge of epithelial cells with *L. pneumophila* induces the expression of human β-defensin-2 (hBD-2), an antimicrobial peptide, MUC5AC, a major mucin protein, and prostaglandin E_2 (PGE_2), which regulates lung surfactant secretion as well as the immune response (N'Guessan et al. 2007; Scharf et al. 2010; Morinaga et al. 2012). hBD-2 production by cells challenged with *L. pneumophila* is decreased in cells preincubated with either JNK or p38 inhibitors (Scharf et al. 2010). In epithelial cells pretreated with ERK or JNK inhibitors, MUC5AC expression is inhibited (Morinaga et al. 2012). PGE_2 induction in response to *L. pneumophila* is also inhibited by both ERK and p38 inhibitors (N'Guessan et al. 2007).

The effect of pharmacokinetic inhibition of MAPKs on *L. pneumophila* intracellular replication is less clear. In monocyte-derived macrophages, inhibition of p38 or JNK, but not ERK, significantly inhibited intracellular growth observed at 48 h post-infection (Welsh et al. 2004). In epithelial cells, the JNK inhibitor SB600125 inhibited replication over a 24 h time period, consistent with treatment of monocyte-derived macrophages with this inhibitor (Morinaga et al. 2012). Another JNK inhibitor, JNK II, did not inhibit *L. pneumophila* replication in epithelial cells (Scharf et al. 2010). Inhibition of ERK or p38 had no effect on replication in epithelial cells (Scharf et al. 2010; Morinaga et al. 2012). Although in some cases, there appears to be reduction of intracellular growth in the presence of pharmacological inhibitors of MAPK, it should be noted that *L. pneumophila* mutants that fail to activate MAPK are still able to grow intracellularly. Therefore, MAPK activation in response to *L. pneumophila* does not appear to be required for intracellular growth.

3.3 Amoebal MAPK Activation

Activation of MAPK signaling is also observed during challenge of *Dictyostelium discoideum*, a natural amoebal host of *L. pneumophila* (Li et al. 2009). *D. discoideum* encodes two MAPKs, ERK1, and ERK2, which are analogous to the mammalian ERK family (Gaskins et al. 1994; Segall et al. 1995). During challenge with *L. pneumophila*, ERK1 is phosphorylated to a maximal level 1 h post-infection and activation continues for at least 4 h (Li et al. 2009). As with mammalian cells, ERK1 activation was observed with similar kinetics when *D. discoideum* was challenged with an *icm/dot*$^-$ mutant (Li et al. 2009).

3.4 Role of Dusps in the MAPK Response

Negative regulation of MAPK signaling in *D. discoideum* is necessary for efficient *L. pneumophila* replication. One mechanism of MAPK modulation is through Dual-specificity protein phosphatases (Dusps) which are upregulated following MAPK signaling. Dusp activation provides a feedback loop whereby MAPKs are dephosphorylated to attenuate the initial response (Patterson et al. 2009). A screen to identify *D. discoideum* mutants defective for *L. pneumophila* intracellular replication identified DupA, a protein which encodes an N-terminal eukaryotic protein kinase domain, as well as a dual-specificity protein phosphatase domain. *dupA* transcript expression is enhanced in *L. pneumophila* infected cells, consistent with a MAPK feedback response. A *dupA* mutant showed constitutive ERK1 activity which may explain the *L. pneumophila* growth defect in these cells and points to the importance of the phosphatase activity of this protein (Li et al. 2009).

Enhanced *Dusp* transcription has also been shown to occur in mammalian cells following *L. pneumophila* challenge (Fig. 1) (Losick and Isberg 2006; Shin et al. 2008). This expression is dependent on maximal MAPK signaling as the *L. pneumophila* strain, lacking five effectors shown to inhibit protein synthesis, does not induce *Dusp* expression (Haenssler and Isberg 2011). However, the importance of this enhanced expression in mammalian cells is unclear, because even though the transcript is increased, Dusp protein levels do not increase. It appears that the inhibition of protein synthesis by *L. pneumophila*, while enhancing Dusp transcript expression through MAPK activation, may be preventing the Dusp feedback response by inhibiting Dusp translation (unpublished results, Asrat and Isberg).

4 The Role of Protein Kinase C in the Response to *L. pneumophila*

The PKC family of serine/threonine kinases responds to varying environmental stimuli allowing for a broad range of responses including cell cycle progression, apoptosis, motility, and gene transcription (Ghayur et al. 1996; Black 2000; Ventura and Maioli 2001; Xiao and Liu 2012). PKCs can be categorized into three subfamilies based on their structural features and the requirement of different upstream signals for activation. Conventional PKCs (α, βI-II, and γ) require diacylglycerol (DAG) and Ca^{2+} to stimulate kinase activity. Novel PKCs (δI-III, ε, η, and θI-II) are activated downstream of DAG but do not require Ca^{2+}. Lastly, atypical PKCs (ζ and ι/λ) require neither DAG or Ca^{2+} (Tan and Parker 2003).

Members of each family of PKCs play important roles in both adaptive and innate immunity in response to bacterial pathogens. PKCs localize to the phagosome during phagocytosis and their pharmacological inhibition has been shown to limit uptake in some systems (Garcia-Garcia and Rosales 2002). Signaling downstream of both cytokine receptors and TLRs is mediated by varying isoforms of PKC. Inhibition, or the absence, of specific isoforms has been shown to block NF-κB and MAPK activation, as well as subsequent cytokine expression (Tan and Parker 2003; Loegering and Lennartz 2011).

4.1 Involvement of PKCs During Uptake

Studies of complement-mediated uptake of *L. pneumophila* have pointed to an important role of PKC in this process. Treatment of human monocytes with either staurosporine or calphostin C, both PKC inhibitors, decreased the ability of these cells to take up C3b/C3bi coated *L. pneumophila*. This was correlated with a decrease in actin polymerization when these cells were challenged, relative to untreated cells. Treatment with these chemicals 2 h post-infection resulted in only a minor reduction in bacterial growth observed at 48 h, implicating that while PKCs are important for CR3 mediated uptake, they may not play a role in intracellular replication (Coxon et al. 1998).

The importance of a conventional PKC, PKC-α, during *L. pneumophila* uptake and intracellular growth has also been studied. RAW 264.7 murine macrophages expressing a dominant negative PKC-α showed no defect in *L. pneumophila* uptake (St-Denis et al. 1999). Interestingly, expression of the dominant negative PKC-α resulted in RAW 264.7 cells which were permissive for *L. pneumophila* intracellular replication. These data are consistent with a model in which PKC-α is not required for bacterial uptake but is necessary for the restriction of bacterial growth exhibited by RAW 264.7 macrophages.

4.2 Innate Immune Signaling Mediated by PKC

Protein kinase C activation has been shown to occur in multiple systems in response to *L. pneumophila*. Early studies revealed that the addition of a *L. pneumophila* heat shock protein, Hsp60, to mouse peritoneal macrophages induced PKC activation (Retzlaff et al. 1996). Challenge of epithelial cells by *L. pneumophila* has also been shown to activate multiple PKC isotypes (N'Guessan et al. 2007; Vardarova et al. 2009). This activation was decreased in cells challenged with strains deficient in flagellin or treated with heat-killed *Legionella*, indicating that multiple signals from the pathogen are required for high-level activation (Vardarova et al. 2009).

Activation of PKC by *L. pneumophila* results in the initiation of innate immune signaling pathways. Enhanced IL-1β transcription, seen during treatment of macrophages with *L. pneumophila* Hsp60, is blocked by chemical inhibition of PKC (Retzlaff et al. 1996). Release of granulocyte macrophage colony-stimulating factor (GM-CSF) and PGE_2 is also limited by PKC inhibition (N'Guessan et al. 2007; Vardarova et al. 2009). Studies using chemical inhibitors of specific PKC isoforms have shown that PKC-α is involved in the activation of NF-κB, while PKC-ε is relevant to c-Jun signaling, in response to *L. pneumophila* challenge (Vardarova et al. 2009).

5 The NF-κB Response to *L. pneumophila* Challenge

Signaling through the regulated transcription factor NF-κB results in changes in the expression of hundreds of genes, including proinflammatory cytokines and regulators of cell survival and differentiation (Natoli 2009). In an unstimulated cell, NF-κB hetero- and homodimers are inhibited by interaction with the inhibitor of κB (IκB) family proteins. NF-κB signaling is initiated when IκB kinases (IKKs) are activated, leading to the phosphorylation of IκB which results in its ubiquitination and subsequent proteasomal degradation. Released NF-κB translocates to, and is maintained in the nucleus where it binds to κB sequences located in the promoter and enhancer regions of target genes (Li and Verma 2002).

NF-κB signaling is an innate immune response to pathogens initiated by both PRRs and cytokine receptors. The TNF-α and IL-1 receptors signal through IKK to activate NF-κB during the response to these extracellular ligands (Kelliher et al. 1998; Verstrepen et al. 2008). Detection of PAMPs by TLR-, NOD-, and NOD-like receptors leads directly to IKK activation and downstream NF-κB induction (Ogura et al. 2001; Fritz et al. 2006; Kawai and Akira 2007).

5.1 The Biphasic Activation of NF-κB

The activation of NF-κB and its regulated genes is strongly induced following *L. pneumophila* host cell challenge (Fig. 2). Activation, as measured by nuclear localization of p65 (RelA), a subunit of the canonical form of NF-κB, is seen within 3 h post-challenge of replication permissive BMDMs from A/J mice, U937 cells, or human monocyte-derived macrophages (Losick and Isberg 2006; Abu-Zant et al. 2007). Inactivation of the Icm/Dot T4SS, through a *dotA* mutation, severely limits this response, as a 10-fold higher MOI is required to elicit a strong NF-κB response. Activation in response to *icm/dot*$^-$ mutants is also transient and maintained for only 2 h post-infection, whereas challenge with a wild-type strain continues to elicit a response up to 14 h post-infection (Losick and Isberg 2006; Abu-Zant et al. 2007; Shin et al. 2008; Bartfeld et al. 2009). Induction is also limited in a Δ*icmS L. pneumophila* host cell challenge, indicating that the translocation of specific effectors, rather than a cellular response to the T4SS apparatus, is responsible (Losick and Isberg 2006).

Studies of TLR signaling responsible for the NF-κB response to *L. pneumophila* have shown that, similar to MAPK activation, this response is biphasic in nature.

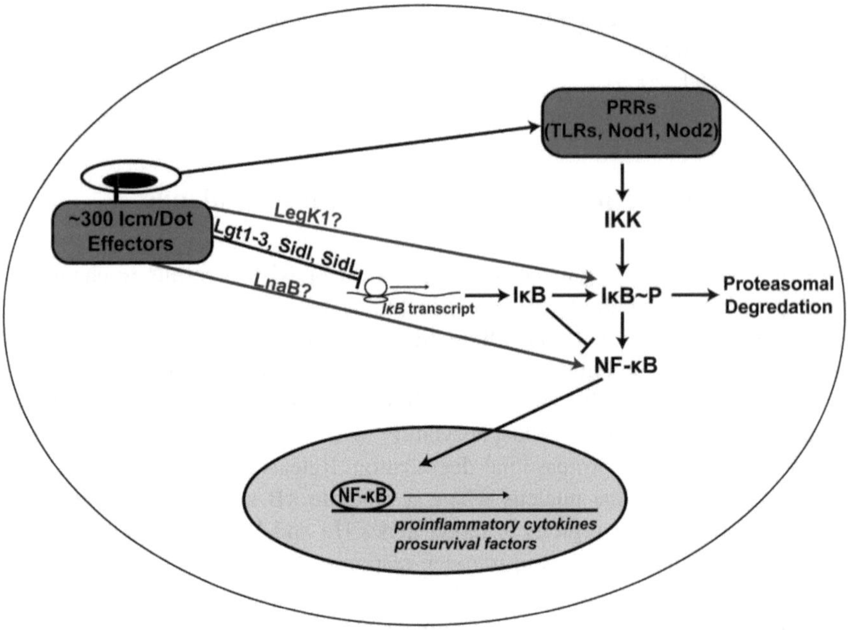

Fig. 2 NF-κB activation in response to *Legionella*. NF-κB activation occurs when the inhibitor of κB (IκB) is phosphorylated by the IκB kinase (IKK), leading to its degradation and the translocation of NF-κB to the nucleus. IKK is activated by *L. pneumophila* in an Icm/Dot independent manner through PRRs, leading to IκB degradation (Shin et al. 2008). High levels of sustained NF-κB activation, in response to *L. pneumophila* challenge, are due to the Icm/Dot mediated translocation of five effectors (Lgt1-3, SidI, and SidL) which inhibit translation of host cell proteins, including IκB (Fontana et al. 2011). LegK1, an Icm/Dot translocated Ser/Thr kinase, is able to phosphorylate IκB in vitro, but its activity in vivo is unclear (Ge et al. 2009). LnaB is another translocated substrate which activates NF-κB by an unknown mechanism (Losick et al. 2010). NF-κB activation leads to enhanced transcription of prosurvival factors, as well as inflammatory cytokines (Losick and Isberg 2006; Abu-Zant et al. 2007)

In epithelial cells depleted for the TLR adaptor protein Myd88, or in *Myd88*$^{-/-}$ BMDMs, there is a dramatic inhibition of the early (1 hpi) NF-κB response (Shin et al. 2008; Bartfeld et al. 2009). At later time points (8 hpi), NF-κB activation is detectable in these hosts when challenged with the wild type, but not with an Icm/Dot deficient strain (Bartfeld et al. 2009). These data pointed to an early Icm/Dot independent response and a later Icm/Dot dependent response.

Recognition of intracellular pathogens by NLRs results in signaling through the adaptor protein Rip2 to activate NF-κB (Chin et al. 2002; Kobayashi et al. 2002; Hayden and Ghosh 2012). Therefore, the role of Rip2 in NF-κB activation in response to *L. pneumophila* was assayed. Unlike Myd88-dependent signaling, Rip2-dependent signaling is dispensable for NF-κB activation by both wild-type and *icm/dot*$^{-}$ *L. pneumophila* (Shin et al. 2008). Interestingly, BMDMs deficient in both Myd88 and Rip2 are defective for NF-κB activation, even at late time points during wild-type *L. pneumophila* challenge (Shin et al. 2008). This indicates that though the late NF-κB response is Icm/Dot dependent, it also requires an additional pattern recognition signal mediated by either TLR or NOD pathways. This is in contrast to MAPK signaling which only requires a single signal for activation.

5.2 Translocated Substrates Implicated in NF-κB Activity

Multiple groups have attempted to identify the translocated substrates of the Icm/Dot T4SS required for the late activation of NF-κB. Ectopic expression of multiple IDTS in HEK293T cells showed moderate (>3-fold) enhancement in NF-κB activity, as measured by a NF-κB-luciferase reporter (Losick et al. 2010). Two substrates, LnaB and LegK1, exhibited greater than 100-fold induction of NF-κB activity (Ge et al. 2009; Losick et al. 2010). The role of these effectors in the induction of NF-κB during *L. pneumophila* challenge is currently unclear as a Δ*lnaB* strain only modestly reduced NF-κB activity while the absence of LegK1 had no effect (Fig. 2) (Losick et al. 2010).

In addition to their role in modulating the host MAPK pathways, the five *L. pneumophila* translocated substrates that inhibit host cell translation also play a role in the NF-κB response (Fig. 2). These proteins inhibit the translation of IκB, which is degraded in response to IKK activation by pattern recognition. In BMDMs challenged with a Δ*dotA* strain, IκB protein levels are decreased shortly after infection, but then return to close to uninfected levels, while in a wild-type infection IκB levels remain low (Shin et al. 2008; Fontana et al. 2011). In cells challenged with a strain lacking the five translation inhibitors, results are similar to a Δ*dotA* infection in which IκB levels return following an early depletion (Fontana et al. 2011). Therefore, in the mutant strain lacking the translation inhibitors, there is a lack of induction of NF-κB activity, as this strain shows lower levels of translocation of the p65 NF-κB subunit into the nucleus, relative to that observed during a wild-type infection (Fontana et al. 2011).

5.3 NF-κB and Host Cell Survival

The outcome of NF-κB activation by *L. pneumophila* appears to be twofold. The first is the enhanced transcription of NF-κB target genes, including cytokines, which play a role in the host response to limit the pathogen (Losick and Isberg 2006; Abu-Zant et al. 2007). The second, which is essential for the ability of *L. pneumophila* to replicate intracellularly, is the activation of prosurvival factors. A/J BMDMs, expressing either a dominant negative IκB, which is inhibitory for NF-κB activation, or treated with caffeic acid phenethyl ester (CAPE), which prevents the nuclear translocation of NF-κB, undergo enhanced cell death in response to *L. pneumophila* (Losick and Isberg 2006). CAPE treatment of human monocyte-derived macrophages, challenged with the AA100 strain, does not induce enhanced cell death, indicating that there may be cell type and strain specificity (Abu-Zant et al. 2007). Regardless of differences in cellular survival, CAPE treatment limits *L. pneumophila* replication in both cell types (Losick and Isberg 2006; Abu-Zant et al. 2007).

6 *L. pneumophila* T4SS Translocated Kinases

L. pneumophila encodes five proteins (LegK1-5) with homology to serine/threonine kinases (de Felipe et al. 2005; Hervet et al. 2011). LegK1-4 are present in all sequenced *L. pneumophila* strains, while LegK5 is found only in the Lens isolate (Hervet et al. 2011). Of these, LegK1-4 have been shown to be translocated by the Icm/Dot T4SS (de Felipe et al. 2008; Shin et al. 2008; Ge et al. 2009; Hervet et al. 2011). Modulation of host signaling through Ser/Thr kinases appears to be a unique mechanism as none of the other identified IDTS have homology to either tyrosine kinases or protein phosphatases (Haenssler and Isberg 2011).

The best studied of these translocated substrates is the previously mentioned LegK1. When expressed in eukaryotic cells, LegK1, but not LegK2 or LegK3, activated NF-κB (Ge et al. 2009). This activation required the kinase activity of LegK1 as a point mutation in its ATP binding site inhibited its ability to activate NF-κB. Recombinant LegK1 is able to phosphorylate the NF-κB inhibitor, IκB, showing that, in vitro, it is able to trigger the canonical NF-κB activation pathway. p100, which is processed into p52 after phosphorylation in the noncanonical pathway of NF-κB activation, was also shown to be phosphorylated by recombinant LegK1 (Ge et al. 2009). The importance of the phosphorylation during intracellular growth is unclear, as a strain lacking *legK1* is able to replicate in both BMDMs and the natural amoebal host *Acanthamoeba castellanii* (Ge et al. 2009; Losick et al. 2010). Furthermore, strains lacking the five translation inhibitors are unable to activate NF-κB above the levels seen in a *dotA*$^-$ strain, calling into question the relevance of the in vitro-demonstrated IκB phosphorylation.

Though specific host targets for LegK2-5 have not been identified, each has begun to be characterized in vitro as well as in vivo. Recombinant LegK2-5 proteins were shown to have autokinase activity, as well as the ability to transfer a phosphate group to the eukaryotic myelin basic protein in vitro (Hervet et al. 2011). In contrast, recombinant LegK1 did not exhibit either of these activities, showing that its ability to phosphorylate NF-κB pathway factors may be specific (Hervet et al. 2011). When *A. castellanii* was challenged with *L. pneumophila* strain Lens lacking each of the LegK proteins, a Δ*legk2* strain exhibited delayed replication and decreased host cell cytotoxicity relative to the wild-type, and other *legK* deletion strains. This may be due to an inability of this strain to form a replication-competent vacuole as *Legionella*-containing vacuoles (LCVs) showed a defect in recruitment of the ER chaperone calnexin (Hervet et al. 2011).

7 Conclusions

The ability of *L. pneumophila* to replicate within a host is dependent on host cell signaling through changes in the phosphorylation state of protein substrates. Modulation of these signaling pathways by *Legionella* is just one example of how this pathogen is able to subvert normal cellular processes for its own benefit. During initial host cell contact, multiple proteins are phosphorylated and may play a role in bacterial uptake. *Legionella* activates both MAPK and NF-κB signaling pathways, when PRRs are engaged, early during infection. These pathways are further activated by the translocation of five IDTS which inhibit host protein synthesis. PKC activation is also involved in these, as well as other, innate immune signaling pathways. Finally, although their targets are currently unknown, *L. pneumophila* translocates at least four effector proteins shown to have protein kinase activity. Further research will elucidate the targets of these proteins and what role they may play in the modulation of the important signaling pathways regulated by protein phosphorylation.

References

Adams SA, Robson SC, Gathiram V, Jackson TF, Pillay TS et al (1993) Immunological similarity between the 170 kD amoebic adherence glycoprotein and human beta 2 integrins. Lancet 341:17–19

Abu-Zant A, Jones S, Asare R, Suttles J, Price C et al (2007) Anti-apoptotic signaling by the Dot/Icm secretion system of *L. pneumophila*. Cell Microbiol 9:246–264

Backert S, Tegtmeyer N, Selbach M (2010) The versatility of *Helicobacter pylori* CagA effector protein functions: the master key hypothesis. Helicobacter 15:163–176

Bardill JP, Miller JL, Vogel JP (2005) IcmS-dependent translocation of SdeA into macrophages by the *Legionella pneumophila* type IV secretion system. Mol Microbiol 56:90–103

Bartfeld S, Engels C, Bauer B, Aurass P, Flieger A et al (2009) Temporal resolution of two-tracked NF-kappaB activation by *Legionella pneumophila*. Cell Microbiol 11:1638–1651

Berger KH, Isberg RR (1993) Two distinct defects in intracellular growth complemented by a single genetic locus in *Legionella pneumophila*. Mol Microbiol 7:7–19

Black JD (2000) Protein kinase C-mediated regulation of the cell cycle. Front Biosci 5:D406–D423

Blom N, Gammeltoft S, Brunak S (1999) Sequence and structure-based prediction of eukaryotic protein phosphorylation sites. J Mol Biol 294:1351–1362

Burstein D, Zusman T, Degtyar E, Viner R, Segal G et al (2009) Genome-scale identification of *Legionella pneumophila* effectors using a machine learning approach. PLoS Pathog 5:e1000508

Caffrey DR, O'Neill LA, Shields DC (1999) The evolution of the MAP kinase pathways: coduplication of interacting proteins leads to new signaling cascades. J Mol Evol 49:567–582

Cambronne ED, Roy CR (2007) The *Legionella pneumophila* IcmSW complex interacts with multiple Dot/Icm effectors to facilitate type IV translocation. PLoS Pathog 3:e188

Cargnello M, Roux PP (2011) Activation and function of the MAPKs and their substrates, the MAPK-activated protein kinases. Microbiol Mol Biol Rev 75:50–83

Charpentier X, Gabay JE, Reyes M, Zhu JW, Weiss A et al (2009) Chemical genetics reveals bacterial and host cell functions critical for type IV effector translocation by *Legionella pneumophila*. PLoS Pathog 5:e1000501

Chin AI, Dempsey PW, Bruhn K, Miller JF, Xu Y et al (2002) Involvement of receptor-interacting protein 2 in innate and adaptive immune responses. Nature 416:190–194

Coers J, Kagan JC, Matthews M, Nagai H, Zuckman DM et al (2000) Identification of Icm protein complexes that play distinct roles in the biogenesis of an organelle permissive for *Legionella pneumophila* intracellular growth. Mol Microbiol 38:719–736

Cohen P (2001) The role of protein phosphorylation in human health and disease. The Sir Hans Krebs Medal Lecture. Eur J Biochem 268:5001–5010

Coxon PY, Summersgill JT, Ramirez JA, Miller RD (1998) Signal transduction during *Legionella pneumophila* entry into human monocytes. Infect Immun 66:2905–2913

de Felipe KS, Glover RT, Charpentier X, Anderson OR, Reyes M et al (2008) *Legionella* eukaryotic-like type IV substrates interfere with organelle trafficking. PLoS Pathog 4:e10000117

de Felipe KS, Pampou S, Jovanovic OS, Pericone CD, Ye SF et al (2005) Evidence for acquisition of *Legionella* type IV secretion substrates via interdomain horizontal gene transfer. J Bacteriol 187:7716–7726

Dorer MS, Kirton D, Bader JS, Isberg RR (2006) RNA interference analysis of *Legionella* in *Drosophila* cells: exploitation of early secretory apparatus dynamics. PLoS Pathog 2:e34

Duesbery NS, Webb CP, Leppla SH, Gordon VM, Klimpel KR et al (1998) Proteolytic inactivation of MAP-kinase-kinase by anthrax lethal factor. Science 280:734–737

Fields BS (1996) The molecular ecology of legionellae. Trends Microbiol 4:286–290

Fields BS, Nerad TA, Sawyer TK, King CH, Barbaree JM et al (1990) Characterization of an axenic strain of *Hartmannella vermiformis* obtained from an investigation of nosocomial legionellosis. J Protozool 37:581–583

Fontana MF, Banga S, Barry KC, Shen X, Tan Y et al (2011) Secreted bacterial effectors that inhibit host protein synthesis are critical for induction of the innate immune response to virulent *Legionella pneumophila*. PLoS Pathog 7:e1001289

Fontana MF, Shin S, Vance RE (2012) Activation of host mitogen-activated protein kinases by secreted *Legionella pneumophila* effectors that inhibit host protein translation. Infect Immun 80:3570–3575

Franco IS, Shohdy N, Shuman HA (2012) The *Legionella pneumophila* effector VipA is an actin nucleator that alters host cell organelle trafficking. PLoS Pathog 8:e1002546

Franke TF, Yang SI, Chan TO, Datta K, Kazlauskas A et al (1995) The protein kinase encoded by the Akt proto-oncogene is a target of the PDGF-activated phosphatidylinositol 3-kinase. Cell 81:727–736

Fritz JH, Ferrero RL, Philpott DJ, Girardin SE (2006) Nod-like proteins in immunity, inflammation and disease. Nat Immunol 7:1250–1257

Garcia-Garcia E, Rosales C (2002) Signal transduction during Fc receptor-mediated phagocytosis. J Leukoc Biol 72:1092–1108

Gaskins C, Maeda M, Firtel RA (1994) Identification and functional analysis of a developmentally regulated extracellular signal-regulated kinase gene in *Dictyostelium discoideum*. Mol Cell Biol 14:6996–7012

Ge J, Xu H, Li T, Zhou Y, Zhang Z et al (2009) A *Legionella* type IV effector activates the NF-kappaB pathway by phosphorylating the IkappaB family of inhibitors. Proc Natl Acad Sci U S A 106:13725–13730

Ghayur T, Hugunin M, Talanian RV, Ratnofsky S, Quinlan C et al (1996) Proteolytic activation of protein kinase C delta by an ICE/CED 3-like protease induces characteristics of apoptosis. J Exp Med 184:2399–2404

Girardin SE, Tournebize R, Mavris M, Page AL, Li X et al (2001) CARD4/Nod1 mediates NF-kappaB and JNK activation by invasive *Shigella flexneri*. EMBO Rep 2:736–742

Guan KL, Dixon JE (1990) Protein tyrosine phosphatase activity of an essential virulence determinant in *Yersinia*. Science 249:553–556

Haenssler E, Isberg RR (2011) Control of host cell phosphorylation by *Legionella pneumophila*. Front Microbiol 2:64

Hayden MS, Ghosh S (2012) NF-kappaB, the first quarter-century: remarkable progress and outstanding questions. Genes Dev 26:203–234

Hervet E, Charpentier X, Vianney A, Lazzaroni JC, Gilbert C et al (2011) Protein kinase LegK2 is a type IV secretion system effector involved in endoplasmic reticulum recruitment and intracellular replication of *Legionella pneumophila*. Infect Immun 79:1936–1950

Huang G, Shi LZ, Chi H (2009) Regulation of JNK and p38 MAPK in the immune system: signal integration, propagation and termination. Cytokine 48:161–169

Huang L, Boyd D, Amyot WM, Hempstead AD, Luo ZQ et al (2011) The E Block motif is associated with *Legionella pneumophila* translocated substrates. Cell Microbiol 13:227–245

Johnson GL, Lapadat R (2002) Mitogen-activated protein kinase pathways mediated by ERK, JNK, and p38 protein kinases. Science 298:1911–1912

Kawai T, Akira S (2007) TLR signaling. Semin Immunol 19:24–32

Kelliher MA, Grimm S, Ishida Y, Kuo F, Stanger BZ et al (1998) The death domain kinase RIP mediates the TNF-induced NF-kappaB signal. Immunity 8:297–303

Khelef N, Shuman HA, Maxfield FR (2001) Phagocytosis of wild-type *Legionella pneumophila* occurs through a wortmannin-insensitive pathway. Infect Immun 69:5157–5161

Kobayashi K, Inohara N, Hernandez LD, Galan JE, Nunez G et al (2002) RICK/Rip2/CARDIAK mediates signalling for receptors of the innate and adaptive immune systems. Nature 416:194–199

Kobayashi KS, Chamaillard M, Ogura Y, Henegariu O, Inohara N et al (2005) Nod2-dependent regulation of innate and adaptive immunity in the intestinal tract. Science 307:731–734

Kubori T, Hyakutake A, Nagai H (2008) *Legionella* translocates an E3 ubiquitin ligase that has multiple U-boxes with distinct functions. Mol Microbiol 67:1307–1319

Laguna RK, Creasey EA, Li Z, Valtz N, Isberg RR (2006) A *Legionella pneumophila*-translocated substrate that is required for growth within macrophages and protection from host cell death. Proc Natl Acad Sci U S A 103:18745–18750

Li H, Xu H, Zhou Y, Zhang J, Long C et al (2007) The phosphothreonine lyase activity of a bacterial type III effector family. Science 315:1000–1003

Li Q, Verma IM (2002) NF-kappaB regulation in the immune system. Nat Rev Immunol 2:725–734

Li Z, Dugan AS, Bloomfield G, Skelton J, Ivens A et al (2009) The amoebal MAP kinase response to *Legionella pneumophila* is regulated by DupA. Cell Host Microbe 6:253–267

Loegering DJ, Lennartz MR (2011) Protein kinase C and toll-like receptor signaling. Enzyme Res 2011:537821

Losick VP, Haenssler E, Moy MY, Isberg RR (2010) LnaB: a *Legionella pneumophila* activator of NF-kappaB. Cell Microbiol 12:1083–1097

Losick VP, Isberg RR (2006) NF-kappaB translocation prevents host cell death after low-dose challenge by *Legionella pneumophila*. J Exp Med 203:2177–2189

Manning G, Plowman GD, Hunter T, Sudarsanam S (2002) Evolution of protein kinase signaling from yeast to man. Trends Biochem Sci 27:514–520

Marra A, Blander SJ, Horwitz MA, Shuman HA (1992) Identification of a *Legionella pneumophila* locus required for intracellular multiplication in human macrophages. Proc Natl Acad Sci U S A 89:9607–9611

Morinaga Y, Yanagihara K, Araki N, Migiyama Y, Nagaoka K et al (2012) Live *Legionella pneumophila* induces MUC5AC production by airway epithelial cells independently of intracellular invasion. Can J Microbiol 58:151–157

N'Guessan PD, Etouem MO, Schmeck B, Hocke AC, Scharf S et al (2007) *Legionella pneumophila*-induced PKCalpha-, MAPK-, and NF-kappaB-dependent COX-2 expression in human lung epithelium. Am J Physiol Lung Cell Mol Physiol 292:L267–L277

Nagai H, Kagan JC, Zhu X, Kahn RA, Roy CR (2002) A bacterial guanine nucleotide exchange factor activates ARF on *Legionella* phagosomes. Science 295:679–682

Natoli G (2009) Control of NF-kappaB-dependent transcriptional responses by chromatin organization. Cold Spring Harb Perspect Biol 1:a000224

Ninio S, Zuckman-Cholon DM, Cambronne ED, Roy CR (2005) The *Legionella* IcmS-IcmW protein complex is important for Dot/Icm-mediated protein translocation. Mol Microbiol 55:912–926

O'Connor TJ, Boyd D, Dorer MS, Isberg RR (2012) Aggravating genetic interactions allow a solution to redundancy in a bacterial pathogen. Science 338:1440–1444

Ogura Y, Inohara N, Benito A, Chen FF, Yamaoka S et al (2001) Nod2, a Nod1/Apaf-1 family member that is restricted to monocytes and activates NF-kappaB. J Biol Chem 276:4812–4818

Patterson KI, Brummer T, O'Brien PM, Daly RJ (2009) Dual-specificity phosphatases: critical regulators with diverse cellular targets. Biochem J 418:475–489

Ray LB, Sturgill TW (1988) Insulin-stimulated microtubule-associated protein kinase is phosphorylated on tyrosine and threonine in vivo. Proc Natl Acad Sci U S A 85:3753–3757

Retzlaff C, Yamamoto Y, Okubo S, Hoffman PS, Friedman H et al (1996) *Legionella pneumophila* heat-shock protein-induced increase of interleukin-1 beta mRNA involves protein kinase C signalling in macrophages. Immunology 89:281–288

Scharf S, Hippenstiel S, Flieger A, Suttorp N, N'Guessan PD (2010) Induction of human beta-defensin-2 in pulmonary epithelial cells by *Legionella pneumophila*: involvement of TLR2 and TLR5, p38 MAPK, JNK, NF-kappaB, and AP-1. Am J Physiol Lung Cell Mol Physiol 298:L687–L695

Segall JE, Kuspa A, Shaulsky G, Ecke M, Maeda M et al (1995) A MAP kinase necessary for receptor-mediated activation of adenylyl cyclase in *Dictyostelium*. J Cell Biol 128:405–413

Shin S, Case CL, Archer KA, Nogueira CV, Kobayashi KS et al (2008) Type IV secretion-dependent activation of host MAP kinases induces an increased proinflammatory cytokine response to *Legionella pneumophila*. PLoS Pathog 4:e1000220

St-Denis A, Caouras V, Gervais F, Descoteaux A (1999) Role of protein kinase C-alpha in the control of infection by intracellular pathogens in macrophages. J Immunol 163:5505–5511

Susa M, Marre R (1999) *Legionella pneumophila* invasion of MRC-5 cells induces tyrosine protein phosphorylation. Infect Immun 67:4490–4498

Swantek JL, Tsen MF, Cobb MH, Thomas JA (2000) IL-1 receptor-associated kinase modulates host responsiveness to endotoxin. J Immunol 164:4301–4306

Tachado SD, Samrakandi MM, Cirillo JD (2008) Non-opsonic phagocytosis of *Legionella pneumophila* by macrophages is mediated by phosphatidylinositol 3-kinase. PLoS ONE 3:e3324

Takamatsu R, Takeshima E, Ishikawa C, Yamamoto K, Teruya H et al (2010) Inhibition of Akt/GSK3beta signalling pathway by *Legionella pneumophila* is involved in induction of T-cell apoptosis. Biochem J 427:57–67

Tan SL, Parker PJ (2003) Emerging and diverse roles of protein kinase C in immune cell signalling. Biochem J 376:545–552

Trosky JE, Mukherjee S, Burdette DL, Roberts M, McCarter L et al (2004) Inhibition of MAPK signaling pathways by VopA from *Vibrio parahaemolyticus*. J Biol Chem 279:51953–51957

Ubersax JA, Ferrell JE Jr (2007) Mechanisms of specificity in protein phosphorylation. Nat Rev Mol Cell Biol 8:530–541

Vardarova K, Scharf S, Lang F, Schmeck B, Opitz B et al (2009) PKC(alpha) and PKC(epsilon) differentially regulate *Legionella pneumophila*-induced GM-CSF. Eur Respir J 34:1171–1179

Venkataraman C, Gao LY, Bondada S, Kwaik YA (1998) Identification of putative cytoskeletal protein homologues in the protozoan host *Hartmannella vermiformis* as substrates for induced tyrosine phosphatase activity upon attachment to the Legionnaires' disease bacterium, *Legionella pneumophila*. J Exp Med 188:505–514

Venkataraman C, Haack BJ, Bondada S, Abu Kwaik Y (1997) Identification of a Gal/GalNAc lectin in the protozoan *Hartmannella vermiformis* as a potential receptor for attachment and invasion by the Legionnaires' disease bacterium. J Exp Med 186:537–547

Venkataraman C, Kwaik YA (2000) Signal transduction in the protozoan host *Hartmannella vermiformis* upon attachment to *Legionella pneumophila*. Microbes Infect 2:867–875

Ventura C, Maioli M (2001) Protein kinase C control of gene expression. Crit Rev Eukaryot Gene Expr 11:243–267

Verstrepen L, Bekaert T, Chau TL, Tavernier J, Chariot A et al (2008) TLR-4, IL-1R and TNF-R signaling to NF-kappaB: variations on a common theme. Cell Mol Life Sci 65:2964–2978

Vitale G, Pellizzari R, Recchi C, Napolitani G, Mock M et al (1998) Anthrax lethal factor cleaves the N-terminus of MAPKKs and induces tyrosine/threonine phosphorylation of MAPKs in cultured macrophages. Biochem Biophys Res Commun 248:706–711

Weinstein SL, Sanghera JS, Lemke K, DeFranco AL, Pelech SL (1992) Bacterial lipopolysaccharide induces tyrosine phosphorylation and activation of mitogen-activated protein kinases in macrophages. J Biol Chem 267:14955–14962

Welsh CT, Summersgill JT, Miller RD (2004) Increases in c-Jun N-terminal kinase/stress-activated protein kinase and p38 activity in monocyte-derived macrophages following the uptake of *Legionella pneumophila*. Infect Immun 72:1512–1518

Xiao H, Liu M (2012) Atypical protein kinase C in cell motility. Cell Mol Life Sci doi:10.1007/s00018-012-1192-1

Zhang ZY, Zhou B, Xie L (2002) Modulation of protein kinase signaling by protein phosphatases and inhibitors. Pharmacol Ther 93:307–317

Zhu JW, Brdicka T, Katsumoto TR, Lin J, Weiss A (2008) Structurally distinct phosphatases CD45 and CD148 both regulate B cell and macrophage immunoreceptor signaling. Immunity 28:183–196

Zhu W, Banga S, Tan Y, Zheng C, Stephenson R et al (2011) Comprehensive identification of protein substrates of the Dot/Icm type IV transporter of *Legionella pneumophila*. PLoS ONE 6:e17638

Zuckman DM, Hung JB, Roy CR (1999) Pore-forming activity is not sufficient for *Legionella pneumophila* phagosome trafficking and intracellular growth. Mol Microbiol 32:990–1001

Mouse Models of Legionnaires' Disease

Andrew S. Brown, Ian R. van Driel and Elizabeth L. Hartland

Abstract *Legionella pneumophila* is an accidental respiratory pathogen of humans that provokes a robust inflammatory response upon infection. While most people exposed to *L. pneumophila* will clear the infection, certain groups with underlying susceptibility will develop Legionnaires' disease. Mice, like most humans, are inherently resistant to *L. pneumophila* and infection of most inbred strains reflects the response of immune competent people to *L. pneumophila* exposure. Hence, the use of mouse models of *L. pneumophila* infection has taught us a great deal about the innate and adaptive factors that lead to successful clearance of the pathogen and avoidance of Legionnaires' disease. At the same time, *L. pneumophila* has provided new insight into innate immunity in general and is now a model pathogen with which to study acute lung inflammation and inflammasome activation. This chapter will explore the history and use of the mouse model of *L. pneumophila* infection and examine what we know about the innate and adaptive factors that contribute to the control of *L. pneumophila* in the mouse lung.

Contents

1 Introduction 272
2 Animal Models of Legionnaires' Disease 273
 2.1 The Guinea Pig Model of Legionnaires' Disease 273

A. S. Brown · I. R. van Driel
Department of Biochemistry and Molecular Biology and the Bio21 Institute, University of Melbourne, Victoria 3010, Australia

E. L. Hartland
Department of Microbiology and Immunology, University of Melbourne, Victoria 3010, Australia

E. L. Hartland (✉)
Murdoch Children's Research Institute, Victoria 3052, Australia
e-mail: hartland@unimelb.edu.au

Current Topics in Microbiology and Immunology (2013) 376: 271–291
DOI: 10.1007/82_2013_349

Published Online: 6 August 2013

2.2 The Mouse Model of Legionnaires' Disease 274
3 The Immune Response to *Legionella pneumophila* in Mice 276
3.1 Molecular-Based Interactions of L. pneumophila with the Innate Immune System 277
3.2 Cellular Responses in the Lung During L. pneumophila Infection 283
3.3 Interactions of L. pneumophila with the Adaptive Immune Response 284
4 Summary 286
References 286

1 Introduction

Legionnaires' disease is largely an infection of immune compromised people, especially those with compromised lung function. Exposure of the lung to water droplets containing *Legionella pneumophila* leads to bacterial replication in resident alveolar macrophages and the induction of an inflammatory response (Rodgers 1979). In the USA, approximately 90 % of reported cases of Legionnaires' disease are accounted for by infection with *L. pneumophila*. Worldwide, other members of the genus such as *L. longbeachae, L. bozemanii,* and *L. micdadei* also contribute significantly to cases of Legionnaires' disease. Infection with *Legionella* is epidemiologically important in both community- and nosocomial-acquired settings, as elderly adults, patients on immunosuppressive therapies, smokers, and other individuals with poor immune function as a result of unrelated illnesses are at an increased risk of severe disease progression and fatality (Fields et al. 2002; Hilbi et al. 2010; Newton et al. 2010). In healthy individuals, *L. pneumophila* infection does not typically progress to productive disease, and patients remain largely asymptomatic. However in susceptible people, the infection may lead to either one of two clinical manifestations, Legionnaires' disease, a severe and potentially fatal pulmonary pneumonia, or Pontiac fever, a less severe, self-limiting nonpneumonic illness with headaches, chills, fever, and myalgia (Fraser et al. 1977; Glick et al. 1978; Fields et al. 2002). Generally, Legionnaires' disease patients will present with lobar consolidation of the lung comprising a strong cellular infiltrate of neutrophils and macrophages into the alveolar septa and spaces alongside a small number of eosinophils (Glavin et al. 1979).

The dramatic appearance of Legionnaires' disease in humans has led to the development of small animal models of *Legionella* infection to understand and study the immune response to infection. Along the way, the mouse infection model in particular has revealed novel aspects of innate immunity relevant to the immune response against many bacterial pathogens.

2 Animal Models of Legionnaires' Disease

In immune competent individuals, bacterial replication is controlled by a robust proinflammatory response involving cell-mediated clearance of *L. pneumophila* from the lung (Ang et al. 2010). To investigate and understand the precise host–pathogen interactions that occur in the mammalian host during *Legionella* infection, and therefore thoroughly understand the immune response initiated against the bacterium, several animal-based infection models of Legionnaires' disease have been developed, most notably the use of guinea pigs and mice.

2.1 The Guinea Pig Model of Legionnaires' Disease

Initially, the isolation of *L. pneumophila* for taxonomic identification was performed using guinea pigs in a protocol otherwise used for the isolation of *Rickettsia* bacteria. Injection of homogenized human lung samples into the peritoneum of guinea pigs induced acute sickness, and guinea pigs became moribund within 3–6 days of infection, displaying severe bacterial dissemination throughout the peritoneum, spleen, and liver (McDade et al. 1977). Subsequent investigations using aerosol-based exposure of guinea pigs to *L. pneumophila*-containing water droplets revealed a high susceptibility of these animals to the pathogen. Infection of guinea pigs generated an illness reminiscent of typical human Legionnaires' disease (Baskerville et al. 1981). While the guinea pig model of Legionnaires' disease has considerable benefit in informing our understanding of *L. pneumophila*-induced pneumonia and disease pathogenesis, much of the research using these animals has focussed on the therapeutic benefit of antimicrobial treatment after aerosol infection (Fraser et al. 1978; Edelstein et al. 1984; Fitzgeorge et al. 1990). The findings from treatment studies have supported those of human clinical trials and the development of current treatment regimens for hospitalized patients suffering Legionnaires' disease (Amsden 2005). Aside from antimicrobial therapy, this model has also been studied in the context of immune responses, focussing primarily on the development of immune-protective vaccines to lethal *L. pneumophila* infection. Indeed, these studies have proven that sub-lethal exposure of guinea pigs to the bacteria induces an effective humoral- and cell-mediated *L. pneumophila*-specific response with protective benefit upon lethal challenge (Breiman and Horwitz 1987; Blander and Horwitz 1989; Weeratna et al. 1994). However, due to the scarce availability of commercially produced guinea pig-specific immunological reagents that are otherwise widely and readily available for mouse studies, the guinea pig model is inadequate for the investigation of the immune response to *L. pneumophila* infection (Brieland et al. 1994; Ang et al. 2010). Consequently, research into the molecular and cellular aspects of the antibacterial immune response has shifted to mouse infection models, which have revealed the complex involvement of innate and adaptive immune cells and molecules in host resistance against *L. pneumophila*.

2.2 The Mouse Model of Legionnaires' Disease

2.2.1 History

Although a proper murine-based model for Legionnaires' disease was not established until 1994, the first reported evidence of *L. pneumophila* infection in mice dates from 1979 after an investigation by Kenneth Hedlund into the efficacy of immunologic protection against the bacterium upon lethal murine challenge (Hedlund et al. 1979). It was quickly discovered that relative to guinea pigs, the majority of inbred mouse strains were resistant to bacterial infection and disease progression (Fitzgeorge et al. 1983; Yoshida and Mizuguchi 1986; Brieland et al. 1994). Mice generally cleared *L. pneumophila* from the lung within a week of infection, and did not display features typical of either Legionnaires' disease patients or *L. pneumophila* infected guinea pigs (Fitzgeorge et al. 1983). Despite these observations, however, inoculation of A strain mice (frequently referred to as A/J) with *L. pneumophila* revealed a strain specific susceptibility to infection, as A mice presented with acute pneumonia within 48 h of infection that resembled human disease (Brieland et al. 1994). These findings supported earlier in vitro-based infections of peritoneal macrophages demonstrating that cells derived from A strain mice were substantially more permissive for replication of *L. pneumophila* in the first 72 h of infection than cells derived from other inbred mouse strains, including C57BL/6 and BALB/c (Yamamoto et al. 1988; Yoshida et al. 1991). Genetic crosses between permissive A mice and resistant C57BL/6 mice mapped this susceptibility to a genetic locus termed *Lgn1* on the distal region of mouse chromosome 13 (Yoshida et al. 1991; Beckers et al. 1995; Dietrich et al. 1995).

The *Lgn1* genetic region contains a series of 80–100 kb polymorphic repeats comprising the neuronal apoptosis-inhibitory proteins (Naip) (otherwise known as baculoviral inhibitor of apoptosis repeat-containing 1, Birc1) (Diez et al. 2002; Wright Jr et al. 2003). Currently, it is believed that missense polymorphisms or alterations in expression of the Naip5/Birc1e protein are responsible for the associated susceptibility of A mice to infection (Lamkanfi and Dixit 2009; Newton et al. 2010). Naip5/Birc1e is a member of the nucleotide-binding and oligomerization domain (NOD)-like receptor (NLR) protein family of innate immune surveillance molecules and is involved in the immune recognition of bacterial flagellin in the cytosol of a host cell (Lightfield et al. 2008; Lamkanfi and Dixit 2009). Upon recognition of flagellin, activation of Naip5 mediates the generation of a large cytosolic multiprotein scaffold complex termed the inflammasome (Molofsky et al. 2006; Ren et al. 2006). Inflammasome generation and activation mediate the activation of caspase-1, which cleaves pro-interleukin (IL)-1b and pro-IL-18 to their active forms, subsequently initiating pyroptosis, a form of proinflammatory cell death (Fink and Cookson 2005; Case et al. 2009). In resistant C57BL/6 mice, flagellin mutants of *L. pneumophila* replicate as efficiently as wild-type *L. pneumophila* in A mice and hence many investigators study immune responses in the C57BL/6 background using *flaA* mutants (Archer et al. 2009, 2010; Berrington et al. 2010).

2.2.2 Testing of *L. pneumophila* Mutants in Mice

The mouse model of infection may be used for examining alterations in virulence of bacterial mutants. In general A strain mice are preferred for pathogenesis studies given their enhanced susceptibility. However, few groups routinely investigate the phenotype of mutant strains of *L. pneumophila* in mice, possibly due to the lack of strong phenotypes in vitro. The Dot/Icm type IV secretion system is a notable exception as the system is absolutely required for bacterial replication in macrophages (Isberg et al. 2008; Newton et al. 2010). The Dot/Icm apparatus is responsible for the secretion and translocation of around 300 effector proteins into the infected macrophage which allow the bacteria to avoid phago-lysosome fusion and establish a specialized *Legionella* containing vacuole (LCV) that resembles rough endoplasmic reticulum and supports bacterial replication. Mutations in *dot* or *icm* genes that abolish the function of proteins necessary for type IV secretion render *Legionella* unable to form a vacuole that supports replication. The vacuoles containing these mutants rapidly mature along the default endocytic pathway (Marra et al. 1992; Berger and Isberg 1993; Berger et al. 1994; Roy et al. 1998). Mice infected with *dotA* mutants of *L. pneumophila* quickly clear the bacteria from the lungs with minimal induction of a proinflammatory response (Archer and Roy 2006).

The lack of strong attenuation of many *L. pneumophila* mutants has led researchers to assess attenuation using mixed or competitive infections. In these experiments, wild-type and mutant *L. pneumophila* are mixed in a ratio of 1:1 and then inoculated into the same animal. After a certain period, bacteria are recovered from the lung and the ratio of wild type to mutant bacteria determined by plating for CFU on differential antibiotic selective media. In general, an output ratio of mutant to wild-type bacteria of 0.5 or less is considered attenuated (meaning less than half of the mutant bacteria were recovered compared to the wild-type strain) (Beuzón and Holden 2001). In this way, more subtle effects on virulence can be examined (Rossier et al. 2004; DebRoy et al. 2006; Newton et al. 2007b, 2008; Lomma et al. 2010).

2.2.3 Other Species of *Legionella* in Mouse Infections

Even though other species of *Legionella* are known to contribute to the incidence of Legionnaires' disease worldwide, other than *L. longbeachae*, few studies have directly compared the virulence of these species with *L. pneumophila* using mouse models. In one study, researchers revealed that intra-tracheal infection of 10^5 CFU *L. longbeachae* serogroup 1 in A mice resulted in over 90–100 % mortality (Gobin et al. 2009). This contrasts with *L. pneumophila* where infection with 10^6 CFU is easily limited and cleared from the lungs within 7 days (Brieland et al. 1994). Furthermore, this observation of enhanced bacterial virulence of *L. longbeachae* was consistent across the typically *L. pneumophila*-'resistant' mouse strains, BALB/c and C57BL/6 (Gobin et al. 2009). In the lung, *L. longbeachae* infections are different in their pathology to *L. pneumophila*, as the inflammatory response (including neutrophil influx) is more focal and localized in the bronchiolar regions rather than the alveolar spaces (Brieland et al. 1994; Gobin et al. 2009). The increased virulence and lack of resistance in non-A mouse strains may be attributed in part to the absence of flagella in *L. longbeachae* (Cazalet et al. 2010). In addition, *L. longbeachae* produces a capsule that may further enhance its virulence (Cazalet et al. 2010). Despite this the Dot/Icm system is still critical for infection in vivo as a *dotA* mutant of *L. longbeachae* is strongly attenuated (Cazalet et al. 2010). Aside from *L. longbeachae,* one study has investigated *L. micdadei* infection of mice. Unlike *L. longbeachae* and *L. pneumophila*, *L. micdadei* did not replicate efficiently in the lung, and was promptly cleared from all tissues within 7 days despite some systemic dissemination (Gao et al. 1999). As the vast majority of in vivo research is centered primarily on *L. pneumophila*, the remainder of this chapter will focus on the host response to *L. pneumophila* initiated in the lung.

3 The Immune Response to *Legionella pneumophila* in Mice

Unlike the guinea pig model of Legionnaires' disease, mouse-based investigations of Legionnaires' disease have focussed primarily on the host innate and adaptive immune responses that lead to eradication of the pathogen from the lung. Indeed, research using this model has revealed a concerted requirement of both host molecular- and cellular-based interactions to effect bacterial clearance (Massis and Zamboni 2011; Schuelein et al. 2011). Although the exact processes taking place from the onset of infection to bacterial clearance are yet to be fully elucidated, it has become an accepted paradigm that multiple pattern recognition receptors (PRRs) of the Toll-like receptor (TLR), Nod-like receptor (NLR), and Rig-like helicase (RLH) families are important for synergistic recognition of the bacterium upon infection in vivo and these contribute to the resultant molecular and cellular immune responses that drive pathogen clearance (Massis and Zamboni 2011; Schuelein et al. 2011).

3.1 *Molecular-Based Interactions of L. pneumophila with the Innate Immune System*

The TLRs comprise a series of PRRs located on either the cell surface or embedded in the membranes of endosomes. With the exception of TLR3, each TLR signals via the adapter protein myeloid differentiation primary response gene-88 (MyD88). TLR-mediated activation of MyD88 initiates a signaling cascade involving the recruitment of adapter and kinase enzymes operating to upregulate transcription and translation of genes encoding proinflammatory cytokines and chemokines (Medzhitov et al. 1998; Kumagai et al. 2008). Upon infection with *L. pneumophila,* mice deficient in MyD88 signaling display severely impaired recruitment of mononuclear and polymorphonuclear leukocytes, in conjunction with reduced secretion of proinflammatory cytokines and chemokines within the first 24 h of infection (Archer and Roy 2006; Hawn et al. 2006; Spörri et al. 2006). MyD88-deficient mice fail to control replication of *L. pneumophila* in the lung, which results in dissemination of the bacteria to the spleen and liver and which becomes fatal within 9 days of infection (Archer and Roy 2006; Hawn et al. 2006; Archer et al. 2009). MyD88 is critical for NK cell production of IFNγ, a key cytokine in the *Legionella* response (Shinozawa et al. 2002; Spörri et al. 2006). When MyD88-deficient mice were infected with a *dotA* mutant strain of *L. pneumophila*, bacteria were cleared from the lung at a comparable rate to wild-type mice. This indicates that while establishment of a functional LCV by *L. pneumophila* is essential for its replication within the lung, the Dot/Icm system is also essential for the initiation of a robust proinflammatory response against the bacterium in vivo (Archer and Roy 2006).

3.1.1 Contribution of TLRs to Controlling *L. pneumophila* Infection

Despite the importance of MyD88 signaling in controlling wild-type *L. pneumophila*, infection of single or multiple TLR deficient mice has not been able to explain the extreme susceptibility of MyD88 knockout mice. Contributions of the TLRs to defense against *L. pneumophila* have focussed primarily on TLR2, TLR4, TLR5, and TLR9, as both in vitro and in vivo-based research has demonstrated recognition of *L. pneumophila* is restricted to these TLRs (Girard et al. 2003; Hawn et al. 2006; Hawn et al. 2007; Newton et al. 2007a; Archer et al. 2009). Early investigations revealed TLR4, a TLR typically involved in the recognition of lipopolysaccharide (LPS) on the outer membrane of Gram-negative bacteria, was not involved in the immune recognition and/or response against *L. pneumophila* in vivo (Lettinga et al. 2002). Rather, in vitro studies indicated TLR2, but not TLR4, recognized *L. pneumophila*-associated LPS, despite TLR2 typically recognizing bacterial peptidoglycan (Girard et al. 2003; Kawai and Akira 2009). Mice deficient in TLR2 exhibited delayed production of proinflammatory cytokines and chemokines and thus recruitment of neutrophils to the site of infection.

Furthermore, these mice displayed increased bacterial load after 5 days of infection relative to wild-type control mice. TLR2-deficient mice did not exhibit disseminated disease, nor did they develop pneumonia in the lung, as seen previously with MyD88 deficiency (Hawn et al. 2006). Similar to TLR2, TLR5 is also involved in the initial stages of bacterial resolution, as TLR5-deficient mice also fail to recruit neutrophils into the lung within hours of *L. pneumophila* infection. However, neutrophil influx is restored by 24 h and these early defects result in only a modest increase bacterial burden in the lungs (Hawn et al. 2007; Archer et al. 2009). Given TLR5 recognizes bacterial flagellin, infection of wild-type C57BL/6 mice with a flagellin-deficient strain of *L. pneumophila* results in a similar defect in neutrophil recruitment into the lung, as seen with TLR5 deficiency. However, unlike TLR5 deficiency, infection of mice with a flagellin-deficient strain is accompanied by a modest increase in bacterial load in the lungs of mice at 3 and 6 days after infection (Hawn et al. 2007). This difference is likely due to the recognition of flagellin by the Naip5/NLRC4 inflammasome in wild-type mice that directs early neutrophil recruitment (see Sect.3.1.3).

The response of TLR9-deficient mice to *L. pneumophila* is less well understood. One study found that TLR9 does not contribute to the secretion of proinflammatory cytokines in the lung within 48–72 h of infection, nor does TLR9 alter bacterial clearance over 7 days of infection (Archer et al. 2009). However, this observation contrasts with another study that found TLR9-deficient mice are delayed in their clearance of *L. pneumophila* from the lung and produce significantly less TNFα and IL-12p70 compared to wild-type controls (Bhan et al. 2008). Evidently, more work needs to be performed to understand the true contribution of TLR9 during *L. pneumophila* resistance. When TLR2 and TLR9 double knockout mice were infected with flagellin-deficient *L. pneumophila ΔflaA*, thus avoiding the activation of TLR2, TLR5, Naip5, and TLR9, only modest increases in bacterial load were observed and these mice were still able to eradicate *L. pneumophila ΔflaA* from the lung over a 7 days period. As a result, these observations have warranted speculation that another TLR and/or yet-to-be-discovered MyD88-dependent PRR may be involved in contributing to the immune control of *L. pneumophila* as the triple TLR/Naip5 deficiency was not comparable to the susceptibility of MyD88 knockout mice to *L. pneumophila* infection (Archer et al. 2009).

3.1.2 Contribution of NOD1 and NOD2 to Limiting *L. pneumophila* Replication In Vivo

As with the TLRs, the NLRs have also been investigated for their role in the innate immune response against *L. pneumophila*. The NLRs are a family of cytosolic pattern recognition receptors that assemble through homo and heteromeric protein–protein domain interactions (Lamkanfi and Dixit 2009). Typically, the NLRs contain an N-terminal interaction domain, a central nucleotide-binding domain, and a variable portion of C-terminal leucine-rich repeats (LRRs) differing in number among the many defined NLRs (Kumagai et al. 2008; Kawai and Akira 2009; Lamkanfi and Dixit 2009). Stimulation of the nucleotide-binding and oligomerization domain protein (NOD)-1 and NOD2 results in the stimulation of an intracellular signaling cascade that involves the activation of the transcription factor, NF-κB, which regulates the expression of many proinflammatory cytokines. NOD signaling is also dependent on 'receptor interacting protein kinase 2' (RIPK2) (Kumagai et al. 2008; Kawai and Akira 2009). Mice deficient in either one of NOD1, NOD2, or RIPK2 exhibit poor neutrophil recruitment into the lung within the first 9 h of infection with *L. pneumophila* but only RIPK2 knockout mice (that have a functional deficiency in NOD1 and NOD2) exhibit a significant reduction in the secretion of proinflammatory cytokines and chemokines. The poor neutrophil recruitment observed in RIPK2 knockout mice was associated with an accompanying increase in bacterial burden up to 4 days after infection (Frutuoso et al. 2010). These observations differed in relation to an earlier study by Archer et al. that demonstrated there was no difference in bacterial burden between RIPK2-deficient and C57BL/6 mice at days 2, 3 or 5 after infection. This latter study only observed a reduction in one chemokine (MCP-1) at day 3 after infection in the RIPK2 deficient mice (Archer et al. 2010).

In these studies, RIPK2 deficiency was used to examine an overall involvement of NOD1 and NOD2. However, a caveat of investigating RIPK2 in this manner is that the enzyme is also known to contribute to the signal transduction cascades of TLR2, IL-1R, and IL-18R (Kobayashi et al. 2002). Therefore the observed effect could be masked by a contribution of these receptors, rather than that of NOD1 and NOD2. Hence, two subsequent studies investigated the contribution of individual NOD1 and NOD2 receptors, rather than RIPK2. However, while one study proposed only NOD1 involvement (Berrington et al. 2010), the other study implicated both NOD1 and NOD2 (Frutuoso et al. 2010). Hence, the involvement of NOD1/NOD2 signaling in the control of *L. pneumophila* infection remains to be resolved.

3.1.3 Inflammasome-Mediated Control of *L. pneumophila* Infection

Many NLRs contribute to the formation of multiprotein inflammasomes that activate caspase-1 (Lamkanfi and Dixit 2009). These include Naip5, NLR family CARD domain-containing protein 4 (NLRC4), also known as IL-1-converting enzyme-protease activation factor (Ipaf), and NACHT-LRR-pyrin domain-containing protein-3 (NLRP3). NLRC4 can heterodimerise with other Naip and NACHT family proteins to influence their functionality, including the activation of NF-κB and caspase-1 (Damiano et al. 2004). Together with Naip5, NLRC4 is involved in the pyroptotic response to bacterial flagellin (Lamkanfi and Dixit 2009). In vitro infections of peritoneal macrophages derived from NLRC4-deficient mice have revealed the importance of flagellin, caspase-1, and the Dot/Icm type IV secretion system in the immune control of *L. pneumophila* (Amer et al. 2006; Molofsky et al. 2006; Ren et al. 2006; Zamboni et al. 2006). Investigation of these phenomena in vivo suggests that NLRC4-deficient mice cannot adequately restrict the replication of *L. pneumophila* in the first 4 days of infection which supports a role for NLRC4, alongside Naip5, in the immune control of *L. pneumophila* (Amer et al. 2006; Molofsky et al. 2006; Ren et al. 2006; Zamboni et al. 2006; Case et al. 2009).

The importance of the inflammasome in the immune response to *L. pneumophila* is further verified in caspase-1-deficient mice, which exhibit a substantially higher bacterial burden than wild-type mice 48 h after infection (Zamboni et al. 2006). Subsequent reports have emphasized a complexity in the cascade of events beyond NLRC4 activation in the restriction of *L. pneumophila* (Pereira et al. 2011). Although NLRC4 is typically involved in directing caspase-1 activation, during *L. pneumophila* infection, NLRC4 restricts bacterial growth via both caspase-1-dependent and independent pathways (Pereira et al. 2011). Wild-type *L. pneumophila* infection worsened in NLRC4-deficient mice over 72 h of infection compared to caspase-1-deficient or wild-type mice (Pereira et al. 2011).

In fact caspase-1 may be activated independently of NLRC4 through the 'apoptosis-associated speck-like protein' (Asc) adaptor (Case et al. 2009), which is required for the secretion of IL1-β and IL-18 during *Legionella* infection (Broz et al. 2010; Case and Roy 2011). Despite this, Asc is dispensable for restricting the replication of *L. pneumophila* in vivo. Asc-deficient mice clear *L. pneumophila* from the lung at the same rate as wild-type mice, unlike NLRC4-deficient mice where the rate of clearance is significantly reduced (Case et al. 2009). Hence, instead of IL1-β and IL-18 processing, the major impact of NLRC4-mediated caspase-1 activation on *L. pneumophila* infection may be the induction of host cell death. Other recent work has shown that NLRC4-induced cell death involves caspase-7, which is cleaved by caspase-1. Caspase-7 drives more efficient phagolysosome fusion in infected macrophages, as well as cell death, and caspase-7 knockout mice are more permissive for *L. pneumophila* replication (Akhter et al. 2009). *L. pneumophila* also induces caspase-11 activation which can stimulate activation of caspase-1, and this alternative caspase-1 activation pathway requires NLRP3/Asc (Case et al. 2013). However, this does not appear to be a major immune requirement during *L. pneumophila* mouse infection as caspase-11 deficient mice are only marginally more susceptible to infection than wild-type mice (Akhter et al. 2012).

In summary, it is undeniable that the larger family of NLRs contribute significantly to the restriction of *L. pneumophila* in vivo. So far the evidence convincingly demonstrates that the Naip5/NLRC4 inflammasome is integral to the restriction of *L. pneumophila* in the lung. Diminished Naip5/NLRC4 function thus serves as a basis for host susceptibility in otherwise restricted inbred strains of mice.

3.1.4 Involvement of the RIG-like Helicases in the Immune Response to *L. pneumophila*

Aside from the TLRs and NLRs, *L. pneumophila* also interacts with a third major PRR group, the cytosolic RLHs. The RLHs comprise the 'retinoic-acid inducible gene-I' (RIG-I) and 'melanoma-differentiation-associated gene-5' (MDA5). RIG-I and MDA5 are traditionally understood to detect double-stranded ribonucleic acid (dsRNA) of viral origin in the cytosolic compartment of a cell (Akira et al. 2006; Kawai and Akira 2009). Activation of RIG-I/MDA5 induces a cellular signaling cascade that incorporates multiple adapter proteins including interferon-β promoter stimulator-1 (IPS-1). IPS-1 regulates the production of interferon regulatory factors (IRFs) and NF-κB, that ultimately mediate the production of type-I interferon (IFN-I) and proinflammatory cytokines, respectively (Akira et al. 2006; Kawai and Akira 2009). *L. pneumophila* can stimulate this pathway through a process that may involve DNA/RNA translocation (Stetson and Medzhitov 2006; Chiu et al. 2009; Monroe et al. 2009). Unlike the wealth of in vivo studies investigating the involvement of the TLRs and the NLRs, the in vivo significance of the RLHs during *L. pneumophila* infection is poorly understood as study of these factors has focussed primarily on in vitro-based interactions. Nevertheless, infection of mice deficient in IPS-1 has revealed a requirement for this adapter protein in the production of IFNβ during infection, although there was no effect on restriction of *L. pneumophila* replication in the lung (Monroe et al. 2009). Indeed, despite evidence that *L. pneumophila* induces IFN-I during lung infection, the functional activity of IFN-I is disputable. Multiple reports have shown that a functional deficiency of IFN-I activity has no effect on *L. pneumophila* replication in the lungs of mice (Monroe et al. 2009; Ang et al. 2010). Mice deficient in the type-I interferon receptor (IFNAR) are no more affected in their ability to restrict *L. pneumophila* than wild-type C57BL/6 mice after 48 h of infection (Ang et al. 2010). This is despite multiple in vitro-based studies using human and murine cells reporting a protective effect of both IFNα and IFNβ prior to and during *L. pneumophila* infection (Schiavoni et al. 2004; Opitz et al. 2006; Plumlee et al. 2009). In the absence of interferon γ (IFNγ), IFNAR deficiency may have some effect on the immune response although the significance of this is uncertain as the requirement for IFNγ dominates (Lippmann et al. 2011). In any case the inconsistency between in vivo and in vitro observations illustrates the importance of validating in vitro-based observations in vivo, especially given host immunity is multifactorial.

3.1.5 Cytokine Responses During *L. pneumophila* Infection

In contrast to IFN-I, other proinflammatory cytokines have a profound impact on resistance to *Legionella* infection in vivo. Mice challenged with *L. pneumophila* induce robust production of IFNγ that peaks in the serum approximately 24 h after infection (Brieland et al. 1994). Mice deficient in the production of IFNγ are unable to restrict bacterial replication in the lung and suffer heavy systemic bacterial dissemination 3 days after infection. IFNγ deficiency frequently results in a fatal infection beyond this time point (Shinozawa et al. 2002). The importance of IFNγ in *L. pneumophila* clearance is further supported by cytokine over-expression in a model performed by Deng et al. Recombinant adenovirus expressing IFNγ, co-administered with *L. pneumophila*, led to enhanced production of IFNγ in the lung as well as faster eradication of *L. pneumophila* (Deng et al. 2001). The production of IFNγ during *L. pneumophila* infection is linked to IL-12 as mice depleted of endogenous IL-12 produce less IFNγ and TNFα (Brieland et al. 1995, 2000), and these changes are associated with increased bacterial burden 5 days after infection (Brieland et al. 1998, 2000). Similar to IFNγ, adenovirus-based over-expression of IL-12p40 and IL-12p35 leads to reduced bacterial load upon *L. pneumophila* infection although the effect is not as great as mice receiving IFNγ-expressing adenovirus (Deng et al. 2001).

Aside from IFNγ and IL-12, mice also produce TNFα upon infection with *L. pneumophila* (Brieland et al. 1995). Treatment of A strain mice with anti-TNFα neutralizing antibodies results in poor clearance of *L. pneumophila* from the lung at 3 and 5 days after infection (Brieland et al. 1995). The role of TNFα in *L. pneumophila* infection has been further assessed using mice deficient in TNFR1 and TNFR2. Both strains display severe lung pathology by 3 days after infection and suffer acute mortality after 7 days upon lethal challenge (Fujita et al. 2008). Despite the similar outcome, the receptors appear to have different roles during infection. TNFR1-deficient mice fail to clear *L. pneumophila* from the lung and show signs of an impaired innate immune response, including severely reduced neutrophil infiltration (Fujita et al. 2008). In contrast, TNFR2-deficient mice suffer a massive accumulation of neutrophils in the lung suggestive of an excessive and fatal inflammatory response.

In addition to IFNγ, IL-12, and TNFα, mice produce significant amounts of IL-17A and IL-17F in the lung during the first 24 h of exposure to *L. pneumophila* (Kimizuka et al. 2012). Although IL-17A/F-deficient mice contain only marginally more bacteria in the lungs 7 days after a sub-lethal infection with *L. pneumophila*, beyond this time, the mice struggle to clear a lethal bacterial challenge, and considerably fewer mice survive by 3 weeks of infection relative to wild-type control animals (Kimizuka et al. 2012). However, despite its clear contribution to host defense, the precise role of IL17A/F during *L. pneumophila* infection is uncertain.

Likewise, both IL-1β and IL-18 have been detected in the lungs and serum of mice infected with *L. pneumophila*, yet the involvement of these cytokines is also not well understood. This is despite intense investigation of inflammasome activation by *L. pneumophila* (Susa et al. 1998; Shinozawa et al. 2002; Hawn et al. 2007; Berrington et al. 2010; Frutuoso et al. 2010). For IL-1β, the immune response appears to be modulated by nonhematopoietic cell production of the cytokine in the lungs and IL-1R deficiency leads to a $\sim$10-fold increase in bacterial burden after 24 h (LeibundGut-Landmann et al. 2011). Most experiments investigating the contribution of IL-18 have utilized either A strain mice or C57BL/6 mice infected with a flagellin-null variant of *L. pneumophila*. Infection of A strain mice induces robust secretion of IL-18 during the acute stage of infection (Brieland et al. 2000), as does infection of C57BL/6 mice with a Δ*flaA* strain of *L. pneumophila*, however the latter is markedly reduced relative to

C57BL/6 mice infected with wild-type *L. pneumophila* (Case et al. 2009). While blockade of IL-18 function, either by neutralizing antibody (Brieland et al. 2000) or IL-18R deficiency (Archer et al. 2010) substantially reduces the amount of IFNγ secreted early during infection, the functional depletion of IL-18 does not alter the bacterial burden or clearance of *L. pneumophila* from the lungs of mice over 7 days of infection (Brieland et al. 2000; Archer et al. 2010). However, given wild-type *L. pneumophila* infection induces substantially more secretion of IL-18 that is reduced in the Naip5-independent infection model, it is not clear whether IL-18 may be one of the primary causes of the resistant phenotype in mice. Thus further studies in IL-18-deficient, Naip5/NLRC4-competent mice using wild-type *L. pneumophila* infection are warranted to understand fully the contribution of IL-18 to resistance.

3.2 Cellular Responses in the Lung During L. pneumophila Infection

Concomitant with the burst of proinflammatory cytokines within the first 72 h of *L. pneumophila* infection in mice, there is a strong influx of polymorphonuclear neutrophils into the lung together with mononuclear phagocytes, including macrophages and a small number of NK cells (Brieland et al. 1994; Susa et al. 1998). Strain mice depleted of neutrophils by RB6-8C5 monoclonal antibody become acutely susceptible to nonlethal infection with *L. pneumophila* within 3 days of infection, and show a skewed cytokine secretion phenotype involving increases in the regulatory cytokines, IL-4, and IL-10, alongside substantially reduced levels of IL-12 and IFNγ (Tateda et al. 2001). Subsequent investigation into NK cell activity during *L. pneumophila* infection shows that NK cells are significant contributors to secreted IFNγ during infection. Their depletion by administration of polyclonal anti-asialo GM1 antibody results in a substantial drop in IFNγ levels in the lungs of *L. pneumophila* infected mice 48 h after infection. Despite this, NK cell deficient mice still clear the bacteria from the lung over 4 days of infection (Archer et al. 2009).

Dendritic cells are also recruited to the lung early in *L. pneumophila* infection. Although conventional dendritic cells have not been well-studied in vivo in response to *L. pneumophila*, work on plasmacytoid dendritic cells (pDCs) suggests that this cell type contributes to host resistance (Ang et al. 2010). pDCs are a rarer subset of dendritic cells that are typified by robust production of IFN-I (Siegal et al. 1999; Honda et al. 2005). Depletion of pDCs from the lung using the monoclonal antibody 120G8 hinders the ability of A strain mice to adequately clear *L. pneumophila* from the lung over 48–72 h of infection and this effect is independent of IFN-I (Ang et al. 2010).

3.3 Interactions of *L. pneumophila* with the Adaptive Immune Response

In contrast to the wealth of research investigating interactions of *L. pneumophila* with the innate immune response, far less is known about the influence of the adaptive immune response on bacterial clearance, possibly because a robust early inflammatory response is critical for controlling *L. pneumophila* infection in humans. In the relative sense, fewer studies have contributed to our understanding of adaptive immune mechanisms in the mouse infection model, and only very recently have we been able to gain insight into the later stages of the infection. Within the first 7 days of infection, $CD4^+$ T cells and cytotoxic lymphocytes (CTLs) infiltrate the lungs of infected mice, which correlates with bacterial clearance. A strain mice depleted of either $CD4^+$ and/or $CD8^+$ T cells by monoclonal antibody administration prior to and during a lethal *L. pneumophila* challenge succumb to the infection at an accelerated rate (Susa et al. 1998). Interestingly, an immunization study against *L. pneumophila* has shown that dendritic cells upregulate CX_3CR1 (also known as fractalkine) on their cell surface in response to the bacterium (Kikuchi et al. 2005). CX_3CR1 serves as a chemoattractant that when cleaved from the cell surface, influences the migration of T cells and monocytes. The transfer of fractalkine expressing dendritic cells pulsed with heat-killed *L. pneumophila* protects mice from a lethal challenge for at least 14 days after infection and this relies on $CD4^+$ T cells, CTLs, and B cells (Kikuchi et al. 2005). From the original observation that $CD4^+$ T cells and CTLs infiltrate the lungs of *L. pneumophila* infected mice, recent work has shown that in the later stages of *L. pneumophila* infection, the $CD4^+$ helper T cell response comprises differentiated Th1 and Th17 helper T cells infiltrating the lung tissue (Trunk and Oxenius 2012). During wild-type infection, *L. pneumophila*-antigen specific T cells are primed in the mediastinal lymph node within the first 3 days during infection. By 6 days, differentiated helper T cells infiltrate the lung parenchyma and begin secreting proinflammatory cytokines, namely IFNγ by Th1 cells and IL-17 by Th17 cells (Trunk and Oxenius 2012). While Th17 restriction of *L. pneumophila* depends on MyD88 and the NLRC4 inflammasome, the Th1 response is still initiated in the absence of MyD88, although appears in conjunction with a Th2 response that is otherwise not seen during wild-type infection (Trunk and Oxenius 2012).

Aside from T cells, B cells also infiltrate the lungs of *L. pneumophila* infected mice (Susa et al. 1998). Mice mount a *L. pneumophila*-specific antibody response by either intranasal inoculation and inhalation into the lung, or by systemic exposure through intravenous injection (Joller et al. 2007). Mice generate IgG and IgA responses in the serum and BAL, respectively, upon primary *L. pneumophila* exposure. The class switching events required for this difference in antibody isotype production depends entirely on $CD4^+$ T cell help. Subsequently, the infection generates a large pool of isotype-switched memory B cells in the lung peaking 7–10 days after infection (Joller et al. 2007). Apart from LPS responses, 30 novel protein-based *L. pneumophila*-specific B cell antigens have been

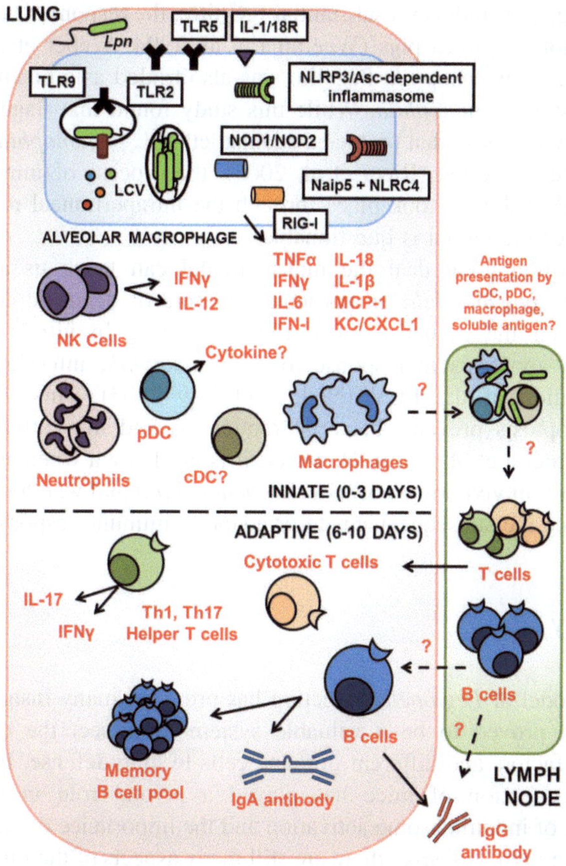

Fig. 1 ***Legionella pneumophila*** **interactions with the host immune system**. During infection *L. pneumophila* is recognized by cell surface, cytosolic and endosomal host pattern recognition systems including the toll-like receptors (TLRs), the Nod-like receptors (NLRs) (which include the Naip5, NLRC4 and NLRP3/Asc-dependent inflammasomes), and the Rig-like helicases (RLHs). This stimulation triggers release of proinflammatory cytokines/chemokines which contribute towards alveolar infiltration of neutrophils, natural killer (NK) cells, macrophages and dendritic cells (cDC and pDC) in the early stages of the immune response. During late stage immunity, the lung is infiltrated with both cytotoxic and helper T cells that have been previously primed in the lung-draining lymph node. Helper T cells have a Th1 and Th17 phenotype, contributing to secretion of IFNγ and IL-17, respectively. B cells infiltrate the lung and generate a memory B cell pool, and also secrete IgA and IgG antibodies into the lung mucosa and serum, respectively

identified, and in most cases these comprise components of the bacterial membrane (Weber et al. 2012). Subcutaneous immunization of mice with purified proteins resulted in increases of serum IgG that appear protective within the first 48 h of infection (Weber et al. 2012). One of the selected candidates was heat shock protein (Hsp)-60, a protein that has been previously investigated in vaccination studies of *L. pneumophila* challenge in guinea pigs. While Hsp60 in both mice and guinea pigs induces a substantial IgG titre, the response is only protective in mice and not in guinea pigs (Weeratna et al. 1994; Weber et al. 2012). In a separate study, *L. pneumophila* flagellin was also trialed as a potential protective antigen against *L. pneumophila*. While this study found that flagellin induced a robust antibody response that protected against lethal *L. pneumophila* challenge up to 30 days after infection (Ricci et al. 2005), the process of immunization and lethal bacterial challenge took place through the intraperitoneal route and hence the relevance of the model is questionable.

There is still a great deal the mouse model can teach us about adaptive immunity to *L. pneumophila*, in particular the role of dendritic cells in antigen presentation and the induction of lasting immunity. In vitro, dendritic cells undergo rapid apoptosis in response to *L. pneumophila* infection despite supporting formation of the LCV (Neild and Roy 2003). This rapid Bak/Bax-dependent apoptosis prevents bacterial replication and may affect antigen presentation (Nogueira et al. 2009). However, it is not known whether dendritic cell apoptosis occurs in vivo in response to *L. pneumophila* and whether this apoptosis would influence the subsequent innate or adaptive immune response (Fig. 1).

4 Summary

The mouse model of *Legionella* infection has provided many insights into innate immunity and proved to be a valuable system to dissect the contribution of inflammatory factors and different immune cells to lung defense. In particular, *L. pneumophila* infection of mice has played a pivotal role in increasing our understanding of inflammasome activation and the importance of NLRC4 and Asc to pyroptosis in vivo. However there are still many aspects of the immune response to explore with this model, in particular the immune cell and cytokine interactions that ultimately lead to clearance of the bacteria from the lung. At the same time, immunologists and microbiologists alike will learn more about the acute inflammatory response during infection and how it may be enhanced to fight bacterial respiratory infections.

Acknowledgments This work was supported by grants awarded to ELH and IVD by the National Australian Health and Medical Research Council (NHMRC). ELH is an Australian Research Council Future Fellow. ASB is supported by an Australian Postgraduate Award.

References

Akhter A, Caution K, Abu Khweek A et al (2012) Caspase-11 promotes the fusion of phagosomes harboring pathogenic bacteria with lysosomes by modulating actin polymerization. Immunity 37:35–47

Akhter A, Gavrilin MA, Frantz L et al (2009) Caspase-7 activation by the Nlrc4/Ipaf inflammasome restricts *Legionella pneumophila* infection. PLoS Pathog 5:e1000361

Akira S, Uematsu S, Takeuchi O (2006) Pathogen recognition and innate immunity. Cell 124:783–801

Amer A, Franchi L, Kanneganti T-D et al (2006) Regulation of *Legionella* phagosome maturation and infection through flagellin and host Ipaf. J Biol Chem 281:35217–35223

Amsden GW (2005) Treatment of Legionnaires' disease. Drugs 65:605–614

Ang DKY, Oates CVL, Schuelein R et al (2010) Cutting edge: pulmonary *Legionella pneumophila* is controlled by plasmacytoid dendritic cells but not type I IFN. J Immunol 184:5429–5433

Archer KA, Ader F, Kobayashi KS, Flavell RA, Roy CR (2010) Cooperation between multiple microbial pattern recognition systems is important for host protection against the intracellular pathogen *Legionella pneumophila*. Infect Immun 78:2477–2487

Archer KA, Alexopoulou L, Flavell RA, Roy CR (2009) Multiple MyD88-dependent responses contribute to pulmonary clearance of *Legionella pneumophila*. Cell Microbiol 11:21–36

Archer KA, Roy CR (2006) MyD88-dependent responses involving toll-like receptor 2 are important for protection and clearance of *Legionella pneumophila* in a mouse model of Legionnaires' disease. Infect Immun 74:3325

Baskerville A, Broster M, Fitzgeorge RB, Hambleton P, Dennis PJ (1981) Experimental transmission of Legionnaires' disease by exposure to aerosols of *Legionella pneumophila*. The Lancet 318:1389–1390

Beckers MC, Si Y, Morgan K, Skamene E, Gros P (1995) Natural resistance to infection with *Legionella pneumophila*: chromosomal localization of the Lgn1 susceptibility gene. Mamm Genome 6:540–545

Berger KH, Isberg RR (1993) Two distinct defects in intracellular growth complemented by a single genetic locus in *Legionella pneumophila*. Mol Microbiol 7:7–19

Berger KH, Merriam JJ, Isberg RR (1994) Altered intracellular targeting properties associated with mutations in the *Legionella pneumophila dotA* gene. Mol Microbiol 14:809–822

Berrington WR, Iyer R, Wells RD, Smith KD, Skerrett SJ, Hawn TR (2010) NOD1 and NOD2 regulation of pulmonary innate immunity to *Legionella pneumophila*. Eur J Immunol 40:3519–3527

Beuzón CR, Holden DW (2001) Use of mixed infections with *Salmonella* strains to study virulence genes and their interactions in vivo. Microbes Infect 3:1345–1352

Bhan U, Trujillo G, Lyn-Kew K et al (2008) Toll-like receptor 9 regulates the lung macrophage phenotype and host immunity in murine pneumonia caused by *Legionella pneumophila*. Infect Immun 76:2895–2904

Blander SJ, Horwitz MA (1989) Vaccination with the major secretory protein of *Legionella pneumophila* induces cell-mediated and protective immunity in a guinea pig model of Legionnaires' disease. J Exp Med 169:691–705

Breiman RF, Horwitz MA (1987) Guinea pigs sublethally infected with aerosolized *Legionella pneumophila* develop humoral and cell-mediated immune responses and are protected against lethal aerosol challenge. A model for studying host defense against lung infections caused by intracellular pathogens. J Exp Med 165:799–811

Brieland J, Freeman P, Kunkel R, Chrisp C, Hurley M, Fantone J, Engleberg C (1994) Replicative *Legionella pneumophila* lung infection in intratracheally inoculated A/J mice. A murine model of human Legionnaires' disease. Amer J Pathol 145:1537

Brieland JK, Jackson C, Hurst S et al (2000) Immunomodulatory role of endogenous interleukin-18 in gamma interferon-mediated resolution of replicative *Legionella pneumophila* lung infection. Infect Immun 68:6567

Brieland JK, Remick DG, Freeman PT, Hurley MC, Fantone JC, Engleberg NC (1995) In vivo regulation of replicative *Legionella pneumophila* lung infection by endogenous tumor necrosis factor alpha and nitric oxide. Infect Immun 63:3253–3258

Brieland JK, Remick DG, LeGendre ML, Engleberg NC, Fantone JC (1998) In vivo regulation of replicative *Legionella pneumophila* lung infection by endogenous interleukin-12. Infect Immun 66:65–69

Broz P, von Moltke J, Jones JW, Vance RE, Monack DM (2010) Differential requirement for caspase-1 autoproteolysis in pathogen-induced cell death and cytokine processing. Cell Host Microbe 8:471–483

Case CL, Kohler LJ, Lima JB, et al. (2013) Caspase-11 stimulates rapid flagellin-independent pyroptosis in response to *Legionella pneumophila*. Proc Natl Acad Sci USA 110:1851–1856

Case CL, Roy CR (2011) Asc modulates the function of NLRC4 in response to infection of macrophages by *Legionella pneumophila*. mBio 2: e00117–111

Case CL, Shin S, Roy CR (2009) Asc and Ipaf inflammasomes direct distinct pathways for caspase-1 activation in response to *Legionella pneumophila*. Infect Immun 77:1981–1991

Cazalet C, Gomez-Valero L, Rusniok C et al (2010) Analysis of the *Legionella longbeachae* genome and transcriptome uncovers unique strategies to cause Legionnaires' disease. PLoS Genet 6:e1000851

Chiu Y-H, MacMillan JB, Chen ZJ (2009) RNA polymerase III detects cytosolic DNA and induces type I interferons through the RIG-I pathway. Cell 138:576–591

Damiano JS, Oliveira V, Welsh K, Reed JC (2004) Heterotypic interactions among NACHT domains: implications for regulation of innate immune responses. Biochem J 381:213–219

DebRoy S, Dao J, Söderberg M, Rossier O, Cianciotto NP (2006) *Legionella pneumophila* type II secretome reveals unique exoproteins and a chitinase that promotes bacterial persistence in the lung. Proc Natl Acad Sci USA 103:19146–19151

Deng JC, Tateda K, Zeng X, Standiford TJ (2001) Transient transgenic expression of gamma interferon promotes *Legionella pneumophila* clearance in immunocompetent hosts. Infect Immun 69:6382–6390

Dietrich WF, Damron DM, Isberg RR, Lander ES, Swanson MS (1995) Lgn1, a gene that determines susceptibility to *Legionella pneumophila,* maps to mouse chromosome 13. Genomics 26:443–450

Diez E, Lee S-H, Gauthier S, Yarghi Z, Tremblay M, Vidal S, Gros P (2002) Birc1e is the gene within the Lgn1 locus associated with resistance to *Legionella pneumophila*. Nat Genet 33:55

Edelstein PH, Calarco K, Yasui VK (1984) Antimicrobial therapy of experimentally induced Legionnaires' disease in guinea pigs. Amer Rev Resp Dis 130:849–856

Fields BS, Benson RF, Besser RE (2002) *Legionella* and Legionnaires' disease: 25 years of investigation. Clin Microbiol Rev 15:506–526

Fink SL, Cookson BT (2005) Apoptosis, pyroptosis, and necrosis: mechanistic description of dead and dying eukaryotic cells. Infect Immun 73:1907–1916

Fitzgeorge RB, Baskerville A, Broster M, Hambleton P, Dennis PJ (1983) Aerosol infection of animals with strains of *Legionella pneumophila* of different virulence: comparison with intraperitoneal and intranasal routes of infection. J Hyg 90:81–89

Fitzgeorge RB, Featherstone ASR, Baskerville A (1990) Efficacy of azithromycin in the treatment of guinea pigs infected with *Legionella pneumophila* by aerosol. J Antimicrob Chem 25:101–108

Fraser DW, Bopp C, Kaye Wachsmuth I, Feeley JC, Tsai TF (1978) Antibiotic treatment of guinea pigs infected with agent of Legionnaires' disease. The Lancet 311:175–178

Fraser DW, Tsai TR, Orenstein W et al (1977) Legionnaires' disease. New Engl J Med 297:1189–1197

Frutuoso MS, Hori JI, Pereira MSF et al (2010) The pattern recognition receptors Nod1 and Nod2 account for neutrophil recruitment to the lungs of mice infected with *Legionella pneumophila*. Microbes Infect 12:819–827

Fujita M, Ikegame S, Harada E et al (2008) TNF receptor 1 and 2 contribute in different ways to resistance to *Legionella pneumophila*-induced mortality in mice. Cytokine 44:298–303

Gao L-Y, Susa M, Ticac B, Kwaik YA (1999) Heterogeneity in intracellular replication and cytopathogenicity of *Legionella pneumophila* and *Legionella micdadei* in mammalian and protozoan cells. Microb Pathog 27:273–287

Girard R, Pedron T, Uematsu S, Balloy V, Chignard M, Akira S, Chaby R (2003) Lipopolysaccharides from *Legionella* and *Rhizobium* stimulate mouse bone marrow granulocytes via toll-like receptor 2. J Cell Sci 116:293–302

Glavin FL, Winn WC, Craighead JE (1979) Ultrastructure of lung in Legionnaires' disease. Observations of three biopsies done during the Vermont epidemic. Ann Int Med 90:555–559

Glick TH, Gregg MB, Berman B, Mallison G, Rhodes WW, Kassanoff I (1978) Pontiac fever. Amer J Epidemiol 107:149–160

Gobin I, Susa M, Begic G, Hartland EL, Doric M (2009) Experimental *Legionella longbeachae* infection in intratracheally inoculated mice. J Med Microbiol 58:723–730

Hawn TR, Berrington WR, Smith IA et al (2007) Altered inflammatory responses in TLR5-deficient mice infected with *Legionella pneumophila*. J Immunol 179:6981

Hawn TR, Smith KD, Aderem A, Skerrett SJ (2006) Myeloid differentiation primary response gene (88)—and toll-like receptor 2—deficient mice are susceptible to infection with aerosolized *Legionella pneumophila*. J Infect Dis 193:1693

Hedlund KW, McGann VG, Copeland DS, Little SF, Allen RG (1979) Immunologic protection against the Legionnaires' disease bacterium in the AKR/J mouse. Ann Int Med 90:676–679

Hilbi H, Jarraud S, Hartland E, Buchrieser C (2010) Update on Legionnaires' disease: pathogenesis, epidemiology, detection and control. Mol Microbiol 76:1–11

Honda K, Yanai H, Negishi H et al (2005) IRF-7 is the master regulator of type-I interferon-dependent immune responses. Nature 434:772–777

Isberg RR, O'Connor TJ, Heidtman M (2008) The *Legionella pneumophila* replication vacuole: making a cosy niche inside host cells. Nat Rev Microbiol 7:13

Joller N, Spörri R, Hilbi H, Oxenius A (2007) Induction and protective role of antibodies in *Legionella pneumophila* infection. Eur J Immunol 37:3414–3423

Kawai T, Akira S (2009) The roles of TLRs, RLRs and NLRs in pathogen recognition. Int Immunol 21:317–337

Kikuchi T, Andarini S, Xin H et al (2005) Involvement of fractalkine/CX3CL1 expression by dendritic cells in the enhancement of host immunity against *Legionella pneumophila*. Infect Immun 73:5350–5357

Kimizuka Y, Kimura S, Saga T et al (2012) Roles of interleukin-17 in an experimental *Legionella pneumophila* pneumonia model. Infect Immun 80:1121–1127

Kobayashi K, Inohara N, Hernandez LD et al (2002) RICK/Rip2/CARDIAK mediates signalling for receptors of the innate and adaptive immune systems. Nature 416:194–199

Kumagai Y, Takeuchi O, Akira S (2008) Pathogen recognition by innate receptors. J Infect Chemo 14:86

Lamkanfi M, Dixit VM (2009) Inflammasomes: guardians of cytosolic sanctity. Immunol Rev 227:95–105

LeibundGut-Landmann S, Weidner K, Hilbi H, Oxenius A (2011) Nonhematopoietic cells are key players in innate control of bacterial airway infection. J Immunol 186:3130–3137

Lettinga KD, Florquin S, Speelman P, van Ketel R, van der Poll T, Verbon A (2002) Toll-like receptor 4 is not involved in host defense against pulmonary *Legionella pneumophila* infection in a mouse model. J Infect Dis 186:570

Lightfield KL, Persson J, Brubaker SW et al (2008) Critical function for Naip5 in inflammasome activation by a conserved carboxy-terminal domain of flagellin. Nat Immunol 9:1171–1178

Lippmann J, Müller HC, Naujoks J et al (2011) Dissection of a type I interferon pathway in controlling bacterial intracellular infection in mice. Cell Microbiol 13:1668–1682

Lomma M, Dervins-Ravault D, Rolando M et al (2010) The *Legionella pneumophila* F-box protein Lpp 2082 (AnkB) modulates ubiquitination of the host protein parvin B and promotes intracellular replication. Cell Microbiol 12:1272–1291

Marra A, Blander SJ, Horwitz MA, Shuman HA (1992) Identification of a *Legionella pneumophila* locus required for intracellular multiplication in human macrophages. Proc Natl Acad Sci USA 89:9607–9611

Massis LM, Zamboni DS (2011) Innate immunity to *Legionella pneumophila*. Front Microbiol 2:109

McDade JE, Shepard CC, Fraser DW, Tsai TR, Redus MA, Dowdle WR (1977) Legionnaires' disease. New Engl J Med 297:1197–1203

Medzhitov R, Preston-Hurlburt P, Kopp E, Stadlen A, Chen C, Ghosh S, Janeway CA (1998) MyD88 is an adaptor protein in the hToll/IL-1 receptor family signaling pathways. Mol Cell 2:253–258

Molofsky AB, Byrne BG, Whitfield NN, Madigan CA, Fuse ET, Tateda K, Swanson MS (2006) Cytosolic recognition of flagellin by mouse macrophages restricts *Legionella pneumophila* infection. J Exp Med 203:1093–1104

Monroe KM, McWhirter SM, Vance RE (2009) Identification of host cytosolic sensors and bacterial factors regulating the type I interferon response to *Legionella pneumophila.* PLoS Pathog 5:e1000665

Neild AL, Roy CR (2003) *Legionella* reveal dendritic cell functions that facilitate selection of antigens for MHC class II presentation. Immunity 18:813–823

Newton CA, Perkins I, Widen RH, Friedman H, Klein TW (2007a) Role of toll-like receptor 9 in *Legionella pneumophila*-induced interleukin-12 p40 production in bone marrow-derived dendritic cells and macrophages from permissive and nonpermissive mice. Infect Immun 75:146

Newton HJ, Ang DKY, van Driel IR, Hartland EL (2010) Molecular pathogenesis of infections caused by *Legionella pneumophila.* Clin Microbiol Rev 23:274

Newton HJ, Sansom FM, Dao J et al (2008) Significant role for *ladC* in initiation of *Legionella pneumophila* infection. Infect Immun 76:3075–3085

Newton HJ, Sansom FM, Dao J, McAlister AD, Sloan J, Cianciotto NP, Hartland EL (2007b) Sel1 repeat protein LpnE is a *Legionella pneumophila* virulence determinant that influences vacuolar trafficking. Infect Immunity 75:5575–5585

Nogueira CV, Lindsten T, Jamieson AM, Case CL, Shin S, Thompson CB, Roy CR (2009) Rapid pathogen-induced apoptosis: a mechanism used by dendritic cells to limit intracellular replication of *Legionella pneumophila.* PLoS Pathog 5:e1000478

Opitz B, Vinzing M, van Laak V et al (2006) *Legionella pneumophila* induces IFNβ in lung epithelial cells via IPS-1 and IRF3, which also control bacterial replication. J Biol Chem 281:36173–36179

Pereira MSF, Morgantetti GF, Massis LM, Horta CV, Hori JI, Zamboni DS (2011) Activation of NLRC4 by flagellated bacteria triggers caspase-1–dependent and –independent responses to restrict *Legionella pneumophila* replication in macrophages and in vivo. J Immunol 187:6447–6455

Plumlee CR, Lee C, Beg AA, Decker T, Shuman HA, Schindler C (2009) Interferons direct an effective innate response to *Legionella pneumophila* infection. J Biol Chem 284:30058–30066

Ren T, Zamboni DS, Roy CR, Dietrich WF, Vance RE (2006) Flagellin-deficient *Legionella* mutants evade caspase-1-and Naip5-mediated macrophage immunity. PLoS Pathog 2:e18

Ricci ML, Torosantucci A, Scaturro M, Chiani P, Baldassarri L, Pastoris MC (2005) Induction of protective immunity by *Legionella pneumophila* flagellum in an A/J mouse model. Vaccine 23:4811–4820

Rodgers FG (1979) Ultrastructure of *Legionella pneumophila.* J Clin Pathol 32:1195–1202

Rossier O, Starkenburg SR, Cianciotto NP (2004) *Legionella pneumophila* type II protein secretion promotes virulence in the A/J mouse model of Legionnaires' disease pneumonia. Infect Immun 72:310–321

Roy CR, Berger KH, Isberg RR (1998) *Legionella pneumophila* DotA protein is required for early phagosome trafficking decisions that occur within minutes of bacterial uptake. Mol Microbiol 28:663–674

Schiavoni G, Mauri C, Carlei D, Belardelli F, Castellani Pastoris M, Proietti E (2004) Type I IFN protects permissive macrophages from *Legionella pneumophila* infection through an IFN-γ-independent pathway. J Immunol 173:1266–1275

Schuelein R, Ang D, van Driel IR, Hartland EL (2011) Immune control of *Legionella* infection: an in vivo perspective. Front Microbiol 2:126

Shinozawa Y, Matsumoto T, Uchida K, Tsujimoto S, Iwakura Y, Yamaguchi K (2002) Role of interferon-gamma in inflammatory responses in murine respiratory infection with *Legionella pneumophila.* J Med Microbiol 51:225–230

Siegal FP, Kadowaki N, Shodell M et al (1999) The nature of the principal type 1 interferon-producing cells in human blood. Science 284:1835–1837

Spörri R, Joller N, Albers U, Hilbi H, Oxenius A (2006) MyD88-dependent IFN-γ production by NK cells is key for control of *Legionella pneumophila* infection. J Immunol 176:6162–6171

Stetson DB, Medzhitov R (2006) Recognition of cytosolic DNA activates an IRF3-dependent innate immune response. Immunity 24:93–103

Susa M, Ticac B, Rukavina T, Doric M, Marre R (1998) *Legionella pneumophila* infection in intratracheally inoculated T cell-depleted or -nondepleted A/J mice. J Immunol 160:316–321

Tateda K, Moore TA, Deng JC et al (2001) Early recruitment of neutrophils determines subsequent T1/T2 host responses in a murine model of *Legionella pneumophila* pneumonia. J Immunol 166:3355–3361

Trunk G, Oxenius A (2012) Innate instruction of CD4+T cell immunity in respiratory bacterial infection. J Immunol 189:616–628

Weber SS, Joller N, Küntzel AB et al (2012) Identification of protective B cell antigens of *Legionella pneumophila*. J Immunol 189:841–849

Weeratna R, Stamler DA, Edelstein PH, Ripley M, Marrie T, Hoskin D, Hoffman PS (1994) Human and guinea pig immune responses to *Legionella pneumophila* protein antigens OmpS and Hsp60. Infect Immun 62:3454–3462

Wright EK Jr, Goodart SA, Growney JD et al (2003) Naip5 Affects host susceptibility to the intracellular pathogen *Legionella pneumophila*. Curr Biol 13:27–36

Yamamoto Y, Klein TW, Newton CA, Widen R, Friedman H (1988) Growth of *Legionella pneumophila* in thioglycolate-elicited peritoneal macrophages from A/J mice. Infect Immun 56:370–375

Yoshida S-I, Mizuguchi Y (1986) Multiplication of *Legionella pneumophila* Philadelphia-1 in cultured peritoneal macrophages and its correlation to susceptibility of animals. Can J Microbiol 32:438–442

Yoshida S, Goto Y, Mizuguchi Y, Nomoto K, Skamene E (1991) Genetic control of natural resistance in mouse macrophages regulating intracellular *Legionella pneumophila* multiplication in vitro. Infect Immun 59:428–432

Zamboni DS, Kobayashi KS, Kohlsdorf T et al (2006) The Birc1e cytosolic pattern-recognition receptor contributes to the detection and control of *Legionella pneumophila* infection. Nat Immunol 7:318–325

Index

A

AAA (ATPases associated with various cellular activities) ATPase, 232
Acanthamoeba castellanii, 264
Adaptive immune response, 284
Adenylylation, 117, 124, 126, 127, 129
ALIS, 238
Amoeba, 1–7, 9, 11, 13, 14, 16–19, 22–24, 85, 90–92, 97
AnkB, 233
AnkX, 121, 124, 126, 127
Arf, 117, 120, 122, 129
Asp-His-His-Cys (DHHC), 140
Autophagy, 240

B

Bacteria, 177–183, 189, 190, 194–196

C

CaaX (Cys—aliphatic—aliphatic—X) motif, 140
CaaX motif proteins (CMPs), 142
cAMP, 144
CD148, 253
CD45, 253
Cellular responses in the lung, 283
Chaperones, 109
Clk1 (Cdc2-like kinase 1), 235
Conjugation, 104
Coupling proteins, 110
CpxR, 37–39, 45, 47
CsrA, 39, 43, 44, 46, 47, 55
Cytokine responses, 281
Cytokines, 95–97

D

DALIS, 238
Diacylglycerol (DAG), 157
Dictyostelium discoideum, 166, 258
Dot/Icm, 142
DrrA, 123, 124, 126–128
Dual-specificity protein phosphatases (Dusps), 256, 259
DupA, 259

E

E3 enzymes, 229
Elongation factor eEF1A, 217–219
Epithelial cells, 81, 89, 94, 95, 97
ER-associated degradation (ERAD), 232

F

Farnesyltransferase, 140
F-box, 233, 259
F-box-containing proteins, 235
FYVE domain, 155

G

Gal/GalNAc lectin receptor, 254
GAP, 119, 121, 122, 124–126, 129
GDI, 119, 122, 124, 126, 128
Geranylgeranyltransferase I, 140
Glycosylation, 221, 222
GTPase, 117–119, 121, 122, 124, 126–130, 157, 158, 162, 164–166, 168
Guanine nucleotide exchange factor (GEF), 119, 121–124, 127, 128, 138, 158
Guinea pig model of Legionnaireś disease, 273

Current Topics in Microbiology and Immunology 376: 293–296
DOI: 10.1007/978-3-642-40591-4

H
Hartmannella vermiformis, 254
Hfq, 57
HLA-B-associated transcript 3 (BAT3), 237
Host response, 18
Hsp60, 260
Hypersensitivity response (HR), 149

I
Infection, 177, 189, 190, 192–196, 198
Inflammasome, 279
Innate immune system, 277
Inositol, 157, 168
Intracellular pathogen, 211
IpgD, 162
Iron uptake, 72

L
LegAU13, 233
LegAU13/AnkB, 235, 238
Legionella, 177, 178, 182, 186, 189, 190, 192, 194, 196
Legionella pneumophila, 1, 3, 8, 10, 11, 81, 82, 211–215, 219–223
Legionella-containing vacuole (LCV), 138, 156
Legionella vacuole association (LVA) domain, 166
LegU1, 233, 235, 237
LegU2, 233
LetA, 38, 42–44
LetA/S, 55
LicA, 233, 235
Lipid, 157, 159, 164, 166–168
LqsR, 38, 39, 44
LubX, 233

M
Macrophages, 1–5, 8–10, 12–15, 18, 22–24, 81, 85, 87, 89, 91, 93–96
Metaeffector, 237
Mitogen-activated protein kinase (MAPK), 252, 255, 256, 258–260
Mouse model of Legionnaireś disease, 274
Mycobacterium, 159

N
NF-κB (nuclear factor kappa-light-chain-enhancer of activated B cells), 252, 260–263
N-myristoyltransferase (NMT), 139
NOD1 and NOD2, 278

O
Oculocerebrorenal syndrome of Lowe (OCRL), 166
Oxidative Stress, 69

P
Palmitoyl acyltransferases (PATs), 140, 148
Phosphatidylinositol (PtdIns), 157
Phosphocholination, 117, 124–127, 129
Phosphoinositide 3-kinase (PI3K), 254
Phosphoinositide (PI), 157
Phospholipases, 178, 180, 181, 183, 184, 186, 194, 198
PI 3-kinase (PI3K), 167
PI 5-phosphatase (PI5P), 165, 166
PmrA, 41, 43, 46, 47
Protein kinase B (PKB/Akt), 254
Protein kinase C (PKC), 252, 259, 260
Protein synthesis inhibition, 221

R
Rab, 117, 119, 120, 122–129
Rab1, 39–42, 45, 117, 121–128, 130
RavZ, 240
Ribosomal binding site, 57
RIG-like helicases, 280
RNA sequencing, 61
RpoS, 71
RsmYZ, 39, 43, 46
RyhB, 72

S
Salmonella, 159, 162
Salmonella-containing vacuole (SCV), 147
SapM, 159
SCF (Skp1-Cullin-F-box protein) complex, 229, 235
Secreted enzymes, 85
Shigella, 159, 162
SidC, 162, 163, 166, 167
SidD, 41, 45, 46
SidH, 237
SidM, 39, 41, 44–46, 48, 163–167
Signaling, 184, 185, 198
Signals, 103
SopB, 159, 162

6S RNA, 64
Surfactant, 88–90, 96

T
Toll-like receptor (TLR), 255, 257, 259, 261, 263, 277
Transporter, 103
Type II protein secretion, 82
Type III secretion systems (T3SS), 147
Type IV secretion system (T4SS), 142

U
Ubiquitin, 228
U-box, 229, 233

V
Vesicular trafficking, 117, 119, 120, 122, 123, 129
Virulence, 4, 10, 24, 177, 178, 183, 190, 194, 195, 198
Virulence factor, 221

MIX
Papier aus verantwortungsvollen Quellen
Paper from responsible sources
FSC® C105338

If you have any concerns about our products,
you can contact us on
ProductSafety@springernature.com

In case Publisher is established outside the EU,
the EU authorized representative is:
Springer Nature Customer Service Center GmbH
Europaplatz 3, 69115 Heidelberg, Germany

Printed by Libri Plureos GmbH
in Hamburg, Germany